Micro Non-Destructive Testing and Evaluation

Micro Non-Destructive Testing and Evaluation

Editor

Giovanni Bruno

MDPI • Basel • Beijing • Wuhan • Barcelona • Belgrade • Manchester • Tokyo • Cluj • Tianjin

Editor
Giovanni Bruno
Bundesanstalt für Materialforschung und–prüfung (BAM),
Berlin, Germany

Editorial Office
MDPI
St. Alban-Anlage 66
4052 Basel, Switzerland

This is a reprint of articles from the Special Issue published online in the open access journal *Materials* (ISSN 1996-1944) (available at: https://www.mdpi.com/journal/materials/special_issues/nondestructive).

For citation purposes, cite each article independently as indicated on the article page online and as indicated below:

LastName, A.A.; LastName, B.B.; LastName, C.C. Article Title. *Journal Name* **Year**, *Volume Number*, Page Range.

ISBN 978-3-0365-6179-0 (Hbk)
ISBN 978-3-0365-6180-6 (PDF)

Cover image courtesy of Prof. Dr. Giovanni Bruno.

Contents

About the Editor

Giovanni Bruno

Giovanni Bruno studied Nuclear Engineering then Physics at the University of Bologna, Italy. He obtained his Ph.D. in Materials Science at the University of Ancona, Italy, working on residual stress analysis in welds of several metallic alloys. He worked as a Post-Doctorate researcher in the UK, in Germany, and in France. He then moved on to industry (Corning Incorporated), where he worked first as the head of the Physical, Mechanical and Structural Characterization Group in France and then as a project leader in the United States.

As of 2012, he became head of the Division 8.5 Micro NDT at BAM (Berlin, D) and professor at the University of Potsdam (D). His research is focused on the use of microstructural investigations (in particular by X-ray computed tomography) and the determination of residual stress and mechanical properties to assess the performance and damage mechanisms of composites, metallic alloys (in particular additively manufactured materials), and ceramics.

materials

MDPI

Editorial

Micro Non-Destructive Testing and Evaluation

Giovanni Bruno

BAM, Bundesanstalt für Materialforschung und -prüfung, Unter den Eichen 87, 12205 Berlin, Germany; giovanni.bruno@bam.de

1. Foreword

What is meant by 'Micro Non-Destructive Testing and Evaluation'? This was the central subject of debate in this Special Issue.

At present, sub-millimeter-size components or even assemblies are pervading the industrial and scientific world. Classic examples are electronic devices and watches (as well as parts thereof), but recent examples encompass additively manufactured lattice structures, stents, or other microparts. Moreover, most assemblies contain micro-components. Testing such components or their miniaturized parts would fit well within the topic of micro non-destructive testing and evaluation.

In all cases, performance and integrity testing, quality control, and dimensional tolerances need to be measured at the sub-millimeter level (ideally with a spatial resolution of about a micron); most of the time, such features and components are embedded in much larger assemblies, which also need to be taken into account. The solution to this dilemma (i.e. measuring large parts with high resolution) depends on the part and on the problem under consideration.

Another possible definition of micro non-destructive testing and evaluation can relate to the characterization of micro-features (e.g., the microstructure) in much larger specimens, such as damage in concrete cores or porosity in additively manufactured components. A further aspect is the use of microscopic probes to evaluate macroscopic properties. This is the case, for instance but not at all exclusively, in the use of diffraction techniques to determine macroscopic stress.

The splits between testing and characterization at the micro-level (or of micro parts) from one side and handling of macroscopic assemblies on the other represent a great challenge for many fields of materials characterization. On top of that, including the use of microscopic methods to test integrity would add a further level of complexity.

Imaging, mechanical testing, non-destructive testing, measurement of properties, structural health monitoring, and dimensional metrology all need to be re-defined if we want to cope with the multi-faceted topic of micro non-destructive testing and evaluation.

The challenge has already been accepted by the scientific and engineering communities for a while but is still far from being universally tackled. This Special Issue yields an interesting answer to the questions posed above. It presents the progress made and the different aspects of the challenge as well as at indicates the paths for the future of NDT&E.

2. Introduction

With the increasing miniaturization of components, performance assessment, quality control, and structural health monitoring have expanded their toolbox of experimental techniques. Classically, non-destructive testing and evaluation (NDT&E) has included macroscopic probes such as radar, X-ray radiography, and ultrasound for structures or large components. Recently, other tools have been used to cope with the challenge of miniaturization. Such tools include not only spatially or temporally resolved techniques such as synchrotron radiation imaging but also investigation techniques, which in the past belonged more to the realm of materials science than to engineering (e.g., diffraction and

Citation: Bruno, G. Micro Non-Destructive Testing and Evaluation. *Materials* **2022**, *15*, 5923. https://doi.org/10.3390/ma15175923

Received: 24 August 2022
Accepted: 25 August 2022
Published: 27 August 2022

laser-induced breakdown spectroscopy). However, such 'new' tools can be and have also been applied to investigate large components: X-ray and neutron diffraction are currently used to determine the residual stress in safety-relevant components (nuclear industry and additive manufacturing) [1,2], and X-ray computed tomography is used to investigate the degradation of concrete cores [3,4]. The meaning of micro-NDT&E (μ-NDT&E) methods has, therefore, been extended from the use of NDT&E techniques on microscopic components to the use of microscopic techniques and to macroscopic components.

3. Summary of the Special Issue

This Special Issue shows that X-ray computed tomography is becoming a major tool for μ-NDT&E, being used for small and large components [3,5,6], and for sensitive materials [7]. Indeed, new methods are also being developed [6–8]. At the same time, optical methods are being perfected to tackle challenging problems at the micro and macro scales [9,10]. Moreover, magnetic methods are still very powerful for detecting defects in several materials and components [11–14], especially when talking about large components. Indeed, such methods are being further developed and extended to new materials such as concrete [15] and to applications such as residual stress determination [16].

From the materials point of view, it is clear that concrete plays an eminent role in the field of NDT&E. Its eternal youth and wide application fields render it always useful, so that new kinds of investigations are paralleled with new compositions and materials designs [17]. However, classic materials such as steels [13,18], novel metallic biomaterials [19], and additively manufactured metallic alloys and structures [5,10] are also at the top of the agenda.

4. Conclusions

From the discussion above, we conclude that, in general, NDT&E methods are of primary importance in the design, performance assessment, quality control, and structural health monitoring of materials and components (metallic or ceramic/cementitious). One could summarize the meaning of the contributions to this Special Issue in a nutshell by stating that, currently, there is no separation between micro-NDT&E and NDT&E, since the latter field already includes the first and the applications of NDT&E methods to miniaturized materials or to the microscopic scale (materials science and characterization) has already been happening for some time.

Funding: This research received no external funding.

Institutional Review Board Statement: Not applicable.

Informed Consent Statement: Not applicable.

Data Availability Statement: Not applicable.

Acknowledgments: All contributing authors and the editorial team of *Materials* are acknowledged for their continued support for this Special Issue.

Conflicts of Interest: The author declares no conflict of interest.

References

1. Schröder, J.; Evans, A.; Mishurova, T.; Ulbricht, A.; Sprengel, M.; Serrano-Munoz, I.; Fritsch, T.; Kromm, A.; Kannengießer, T.; Bruno, G. Diffraction-Based Residual Stress Characterization in Laser Additive Manufacturing of Metals. *Metals* **2021**, *11*, 1830. [CrossRef]
2. Bruno, G. Relaxation of Residual Stress in as-welded and aged AISI 347 pipe by means of time-of-flight neutron diffraction. *Z. Met.* **2002**, *93*, 33–41.
3. Oesch, T.; Weise, F.; Bruno, G. Detection and Quantification of Cracking in Concrete Aggregate through Virtual Data Fusion of X-ray Computed Tomography Images. *Materials* **2020**, *13*, 3921. [CrossRef] [PubMed]
4. Powierza, B.; Stelzner, L.; Oesch, T.; Gollwitzer, C.; Weise, F.; Bruno, G. Water Migration in One-Side Heated Concrete: 4D In-Situ CT Monitoring of the Moisture-Clog-Effect. *J. Nondestruct. Eval.* **2019**, *38*, 15. [CrossRef]

5. Khrapov, D.; Kozadayeva, M.; Manabaev, K.; Panin, A.; Sjöström, W.; Koptyug, A.; Mishurova, T.; Evsevleev, S.; Meinel, D.; Bruno, G.; et al. Different Approaches for Manufacturing Ti-6Al-4V Alloy with Triply Periodic Minimal Surface Sheet-Based Structures by Electron Beam Melting. *Materials* **2021**, *14*, 4912. [CrossRef] [PubMed]
6. Evsevleev, S.; Mishurova, T.; Khrapov, D.; Paveleva, A.; Meinel, D.; Surmenev, R.; Surmeneva, M.; Koptyug, A.; Bruno, G. X-ray Computed Tomography Procedures to Quantitatively Characterize the Morphological Features of Triply Periodic Minimal Surface Structures. *Materials* **2021**, *14*, 3002. [CrossRef] [PubMed]
7. Léonard, F.; Zhang, Z.; Krebs, H.; Bruno, G. Structural and Morphological Quantitative 3D Characterisation of Ammonium Nitrate Prills by X-Ray Computed Tomography. *Materials* **2020**, *13*, 1230. [CrossRef] [PubMed]
8. Linardatos, D.; Koukou, V.; Martini, N.; Konstantinidis, A.; Bakas, A.; Fountos, G.; Valais, I.; Michail, C. On the Response of a Micro Non-Destructive Testing X-ray Detector. *Materials* **2021**, *14*, 888. [CrossRef] [PubMed]
9. Özcan, B.; Schwermann, R.; Blankenbach, J. A Novel Camera-Based Measurement System for Roughness Determination of Concrete Surfaces. *Materials* **2021**, *14*, 158. [CrossRef] [PubMed]
10. Mahmood, M.A.; Chioibasu, D.; Mihai, S.; Iovea, M.; Mihailescu, I.N.; Popescu, A.C. Non-Destructive X-ray Characterization of a Novel Joining Method Based on Laser-Melting Deposition for AISI 304 Stainless Steel. *Materials* **2021**, *14*, 7796. [CrossRef] [PubMed]
11. Chady, T.; Okarma, K.; Mikołajczyk, R.; Dziendzikowski, M.; Synaszko, P.; Dragan, K. Extended Damage Detection and Identification in Aircraft Structure Based on Multifrequency Eddy Current Method and Mutual Image Similarity Assessment. *Materials* **2021**, *14*, 4452. [CrossRef] [PubMed]
12. Chady, T.; Łukaszuk, R.D.; Gorący, K.; Żwir, M.J. Magnetic Recording Method (MRM) for Nondestructive Evaluation of Ferromagnetic Materials. *Materials* **2022**, *15*, 630. [CrossRef] [PubMed]
13. Chady, T.; Łukaszuk, R. Examining Ferromagnetic Materials Subjected to a Static Stress Load Using the Magnetic Method. *Materials* **2021**, *14*, 3455. [CrossRef] [PubMed]
14. Vértesy, G.; Gasparics, A.; Szenthe, I.; Bilicz, S. Magnetic Investigation of Cladded Nuclear Reactor Blocks. *Materials* **2022**, *15*, 1425. [CrossRef] [PubMed]
15. Frankowski, P.K.; Chady, T. Impact of Magnetization on the Evaluation of Reinforced Concrete Structures Using DC Magnetic Methods. *Materials* **2022**, *15*, 857. [CrossRef] [PubMed]
16. Mishurova, T.; Stegemann, R.; Lyamkin, V.; Cabeza, S.; Evsevleev, S.; Pelkner, M.; Bruno, G. Subsurface and Bulk Residual Stress Analysis of S235JRC + C Steel TIG Weld by Diffraction and Magnetic Stray Field Measurements. *Exp. Mech.* **2022**, *62*, 1017–1025. [CrossRef]
17. Mishurova, T.; Rachmatulin, N.; Fontana, P.; Oesch, T.; Bruno, G.; Radi, E.; Sevostianov, I. Evaluation of the probability density of inhomogeneous fiber orientations by computed tomography and its application to the calculation of the effective properties of a fiber-reinforced composite. *Int. J. Eng. Sci.* **2018**, *122*, 14–29. [CrossRef]
18. Vértesy, G.; Gasparics, A.; Szenthe, I.; Rabung, M.; Kopp, M.; Griffin, J.M. Analysis of Magnetic Nondestructive Measurement Methods for Determination of the Degradation of Reactor Pressure Vessel Steel. *Materials* **2021**, *14*, 5256. [CrossRef] [PubMed]
19. Savin, A.; Craus, M.L.; Bruma, A.; Novy, F.; Malo, S.; Chlada, M.; Steigmann, R.; Vizureanu, P.; Harnois, C.; Turchenko, V.; et al. Microstructural Analysis and Mechanical Properties of TiMo20Zr7Ta15Six Alloys as Biomaterials. *Materials* **2020**, *13*, 4808. [CrossRef] [PubMed]

materials

MDPI

Article

Structural and Morphological Quantitative 3D Characterisation of Ammonium Nitrate Prills by X-ray Computed Tomography

Fabien Léonard [1,*], Zhen Zhang [1], Holger Krebs [1] and Giovanni Bruno [1,2]

1 Bundesanstalt für Materialforschung und -prüfung, Unter den Eichen 87, 12205 Berlin, Germany; zhen.zhang@bam.de (Z.Z.); holger.krebs@bam.de (H.K.); giovanni.bruno@bam.de (G.B.)
2 Institute of Physics and Astronomy, University of Potsdam, Karl-Liebknecht-str.24-25, 14476 Potsdam, Germany
* Correspondence: fabien.leonard@bam.de; Tel.: +49-30-8104-4627

Received: 23 January 2020; Accepted: 27 February 2020; Published: 9 March 2020

Abstract: The mixture of ammonium nitrate (AN) prills and fuel oil (FO), usually referred to as ANFO, is extensively used in the mining industry as a bulk explosive. One of the major performance predictors of ANFO mixtures is the fuel oil retention, which is itself governed by the complex pore structure of the AN prills. In this study, we present how X-ray computed tomography (XCT), and the associated advanced data processing workflow, can be used to fully characterise the structure and morphology of AN prills. We show that structural parameters such as volume fraction of the different phases and morphological parameters such as specific surface area and shape factor can be reliably extracted from the XCT data, and that there is a good agreement with the measured oil retention values. Importantly, oil retention measurements (qualifying the efficiency of ANFO as explosives) correlate well with the specific surface area determined by XCT. XCT can therefore be employed non-destructively; it can accurately evaluate and characterise porosity in ammonium nitrate prills, and even predict their efficiency.

Keywords: ANFO; explosives; specific surface area; porosity; XCT; data processing

---

1. Introduction

Prilled/granulated ammonium nitrate (AN) is extensively used in the farming industry as a fertilizer and in the mining industry as an explosive. In particular, the term ANFO, for ammonium nitrate/fuel oil, specifically describes a mixture of solid ammonium nitrate prills (see Figure 1a) and diesel fuel. Thanks to its simplicity, safety, low cost, simple manufacturing process, and high blasting efficiency, ANFO represents the majority of the explosives used worldwide [1]. To quantify the importance of ammonium nitrate product usage internationally, the world's total AN production was estimated at 15.3 million tons in 2017 [2], while the global AN market is expected to reach $6.18 billion by 2025 [3].

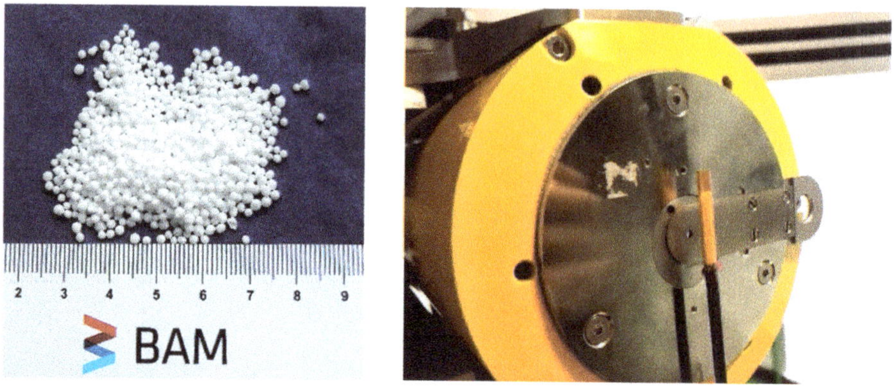

(**a**) (**b**)

Figure 1. Photography of ammonium nitrate (AN) prills type E1: (**a**) loose AN prills; (**b**) AN prills paced into a polyimide tube glued onto a carbon fibre rod, prior to X-ray computed tomography (XCT) scanning.

Ammonium nitrate prills are solid spherical pellets, ranging in diameter from tenths of millimeters to up to 2 mm, manufactured by dropping a supersaturated ammonium nitrate solution through a cooling tower 30 to 60 m in height. AN molten droplets are sprayed at the top of the tower; solidification occurs by water loss during the fall of the prill through a heated air flow. During drying, the water molecules on the surface of the AN drops evaporate, causing the initiation of crystallization on the prill surface. The water concentration gradient created between the surface and the inner part of the prill forces the internal water to migrate to the surface. The space originally occupied by the water molecules is thus replaced by air, resulting in the formation of pore channels with a complex geometry. In addition, the shrinkage in volume leads to stresses between the surface and the interior of the prill, generating cracks and flaws during solidification [4].

For the fertilizing industry, highly concentrated AN solutions (99.7% to 99.8%) are used to produce high density prills. For the mining industry, porous prills are required to promote propagation of detonation and allow a higher loading of fuel oil in ANFO mixtures: low density prills are produced for the mining industry with 96%–97% AN solutions [5]. The porosity (without packing porosity) of AN prills is approximately 20% and high quality explosive-grade prills can absorb up to about 15% fuel oil by weight [6]. The common ANFO mixture composition consists of 94.5% of AN and 5.5% of FO in weight [7]. One of the major performance predictors of the ANFO prills is the fuel oil retention, which is itself governed by the porosity of the AN prills. Presently, the oil retention capacity of ammonium nitrate in prilled and granulated forms is determined in the European Union by means of a standardised test [8]. However, this method faces technical difficulties, mainly because the porosity from different types of ammonium nitrate prills varies significantly. The porosity connected to the prill surface, i.e., open porosity, is available for oil retention. The pores not connected to the prill surface, i.e., closed porosity, are not available for oil retention but are, however, important for the explosive sensitivity to detonation [9], and the current test methods cannot account for the closed porosity.

One way to investigate and characterize the porosity of AN prills in a more accurate way is to use X-ray computed tomography (XCT). In fact, XCT allows the quantitative determination of both the open and the closed porosity, and can therefore yield a measure of the total surface area of the pore structure [10]. The main drawback of XCT is the limited resolution, when very small porosity is present. However, several strategies have been found to extract quantitative information, even in cases where single objects were not imaged [11,12]. In this paper the suitability of XCT to the non-destructive evaluation of AN prill materials is assessed. A data processing workflow is developed to extract relevant structural and morphological parameters from XCT scans of AN prills for each individual prill grain present in the XCT volume, whilst performing the segmentation on the entire 3D reconstructed volume only once. We demonstrate that differences between fertilisers and explosives and among different AN prills can be quantitatively determined from the XCT data.

2. Materials and Methods

2.1. Ammonium Nitrate Prills

Two types of AN prills were XCT scanned, the type labelled hereafter "type E" used in the mining industry as a constituent in ANFO mixtures (see Figure 1a), and the "type F", used as a fertiliser in farming. Specimens E1 and E2 came from the same manufacturer, with similar manufacturing processes, however, E2 was expected to have higher porosity content than E1. Specimens E3 and F1 were supplied by different manufacturers than E1 and E2, and there were no available data as to their respective porosity contents. The prills, around 1.3 mm in diameter for type E and 2.4 mm in diameter for type F, were placed into a polyimide tube of 4.2 mm diameter glued onto a carbon fibre rod (see Figure 1b) and then scanned using laboratory X-ray computed tomography. All E type prills contained more than 95% pure ammonium nitrate, whilst type F contained about 40 % AN, and other potassium and ammonium salts. Typical additives were used to make the surface of the prills hydrophobic. Their production route is described in the introduction whilst further information about the products cannot be released due to confidentiality.

2.2. Laboratory X-Ray Computed Tomography

The AN prills were scanned on a GE V|Tome|x L 180/300 system [13] equipped with a 180 kV source, a tungsten transmission target (actual focal spot size below 2 µm as determined with JIMA test pattern RTC02), a diamond window, and a GE 2000 × 2000 pixel DXR-250 detector (GE Sensing & Inspection Technologies GmbH, Wunstorf, Deutschland). The sample projections were taken at 1800 angular positions over the 360° rotation, with increments of 0.2°, and an exposure time of 5 s per projection. The samples were placed 11.25 mm downstream from the source, with a source-detector distance of 450 mm, so that the effective magnification of 40 was obtained. The resulting pixel size was around 5.0 µm and the total scan time was 2 h 30 min. The voxel size was calibrated after the scans were performed by scanning a ball bar consisting of 2 ruby spheres glued onto a carbon fibre rod and separated by a calibrated distance of 2.273 mm ± 0.001 mm. The calibrated voxel size was determined by comparing the calibrated distance to the distance between the 2 spheres in the volumetric XCT data using VGstudio MAX version 3.2 [14]. Surface determination was performed, then 2 spheres were generated by fitting 25 points to the surface of each ruby spheres, and then the distance between the sphere centres was measured. The calculated value (5.064 µm) was then employed as voxel size for each CT scan during the data processing step. The data visualisation, processing and quantification was performed using Amira ZIB Edition version 2019.03 [15].

2.3. Oil Retention Tests

The oil retention tests were performed according to the Regulation (EC) 2003R2003 [8] (pp. 80–81). The measurement principle prescribes total immersion of the test sample in gas oil for a specified period of time, followed by the draining of the surplus oil under specified conditions, and finally by measuring the increase in mass of the test sample. Such tests are difficult in practice, as many parameters can influence the final increase of mass value: oil viscosity and density (which are themselves dependent on temperature), oil composition and impurity content, amount of time prills are immersed in oil, filter paper type and grade, rolling procedure for draining excess oil, etc. Since such test results are difficult to compare (e.g., between labs), it is of high importance to establish alternative techniques to quantitatively assess the explosion efficiency. We will see that XCT provides a robust quantitative assessment of the specific surface area, and of explosive efficiency.

3. XCT Data Processing Workflow

The data processing workflow was aimed at extracting the most relevant structural parameters of the AN prills, both on a global and a local scale, and in a quantitative fashion. The entire workflow is presented in Figure 2, each step is discussed in detail below and illustrated in Video S1 (supplementary material); further details as to the Amira functions used can be found in Table A1 Appendix A.

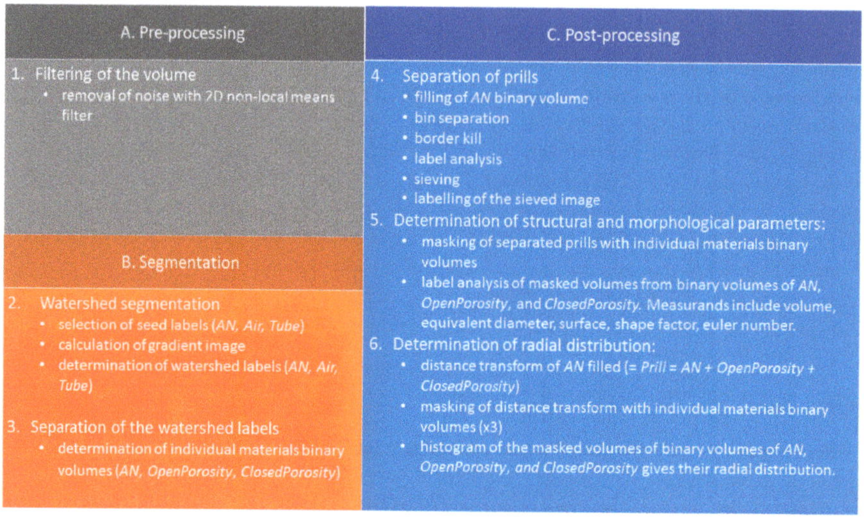

Figure 2. Overview of the data processing workflow.

The first step is the filtering of the data, in order to remove the noise present, whilst conserving the features of interest. For the AN prills, several filters were tested; the best filtering results were obtained with a non-local means filter (employed on all 2D slices, with a search window of 21 pixels, a local neighborhood value of 5 and a similarity value of 0.6). An example of the filtering process is presented in Figure 3: Figure 3a shows a slice of the 3D reconstruction of E2 prills, while Figure 3b represents the same slice after filtering (of the whole reconstructed volume). It can be seen on the plot (Figure 3c) that the peaks corresponding to the AN and the air appear sharper and of higher intensity after the filtering. In addition, two smaller peaks, that were previously combined into the AN peak, are now better resolved. These peaks correspond to the outer polyimide tube and the inner polyimide tube material, respectively.

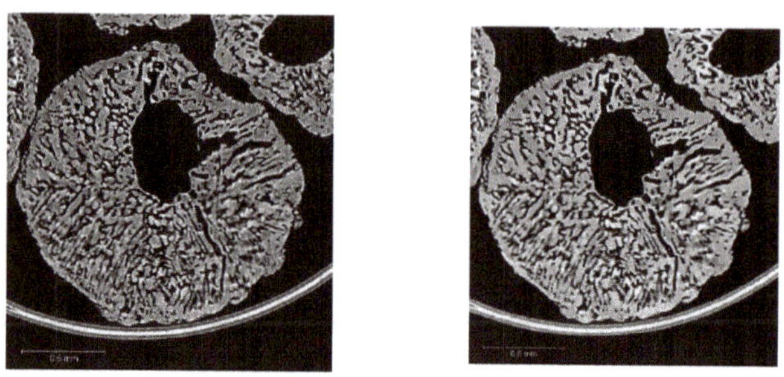

(a) (b)

Figure 3. *Cont.*

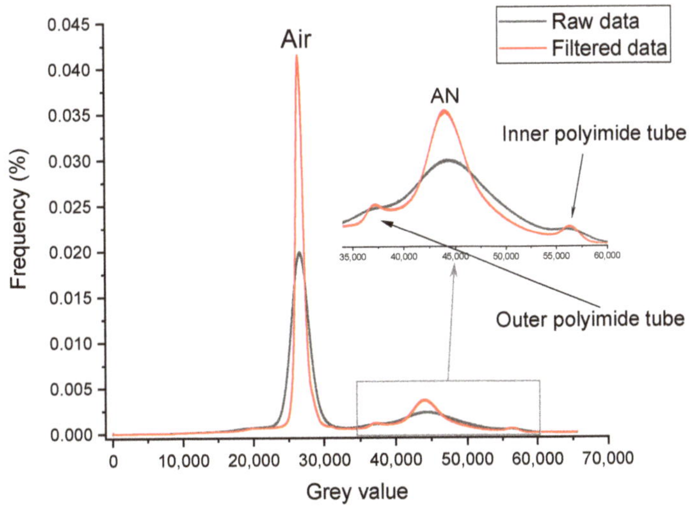

(c)

Figure 3. Example of 2D non-local means filtering on E2 AN prill XCT data (step 2 from Figure 2): (**a**) raw CT data; (**b**) filtered CT data; (**c**) greyscale histograms of raw and filtered XCT volumes (16 bit data).

The segmentation (Figure 2, Step 2) was performed using a seed-based watershed algorithm [16], as it allows a better reproducibility between the different AN material types and lessens the impact of the threshold definition on the surface determination compared to a global threshold or ISOXX approach (see example given in Appendix B, and corresponding results in Table A2). By convention, labels and binary volumes as defined in Amira will be referred to with a name in italic, whilst a name in standard characters will be used to refer to the generic materials. Seeds of three materials were generated, namely *Tube* for the polyimide tube, *AN* for the AN prill solid material only, and *Air* for the air present in the volume, both within the prills (as closed and open porosity), and outside the prills (as air in-between the prills). An overview of the watershed segmentation process is given in Figure 4.

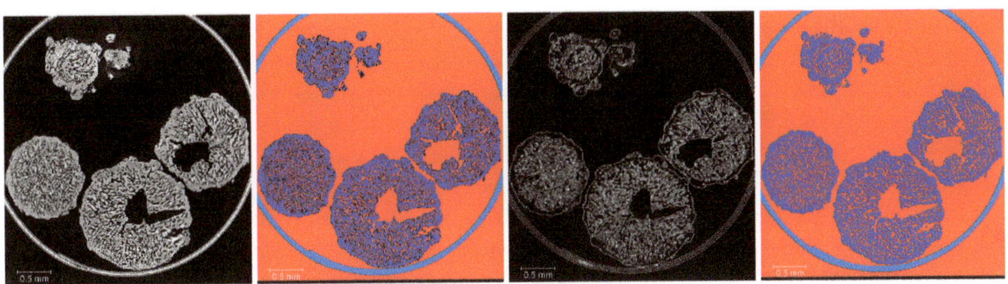

(**a**) (**b**) (**c**) (**d**)

Figure 4. Overview of the watershed segmentation process (Figure 2 step 2): (**a**) filtered data; (**b**) label seeds (*Tube* is light blue, *AN* is dark blue, *Air* is red); (**c**) gradient image from filtered data; (**d**) labels resulting from watershed segmentation (*Tube* is light blue, *AN* is dark blue, *Air* is red), where the label seeds are taken as starting point and grown in combination with the gradient image so that all voxels in the volume get assigned to a label.

Once the watershed segmentation is performed, the following individual materials binary volumes must be defined (Figure 2, Step 3): *AN*, *OpenPororisty*, and *ClosedPorosity* (as shown in Figure 5). This step is necessary for the analysis to be performed on these labels. In particular, the packing porosity (gaps between the grains) must be removed and the open and closed porosity must be separated. The *AN* binary volume is obtained directly from the watershed segmentation labels. The *ClosedPorosity* is obtained by using a *3D_fill* operation on the *AN* selection, then by subtracting the original *AN* binary volume. The *OpenPorosity* is the *Air* label minus the *ClosedPorosity* selection. Alternatively, the *OpenPorosity* selection can be obtained by selecting all the *Air* voxels connected to an *Air* voxel situated between the prills. The *ClosedPorosity* selection would then be the *Air* label minus the *OpenPorosity* selection. These two approaches are identical and yield the same results.

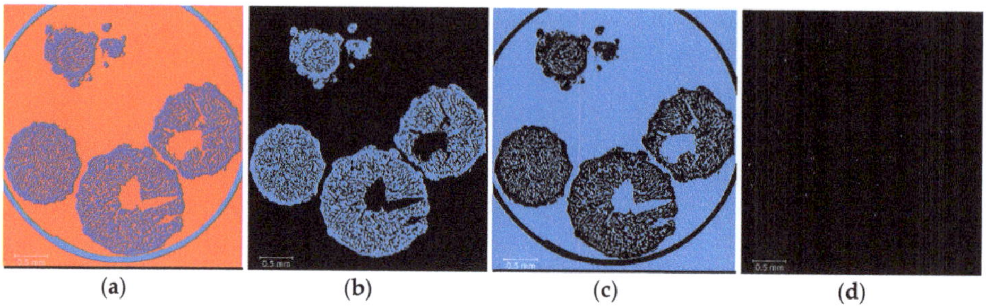

Figure 5. Overview of the separation of the watershed labels (Figure 2, Step 3): (**a**) labels from watershed segmentation (*Tube* is light blue, *AN* is dark blue, *Air* is red); (**b**) *AN* binary volume; (**c**) *OpenPorosity* binary volume; (**d**) *ClosedPorosity* binary volume.

At this stage, it is important to bear in mind that the segmentation has been performed once on the entire volume, which means that if a label analysis was performed, it would yield a single value for each metric of interest for each binary volume (e.g., total surface area of *AN*, or total volume of *OpenPorosity* binary volumes). Although it may be correct for some quantities of some of the binary volumes, it is evidently wrong for the *OpenPorosity* binary volume, as the air in between the prills is included in the binary volume. Further processing is required to obtain the binary volumes with the correct data. To do so, the prills not fully in the 3D image must be removed, and the open porosity must be the one included within the prills, only. In addition, the data processing must be able to separate each binary volume on a prill-by-prill basis, so that the average value and standard deviation of the quantities of interest for each AN material can be determined. This is the key aspect of the work presented here, the segmentation is performed on the entire volume, then the post-processing allows extracting automatically the metrics of interest for each prill present in the 3D image, regardless of the number of prills. The workflow introduced here is an elegant alternative to a painstaking but possible (given enough time) manual segmentation and quantification of each individual prill.

The next step in the workflow is the separation and identification of the individual prills (Figure 2, Step 4), for which a series of six steps is necessary. First, the *AN* binary volume is selected and filled by a succession of a *3D_fill* operation, followed by three *2D_fill* operations along the three orthogonal planes of the CT volume. Then, a *Bin_Separate* operation (3D interpretation considering 26 neighbours, marker extent value of 4, repeatable algorithm) is used to separate each individual prill, followed by a *Border_kill* operation to remove prills not fully contained in the 3D image (i.e., prills that touch the borders of the 3D volume). A *Label_analysis* is performed on the resulting volume, so that each individual prill can be characterised independently. The resulting volume is sieved (*Label_sieve*) based on the volume of the prills, so that any fragment present can be removed. Finally, the prill selection obtained is labelled (*Labeling*) to give each prill a unique identifier (Figure 6d). This is required so that the material labels (*AN*, *OpenPorosity*, and *ClosedPorosity*) can be analysed on an individual prill basis and therefore some statistically relevant structural analysis performed, to highlight the differences between the AN material types.

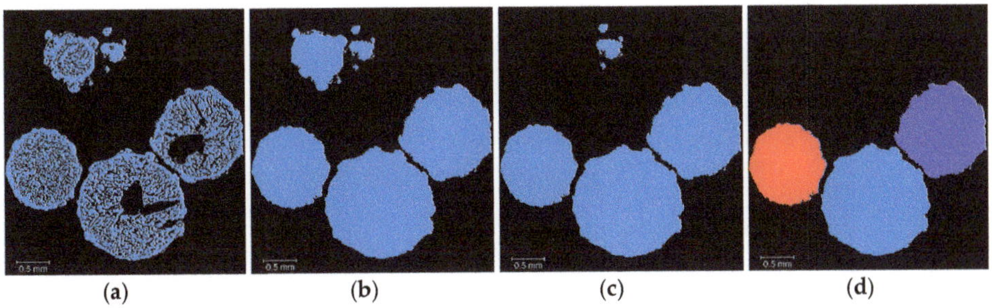

Figure 6. Overview of the prill separation process: (**a**) *AN* binary image; (**b**) *Prill* binary image obtained by filling operation of the AN binary image; (**c**) *Prill* binary image after a bin separation and a border kill operators; (**d**) labelled prill selection (previous *Prill* binary image after further analysis, sieving by volume to remove noise and small prill fragments, and labelling to give a volume where each prill has a unique ID).

The next step in the post-processing stage is the determination of the structural and morphological parameters from the individual materials labels (Figure 2, Step 5). Each of the individual materials binary volumes is successively masked with the labelled prill volume, as shown in Figure 7, and the following parameters are measured for each resulting labelled volume (using a *Label_analysis* operator): volume, surface area, equivalent diameter, shape factor, and Euler characteristic χ. For each parameter, the mean value is given (average of the value for each individual prill), in addition to a confidence interval corresponding to the standard deviation of the parameter.

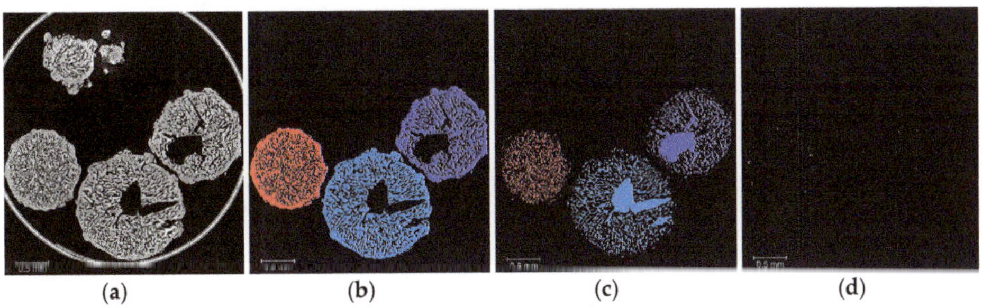

Figure 7. Overview of the determination of structural and morphological parameters: (**a**) filtered data; (**b**) masking of labelled prills with AN binary volume (Figures 5b and 6d); (**c**) masking of labelled prills with *OpenPorosity* binary volume (Figures 5c and 6d); (**d**) masking of labelled prills with *ClosedPorosity* binary volume (Figures 5d and 6d).

The shape factor is defined in Equation (1),

$$Shape\ Factor = \frac{Surface\ Area^3}{(36 \times \pi \times Volume^2)} \quad (1)$$

where as the Euler characteristic χ (also called Euler-Poincaré characteristic or Euler number) [17] is an indicator of the connectivity of a 3D complex structure [18,19]. The Euler characteristic measures what might be called "redundant connectivity", the degree to which parts of an object are multiply connected [20]. It is a measure of how many connections in a structure can be severed before the structure falls into two separate pieces. The Euler-Poincaré formula for a 3D object X is given in Equation (2):

$$\chi(X) = \beta_0 - \beta_1 + \beta_2 \quad (2)$$

where β_0 is the number of objects (the number of connected components), β_1 the connectivity, and β_2 the number of enclosed cavities. The Euler characteristic is used here as an indicator of the complexity of the topology of the AN prill and associated open porosity. The specific surface value was calculated for each prill from the ratio between the surface area of the *AN* (strictly speaking, the surface area of *AN* minus the surface area of *ClosedPorosity* as only open porosity is here of interest, since it provides the surface available for reaction with the oil) and the AN material volume, hence the unit in mm^2/mm^3 (unit of surface area per unit of volume). It is here relevant to mention that it is not possible to convert this into a more common mm^2/g (unit of surface area per unit of mass), as the density of the bulk material is not well known and cannot be accurately defined from the XCT data for each individual prill.

The final post-processing step is dedicated to obtaining the radial distribution of the different phases within the prills (Figure 2 step 6). This is useful as it can help distinguishing between two prill microstructures that could have an overall similar open porosity content, whilst having the pore network radially distributed in a different manner. As shown in Figure 8, a *Distance_transform* operation (Euclid type) is used on the *Prill* selection, then the output volume is masked with the respective *AN*, *OpenPorosity*, and *ClosedPorosity* binary volumes. The distance transform measures the distance of each object point from the nearest boundary [21,22]. The radial distributions (i.e., radial volume fractions) are given by the histograms (voxel count versus distance) of the corresponding masked volumes. In this case however, all the prills present in the volume are processed at the same time, thus only a mean value of the radial distributions averaged over all the prills can be obtained. It is not possible with the actual data processing software to perform the distance transform onto the labelled AN prills: the information contained in the voxels of the 3D image are either the distance to the nearest voxel selected as background (i.e., distance transform from the respective binary volumes) or a label number (i.e., labelled prill volume); these two data fields are mutually exclusive and cannot be combined. As the grain sizes of the different types of AN prills (from F1 through E3) differ slightly, the distances to the outer surface of the prills were normalized for easier comparison between the different materials.

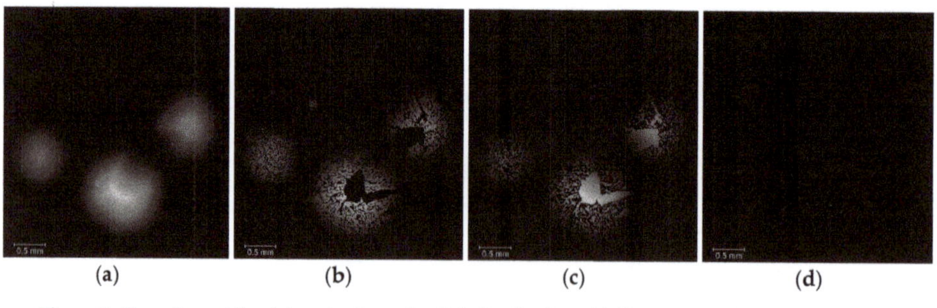

Figure 8. Overview of the determination of radial distribution: (**a**) distance transform of prills; (**b**) masking of prill distance transform with AN binary volume; (**c**) masking of prill distance transform with *OpenPorosity* binary volume; (**d**) masking of prill distance transform with *ClosedPorosity* binary volume. The radial distributions are obtained by plotting the histogram of the respective volumes.

4. Experimental Results and Quantitative Analysis

As shown in Figure 9, representative 2D slices taken in the middle plane of each specimen give a good qualitative overview of the structure of the different types of AN prills. For all type E samples, there is a large cavity close to the centre of the prill, with finer pores located throughout the remainder of the prill. E1 has a pore network with more rounded edges, whilst E2 and E3 have a more dendritic-like pore structure expanding radially. Sample F1 displays a totally different structure, with several large (500 µm to 1 mm) high density inclusions and few individual pores dispersed throughout the prill. Even though the porosity can be seen from a 2D XCT slice, similarly to a SEM image, it is not possible to determine if the visible porosity is open (connected to the exterior and therefore susceptible to interact with oil) or closed (no reaction with oil possible but influence on detonation susceptibility). This is a prime example to showcase the need for 3D visualisation and associated qualitative/quantitative measurements.

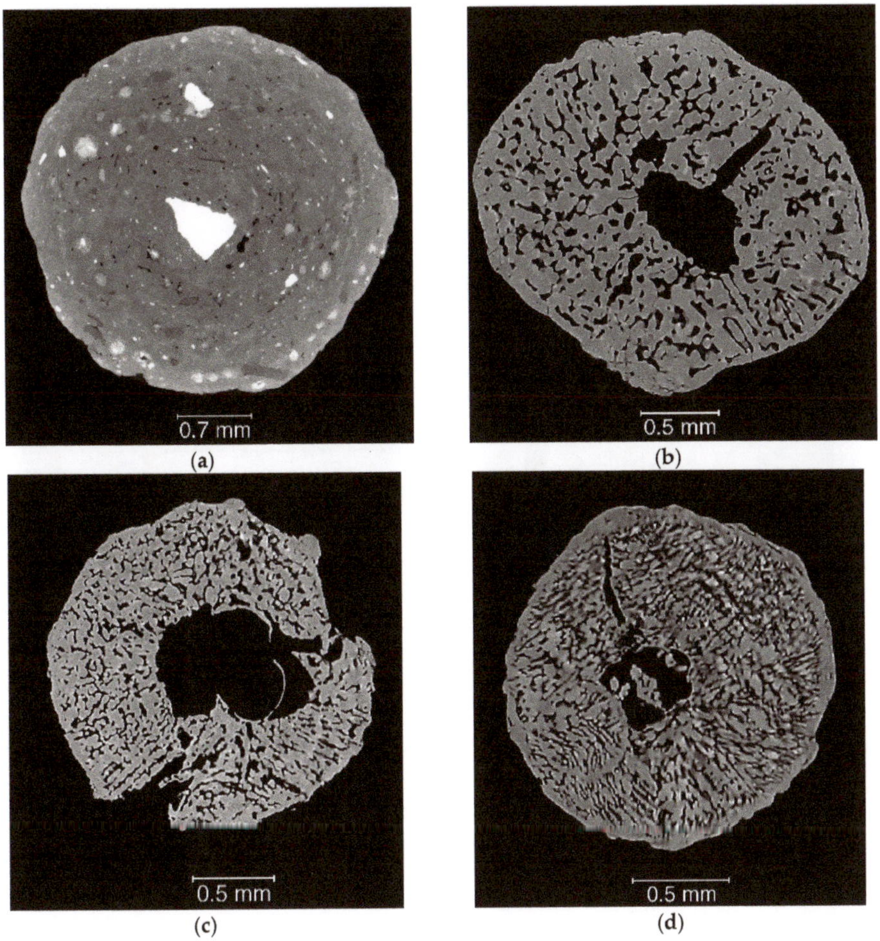

Figure 9. Example of CT slice taken in the middle plane of a single prill for each specimen under investigation: (**a**) specimen F1 (scale bar is different from other samples); (**b**) specimen E1; (**c**) specimen E2; (**d**) specimen E3.

The structural and morphological analysis (Figure 2, Step 6) provides measurements of the volume fraction, surface area, shape factor, Euler characteristic, and specific surface area. An overview of the results is gathered in Figure 10. In terms of volume fraction (Figure 10a), the fertiliser sample (sample F1) is the sample with the highest volume fraction of AN (96.8%) and the lowest volume fraction of open porosity (below 0.02% ± 0.01%). It is the only sample displaying high density inclusions (2.2%) and has the highest volume fraction of closed porosity (1.0%), with a value more than twice higher than that of the other materials. The AN used for explosives (samples E1 through E3) have significantly lower volume fractions of AN, with values decreasing for sample E1 to E3 (see Figure 10a). Conversely, the volume fractions of open porosity are much higher, to allow the retention of the fuel oil: the open porosity increases from E1 to E3. Regarding the closed porosity, the volume fractions for all E samples are well below 1%.

For the morphological analysis (Figure 10b), either the AN or the open porosity can be selected. The analysis of the AN yields more representative results, but it is also interesting to consider the open porosity itself, as it is a true representation of the volume available for the fuel oil to soak into the AN prill. The AN shape factor values increase from F1 to E3 materials (see Figure 10b). The value for F1 is low (1.25), indicating that the prill has a shape close to that of a sphere. For the explosive materials, the values increase to very large values, indicating a dramatic increase in complexity of the AN prill shape. The increase in shape complexity from sample E1 through to E3 is further demonstrated by the open porosity shape factors, which have a similar trend and higher values. Another way to describe the morphology of the AN prills is to consider the Euler characteristic χ. For both the AN and the open porosity, the Euler characteristic values decrease from sample F1 to sample E3, with both values being similar. This evolution is consistent with an increase in complexity of the open porosity network, in good agreement with the open porosity contents and shape factors measured.

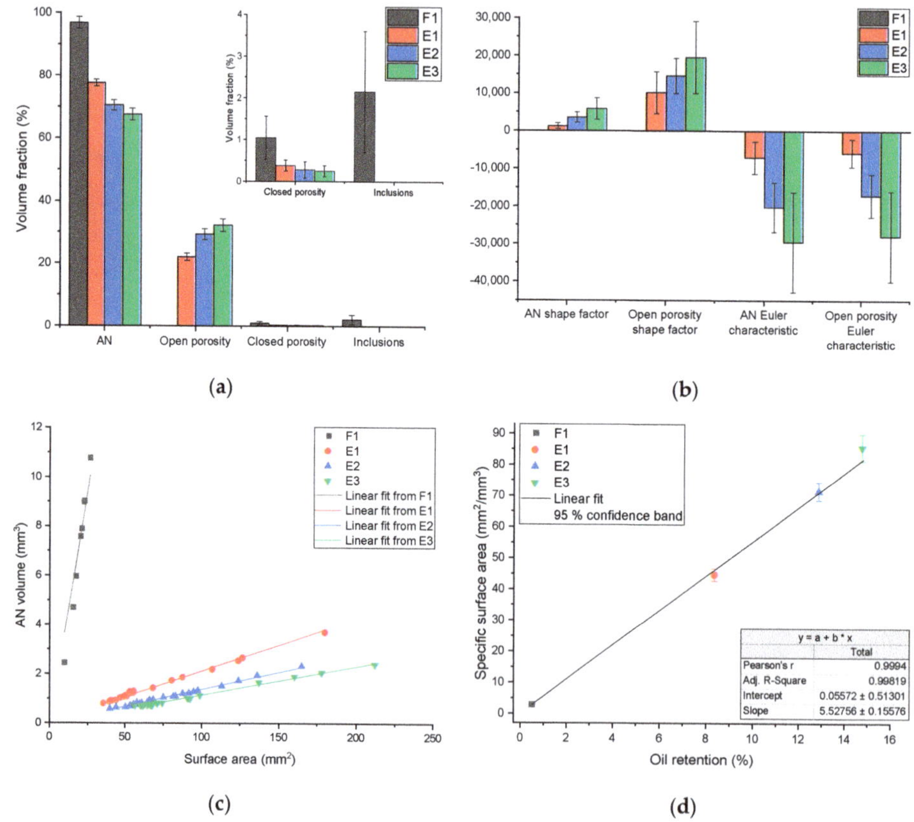

Figure 10. Structural and morphological results: (**a**) volume fraction; (**b**) shape factor and Euler characteristic χ for *AN* and *OpenPorosity* (no *OpenPorosity* parameters calculated for F1 as the open porosity is too small to be significant from a morphological point of view, F1 AN shape factor is 1.25 and F1 Euler characteristic is −5.88); (**c**) correlation between *AN* volume and surface area; (**d**) correlation between specific surface area and oil retention.

Figure 10c shows the correlation between *AN* volume and surface area for each prill of the XCT volume. It is clear that the F1 material has a much greater AN volume than the explosive materials, but with an associated low surface area. The explosive materials have greater surface area (35 mm^2 up to 210 mm^2), but a lower AN volume, due to the lower prill size. As the porosity contents vary somewhat significantly among the explosive materials (10% difference between E1 and E3), whilst the grain size remains comparable, a better metric to compare the materials is the specific surface, defined as the surface area of *AN* per unit of *AN* volume. Figure 10d shows the linear relationship between the specific surface values determined by X-ray computed tomography and the oil retention values determined according to the European regulation [8]. With this result, we demonstrate that XCT could be used to predict the performance of explosives over a very wide range of porosity content.

Based on a distance transform operator, the radial volume fraction of each phase can be determined (Figure 2 step 6), for each prill material under investigation. The plots are gathered in Figure 11. For sample F1, the AN content decreases when moving the grain radially inwards. The closed porosity increases, particularly in the innermost 20 % of the grains, to reach a volume fraction as high as 15%, whilst the average closed porosity content is only 1%. The high-density inclusions are mostly present in the outermost 10 % and innermost 50 %, but are relatively well distributed. All the explosive materials exhibit similar radial distribution profiles for each of their phases. The AN content drops rapidly in the outermost 10 % of the prills, then decreases slowly on the central 10 % to 80 %, and finally drops rapidly over the innermost 20 %. The evolution of the open porosity content is opposite to that of the AN, whilst most of the closed porosity is located in the first 10% of the prills, which corresponds to roughly the outermost 100 μm layer.

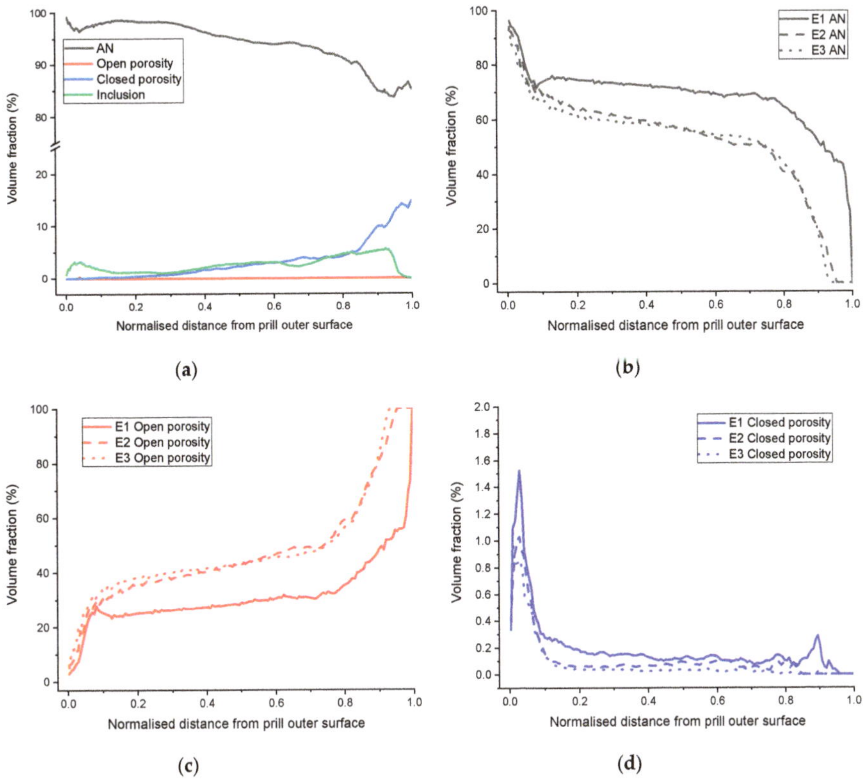

Figure 11. Plots from radial distribution: (**a**) specimen F1; (**b**) AN for type E specimens; (**c**) open porosity for type E specimens; (**d**) Closed porosity for type E specimens.

5. Discussion and Conclusions

As we have seen in the preceding sections and as has been reported in the literature, X-ray computed tomography is a method with great potential and is being increasingly used for non-destructive materials evaluation, in particular for characterising porosity [23–25]. However, one of the limitations to a more wide-spread use of XCT is the relative complexity of the data processing. Whilst simple volume fractions or porosity contents can be easily obtained by standard segmentation (although the surface determination is always a critical step [26]), more complex data processing workflows including pre- and post-processing operations complicate the analysis. Another limitation is the difficulty in assessing the robustness, repeatability, and exactitude of the XCT data processing workflows. Here, we have first demonstrated that the filtering employed (pre-processing) helped increase the separation between the peaks in the grey scale histogram (Figure 3c), hence increasing the separation between the materials in the 3D volume. Then, the surface determination strategy based on a watershed segmentation (seed-based) was proven to be repeatable. In fact, upon changing the threshold values used for defining the watershed seeds, we observed a negligible impact on the measured characteristics of the AN prills (details in Table A2 Appendix B). Finally, the assessment of two different specimens coming from the same AN prill batch showed similar results, with values falling within the standard deviations of the respective parameters determined (details in Table A3 Appendix C), demonstrating the overall robustness and repeatability of the data processing workflow proposed here.

In terms of output, XCT data can provide qualitative information, such as that extracted from the unprocessed 3D reconstructions (Figure 9), where the structure of the prills can be observed. Volume renderings of any of the segmented materials (seen separately, Figure 12) can also provide a good general overview of the complexity of the prill structure in 3D. The qualitative information can be further supported by quantitative information such as volume fraction: fertilizer F1 is the only prill material that has high density inclusions and exhibits very low open porosity (0.02%). On the other hand, the explosive materials have much higher open porosity values, with contents ranging from 20% to 30%. Beyond the structural information, relevant morphological information can be obtained, namely shape factor, Euler characteristic χ and specific surface area. Shape factor values increase and Euler characteristic values decrease with greater open porosity contents, demonstrating that as the open porosity content increases, the whole open porosity network becomes more geometrically complex, i.e., the surface area increases much faster than the volume. This is also quantitatively supported by the specific surface area values: as the oil retention values nearly double between specimen E1 and E3 (8.4% and 14.8%, respectively), there is a doubling of the associated specific surface area values (44.9 mm^2/mm^3 and 85.7 mm^2/mm^3, respectively). The linear relationship (Figure 10d) reported in this work between the oil retention and the specific surface area determined by XCT, demonstrate that XCT can also be used to predict the performance of AN prills, in addition to giving a full structural and morphological characterisation of the prills. Such structural information can be linked to other relevant parameters, such as explosive sensitivity and velocity of detonation. It was shown in [27] that an optimum porosity content exists, and that above and below that content, velocity of detonation and explosive efficiency are reduced. In this study the optimum porosity was determined to be 30%, related to the total porosity which included both open and closed. In fact, the pore size also plays a role in the detonation capability of the prills [28]. However, in the case of open porosity (the most relevant one), the definition of a "pore size" is still controversial in the literature (see e.g., [11,29]), as it depends on the "pore form" attached to the pore channels, characterizing an open porosity structure. In fact, a typical approach to evaluate "pore sizes" is through the skeletonization of the open porosity structure [30] and then through the calculation of the volume of the spheres with centers in the skeleton and fully filling the channels.

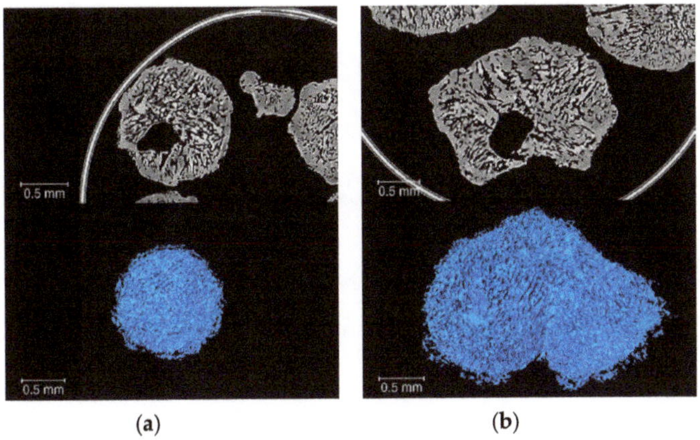

Figure 12. 3D renderings: (**a**) open porosity of one prill from specimen E1, with a specific surface area of 139 mm^2/mm^3, a shape factor of 6559 and a Euler characteristic χ of −7199; (**b**) open porosity of one prill from specimen E1, with a specific surface area of 139 mm^2/mm^3, a shape factor of 25,240 and a Euler characteristic χ of −25,984.

Another important parameter that XCT can deliver is the radial distribution of the different phases. To the best of the authors' knowledge, this is the first time these types of results are presented. For the explosive materials, it appears that the profiles from E1 through E3 are similar: first a sharp increase in open porosity over the outer 10% and inner 15%–20% of the prill radius, and a much slower increase in-between. In addition, the open porosity content increase between E1 through E3 can be seen as being homogeneously distributed through the radius of the prills, as the open porosity line profiles are globally shifted towards higher volume fractions (between E1, E2 and E3). Finally, another important result for the explosive materials community is the fact that most of the closed porosity in the explosive materials is concentrated within the outermost 100 µm material layer (outer 10% of the prills' radius. This result could have significant implications, in particular for the sensitivity to explosion and explosion mechanism (hot spot mechanism [9]).

Overall, the results presented here demonstrate that XCT can be successfully applied to the thorough structural and morphological characterisation of AN prills in a non-destructive manner. As XCT can be used to scan virtually any material, it is important to mention that the data processing workflow developed here can be applicable to a broad range of small porous parts and granular porous materials. An advanced data processing workflow was developed, so that both structural and morphological prill parameters could be extracted for each individual prill, whilst performing the segmentation only once for the entire scanned volume. Future work will focus on further investigating qualitative and quantitative morphological data (such as curvature of open porosity (both local and global), local pore/throat size/diameter, and radial evolution of specific surface area), possibly comparing the XCT results to those of conventional techniques, such as BET and mercury porosimetry. This will provide the most relevant metrics for the explosives community to better understand all aspects of the detonation process.

Supplementary Materials: The following are available online, Video S1: Overview of XCT data processing workflow for ammonium nitrate prills quantitative analysis (uploaded to Zenodo, DOI 10.5281/zenodo.3611339).

Author Contributions: F.L. conceived and designed the experiments; H.K. provided the samples; F.L. and Z.Z. performed the experiments and the data processing; F.L. developed the XCT data processing workflow; F.L. did the original draft preparation; F.L., G.B. and H.K. contributed to subsequent draft reviews and editing. All authors have read and agreed to the published version of the manuscript.

Funding: This research received no external funding.

Acknowledgments: The authors would like to acknowledge Marcus Malow from BAM, Berlin for the fruitful discussions on ammonium nitrate samples and his inputs towards these samples.

Conflicts of Interest: The authors declare no conflict of interest.

Appendix A

Table A1. Overview of the Amira function used in the developed workflow.

Amira Function Name	Operation	Comment
3D_fill	Closes holes fully comprised within a selection in 3D	-
2D_fill	Closes holes fully comprised within a selection in 2D	For high packing density, over filling can occur.
Bin_Separate	Separates individual elements that can be touching each other	Based on a watershed and distance transform
Border_kill	Removes the selected elements that are touching a border of the 3D volume	Required so that prills not fully contained within the analysis volume are removed and thus do not skew the values of the measured parameters.
Label_analysis	Performs an analysis of the labels to extracts user defined parameters	Parameter of interest are volume, surface area, equivalent diameter, shape factor, and Euler characteristic
Label_sieve	Sieves the individual labeled prills based on a discriminating parameter (in this case volume)	Required to remove very small prill fragments
Labeling	Gives a unique label to each individual of the binary volume	Required to get a mean value and associated standard deviation
Distance_transform	Calculates the distance to the closest boundary from each selected voxel	Required to obtain the radial distribution of prill phases

Appendix B

An assessment of the sensitivity of the metrics characterised in the present study to changes in threshold values has been performed for sample E1. The exact same data processing workflow was followed apart from the initial label seed definition. For the first test, a "normal" threshold range was selected for each label seed. For the second test, much smaller threshold ranges were selected for each of the label seeds, resulting in more of the voxels remaining to be assigned to a specific label by the watershed algorithm. The results from the two tests and their comparison are presented in Table A2. For the volume fraction values, the relative deviations observed are very small, below 1% of the measured values for AN and open porosity. For the morphological parameters, the relative deviations observed are greater, with values around 10%. However, as the standard deviations of the morphological parameters are generally high (~50%), relative deviations of only 10% of the measured values are deemed satisfactory. These results clearly demonstrate that the watershed segmentation of the filtered XCT data from AN prills is repeatable and that the associated data processing workflow developed here is robust.

Table A2. Results of the structural and morphological analysis of sample E1 based on two different segmentation strategies.

Parameter	Segmentation 1	Segmentation 2	Relative Deviation [1]
Volume fraction AN (%)	77.6 ± 1.1	77.8 ± 1.2	−0.2%
Volume fraction open porosity (%)	22.0 ± 1.2	21.9 ± 1.2	0.8%
Volume fraction closed porosity (%)	0.4 ± 0.1	0.4 ± 0.1	−2.6%
Specific surface area (mm^2/mm^3)	44.9 ± 1.9	44.8 ± 2.2	0.2%
AN Euler characteristic (-)	−4589 ± 3327	−4134 ± 3321	9.9%
AN shape factor (-)	1364 ± 811	1241 ± 821	9.0%
Open porosity Euler number (-)	−5807 ± 3688	−5258 ± 3720	9.5%
Open porosity shape factor (-)	10,138 ± 5535	9219 ± 5685	9.1%

[1] Deviation between Test 1 and test 2, with test 1 taken as reference.

Appendix C

Three different samples coming from the prill batch E3 were scanned to assess the deviation that can be expected when scanning different samples coming from the same powder. The exact same data processing workflow was followed. Results are shown exemplarily for two scans for the sake of brevity. They all agree within a reasonable interval, which can also be ascribed to genuine materials variability.

Table A3. Results of the structural and morphological analysis of sample E3 based on two different samples coming from the same prill batch.

Parameter	Scan 1	Scan 2	Relative Deviation [1]
Volume fraction AN (%)	67.5 ± 1.9	68.3 ± 2.6	−1.1%
Volume fraction open porosity (%)	32.2 ± 2.0	31.4 ± 2.8	2.5%
Volume fraction closed porosity (%)	0.3 ± 0.1	0.3 ± 0.2	−19.0%
Specific surface area (mm^2/mm^3)	85.7 ± 4.2	85.8 ± 3.8	−0.1%
AN Euler characteristic (-)	−29,502 ± 13,257	−37,519 ± 19,229	−27.2%
AN shape factor (-)	5813 ± 2855	6639 ± 3591	−14.2%
Open porosity Euler number (-)	−27,962 ± 12,018	−33,545 ± 15,630	−20.0%
Open porosity shape factor (-)	19,515 ± 9602	23,687 ± 11,076	−21.4%

[1] Deviation between Test 1 and test 2, with test 1 taken as reference.

References

1. Salyer, T.R.; Short, M.; Kiyanda, C.B.; Morris, J.S.; Zimmerly, T. *Effect of Prill Structure on Detonation Performance of ANFO*; United States Department of Energy: Washington, DC, USA, 2010.
2. Knoema website. Fertilizers by products. Available online: https://knoema.de/FAORFBFP/faostat-fertilizers-by-product. (accessed on 1 September 2019).
3. Grand View Research, Inc. *Ammonium Nitrate Market Analysis by Application (Fertilizers, Explosives), By Region (North America, Europe, Asia Pacific, CSA, MEA), Competitive Landscape, and Segment Forecasts, 2014–2025*; Published by Grand View Research, Inc.: San Francisco, CA, USA, 2017.
4. Chattopadhyay, A.K.; Knight, J.; Kapadia, R.D.; Sarkar, S.K. 3-D NMR microscopy study of the structural characterization of polycrystalline materials. *J. Mater. Sci. Lett.* **1994**, *13*, 983–984. [CrossRef]
5. Oommen, C.; Jain, S.R. Ammonium nitrate: A promising rocket propellant oxidizer. *J. Hazard. Mater.* **1999**, *67*, 253–281. [CrossRef]
6. Hurley, C. *Development of Ammonium Nitrate Based Explosives to Optimize Explosive Properties and Explosive Welding Parameters Used during Explosion Cladding*; Colorado School of Mines: Golden State, CO, USA, 2013.
7. Zygmunt, B.; Buczkowski, D. Influence of Ammonium Nitrate Prills' Properties on Detonation Velocity of ANFO. *PropellantsExplos. Pyrotech.* **2007**, *32*, 411–414. [CrossRef]
8. European Parliament and Council of the European Union. *Regulation (EC) No 2003/2003 Relating to Fertilisers—Technical Provisions for Ammonium nitrate Fertilisers of High Nitrogen Content, in Determination of Oil Retention*; European Parliament and Council of the European Union: Brussels, Belgium, 2003; p. 80.

9. Johansson, C.H.; Persson, P.A. *Detonics of High Explosives*; Academic Press: Cambridge, MA, USA, 1970.
10. Müller, B.R.; Cooper, R.C.; Lange, A.; Kupsch, A.; Wheeler, M.; Hentschel, M.P.; Staude, A.; Pandey, A.; Shyam, A.; Bruno, G. Stress-induced microcrack density evolution in β-eucryptite ceramics: Experimental observations and possible route to strain hardening. *Acta Mater.* **2018**, *144*, 627–641. [CrossRef]
11. Onel, Y.; Lange, A.; Staude, A.; Ehrig, K.; Kupsch, A.; Hentschel, M.P.; Wolk, T.; Müller, B.R.; Bruno, G. Evaluating Porosity in Cordierite Diesel Particulate Filter Materials, Part 2 Statistical Analysis of Computed Tomography Data. *J. Ceram. Sci. Technol.* **2014**, *5*, 13–22.
12. Powierza, B.; Stelzner, L.; Oesch, T.; Gollwitzer, C.; Weise, F.; Bruno, G. Water Migration in One-Side Heated Concrete: 4D In-Situ CT Monitoring of the Moisture-Clog-Effect. *J. Nondestruct. Eval.* **2018**, *38*, 15. [CrossRef]
13. General Electrics Website. Phoenix v|tome|x L 300 CT System. Available online: https://www.industrial.ai/inspection-technologies/radiography-ct/x-ray-computed-tomography/phoenix-vtomex-l-300 (accessed on 3 September 2019).
14. Volume Graphics Website. VGStudio Max Software. Available online: https://www.volumegraphics.com/en/products/vgstudio-max.html (accessed on 3 September 2019).
15. ZIB. Amira Software. Available online: https://www.zib.de/software/amira (accessed on 3 September 2019).
16. Preim, B.; Botha, C. Chapter 4—Image Analysis for Medical Visualization. In *Visual Computing for Medicine*, 2nd ed.; Preim, B., Botha, C., Eds.; Morgan Kaufmann: Boston, MA, USA, 2014; pp. 111–175.
17. Marjanovic, M.M. Euler-Poincaré characteristic—A case of topological self-convincing. *Teach. Math.* **2014**, *XVII*, 21–33.
18. Okuma, G.; Kadowaki, D.; Hondo, T.; Tanaka, S.; Wakai, F. Interface topology for distinguishing stages of sintering. *Sci. Rep.* **2017**, *7*, 11106. [CrossRef]
19. Schmitt, M.; Halisch, M.; Müller, C.; Fernandes, C.P. Classification and quantification of pore shapes in sandstone reservoir rocks with 3-D X-ray micro-computed tomography. *Solid Earth* **2016**, *7*, 285–300. [CrossRef]
20. Odgaard, A.; Gundersen, H.J.G. Quantification of connectivity in cancellous bone, with special emphasis on 3-D reconstructions. *Bone* **1993**, *14*, 173–182. [CrossRef]
21. Danielsson, P.-E. Euclidean distance mapping. *Comput. Graph. Image Process.* **1980**, *14*, 227–248. [CrossRef]
22. Rosenfeld, A.; Pfaltz, J.L. Distance functions on digital pictures. *Pattern Recognit.* **1968**, *1*, 33–61. [CrossRef]
23. Augusto, K.S.; Paciornik, S. Porosity Characterization of Iron Ore Pellets by X-Ray Microtomography. *Mater. Res.* **2018**, *21*, e20170621. [CrossRef]
24. Maire, E. X-Ray Tomography Applied to the Characterization of Highly Porous Materials. *Annu. Rev. Mater. Res.* **2012**, *42*, 163–178. [CrossRef]
25. du Plessis, A.; Yadroitsev, I.; Yadroitsava, I.; Le Roux, S.G. X-Ray Microcomputed Tomography in Additive Manufacturing: A Review of the Current Technology and Applications. *3D Print. Addit. Manuf.* **2018**, *5*, 227–247. [CrossRef]
26. Moroni, G.; Petrò, S. Impact of the Threshold on the Performance Verification of Computerized Tomography Scanners. *Procedia Cirp* **2016**, *43*, 345–350. [CrossRef]
27. Krebs, H. *Sprengtechnische und Sprengtechnologiesche Untersuchungen zum Mischladen von Losen Sprengstoffen*; TU Bergakademie Freiberg: Freiberg, Germany, 1986.
28. Parker, G.R.; Rae, P.J. *Mechanical and Thermal Damage in Non-Shock Initiation of Explosives*; Asay, B.W., Ed.; Springer: Berlin, Germany, 2010; pp. 294–308.
29. Kupsch, A.; Lange, A.D.P.; Hentschel, M.P.; Onel, Y.; Wolk, T.; Staude, A.; Ehrig, K.; Müller, B.R.; Bruno, G. Evaluating porosity in cordierite diesel particulate filter materials, part 1 X-ray refraction. *J. Ceram. Sci. Tech.* **2013**, *4*, 169–176.
30. Pothuaud, L.; Porion, P.; Lespessailles, E.; Benhamou, C.L.; Levitz, P. A new method for three-dimensional skeleton graph analysis of porous media: Application to trabecular bone microarchitecture. *J. Microsc.* **2000**, *199 Pt 2*, 149–161. [CrossRef]

materials

MDPI

Article

A Novel Camera-Based Measurement System for Roughness Determination of Concrete Surfaces

Barış Özcan *, Raimund Schwermann and Jörg Blankenbach

Geodetic Institute and Chair for Computing in Civil Engineering & Geo Information Systems, RWTH Aachen University, Mies-van-der-Rohe-Str. 1, 52074 Aachen, Germany; schwermann@gia.rwth-aachen.de (R.S.); blankenbach@gia.rwth-aachen.de (J.B.)

* Correspondence: oezcan@gia.rwth-aachen.de

Abstract: Determining the roughness of technical surfaces is an important task in many engineering disciplines. In civil engineering, for instance, the repair and reinforcement of building component parts (such as concrete structures) requires a certain surface roughness in order to ensure the bond between a coating material and base concrete. The sand patch method is so far the state-of-the-art for the roughness measurement of concrete structures. Although the method is easy to perform, it suffers from considerable drawbacks. Consequently, more sophisticated measurement systems are required. In a research project, we developed a novel camera-based alternative, which comes with several advantages. The measurement system consists of a mechanical cross slide that guides an industrial camera over a surface to be measured. Images taken by the camera are used for 3D reconstruction. Finally, the reconstructed point clouds are used to estimate roughness. In this article, we present our measurement system (including the hardware and the self-developed software for 3D reconstruction). We further provide experiments to camera calibration and evaluation of our system on concrete specimens. The resulting roughness estimates for the concrete specimens show a strong linear correlation to reference values obtained by the sand patch method.

Keywords: concrete surface roughness; 3D reconstruction; digital photogrammetry; non-destructive testing

Citation: Özcan, B.; Schwermann, R.; Blankenbach, J. A Novel Camera-Based Measurement System for Roughness Determination of Concrete Surfaces. *Materials* **2021**, *14*, 158. https://doi.org/10.3390/ma14010158

Received: 20 October 2020
Accepted: 28 December 2020
Published: 31 December 2020

1. Introduction

The roughness of building materials plays an important role in the field of civil engineering. For example, for the renovation of concrete components, a certain roughness is needed to ensure bonding between the coating material and base concrete [1–4]. Besides specific adhesion—which involves the physical, chemical, and thermodynamic interactions of surfaces—mechanical adhesion is also crucial for a good adhesive bond [5]. Mechanical adhesion supposes that the applied liquid coating material flows into the holes and gaps of the base concrete, hardens, and anchors like dowels or snap fasteners. Moreover, a higher roughness of the base concrete causes a larger composite surface, which leads to a stronger bond as well. Another benefit of determining roughness is the facilitation of estimating the required amount of coating material. The rougher the surface of a concrete component, the more coating material is required to cover the entire surface.

The most widely used and standardized method for the determination of concrete surface roughness is the sand patch method introduced by Kaufmann [6]. Here, the user applies a pre-defined amount of sand with a standardized grain size onto the surface and distributes it evenly in circular movements. The diameter of the resulting sand patch relates to surface roughness. Despite its simplicity, this method has its shortcomings. For instance, the manual execution requires some experience on the part of the user. Studies have shown that the measured roughness can vary by around 20% depending on the user [7]. In addition, this method does not provide reproducible results. Specifically, the repeated execution of the sand patch method on the same area results in varying diameters of sand patches thus leading to a range of different estimates for roughness. Furthermore, it requires direct contact with the surface to be measured. However, direct contact can lead to wear and modify the original structure of the surface. This in turn can lead to falsified results. Another crucial limitation is the lack of applicability on highly slanted surfaces or ceilings. For these reasons, it would be helpful for many applications to develop more sophisticated and contactless measurement systems for the topographic analysis of technical surfaces.

For application in the construction industry, the system must be deployable on construction sites. Thus, the measurement device has to fulfill a range of requirements. In practical use, building elements usually have to be measured that are already installed and in service. Thus, it is not feasible to detach the elements in order to measure the surface roughness at a stationary system (e.g., in a lab). Consequently, the mobility of the measurement device is crucial. Furthermore, the measurement of strongly slanted surfaces or ceilings requires the device to be hold on the surface by the operator during the measurement for a certain period. Hence, the device also has to be as lightweight as possible. In addition, the simplicity of the commonly used sand patch method has made it very popular and widely used in the construction industry. Accordingly, for progressive measurement methods it is a huge challenge to become established in practice. In order to achieve acceptance, the system must be usable without major technical instructions such as the sand patch method. Finally, to make the measurement device as attractive as possible for users, the costs need to be kept low.

Although there are already many systems and recent developments regarding area-based surface reconstruction (see Section 2), they often do not satisfy the aforementioned requirements. Most of these systems are based on optical methods (such as white-light interferometry, confocal 3D laser scanning microscopy, or focus variation microscopy) and rely on a complex and costly measurement setup or equipment. Further, they are not portable and hence not applicable on construction sites. Our objective is to tackle this gap.

In this paper, we introduce a novel camera-based measurement system for the roughness determination of technical materials, such as concrete elements. The hardware of the system primarily consists of a cross slide with a controlling unit and propulsion, which guides a monocular industrial camera over the surface to be measured. Images taken by the camera are used for digital 3D reconstruction. The self-developed software for 3D reconstruction mainly involves a two-step image matching algorithm: Structure from Motion (SfM) and Dense Image Matching (DIM). Finally, the reconstructed dense point clouds are used for the estimation of roughness.

Our measurement system is designed to satisfy the aforementioned requirements and provides additional advantages compared to other methods. The novelty of our approach can be related to the combination of the following features, which is unique in this form:

- Fully digital measurement system and reproducibility of results.
- Contactless and area-based measurement.
- Deployable on construction sites and high mobility.
- Applicability on arbitrary oriented surfaces.
- Easy to use.
- Lightweight.
- Low-cost.

Following this introduction, this paper is organized as follows: Section 2 gives an overview of recent developments regarding methods for measurements on concrete surfaces. In Section 3, we provide some necessary fundamentals, in particular a brief definition of roughness and the basics of digital photogrammetry. Subsequently, in Section 4, the developed measurement system is introduced. This includes the concept for capturing the images of object surfaces, the hardware used to build the prototype, and the custom-built 3D calibration test-field. In Section 5, the methodology for 3D reconstruction and roughness estimation is provided. Following this, Section 6 covers our investigations into camera calibration and trials at estimating the roughness of 18 concrete specimens. Section 7 presents and discusses our results. Finally, in Section 8, we present our conclusions with an additional insight into future work.

2. State of the Art and Related Work

A simple modification of the sand patch method was introduced in [8], which enables the measurement procedure to be performed on arbitrarily oriented surfaces. Instead of pure sand, they propose a paste consisting of two parts of sand and one-part medical ultrasound gel. The measurement procedure is carried out in a similar way to the traditional sand patch method. Even though it is applicable to arbitrarily oriented surfaces due to the stickiness of the paste, it still suffers from the aforementioned disadvantages. Furthermore, after the measurement, additional effort must be expended to remove the paste from the surface.

Besides the frequently used sand patch method, there is also the Digital Surface Roughness Meter (DSRM) [4,9] widely used in practice for roughness determination. These devices are either laser-based or based on mechanical stylus and capture the surface in both cases in a line-based manner. Although they are usually small, handy, and low-cost, they also hold some disadvantages. The stylus-based ones measure by a stylus tip tracing the surface, which can lead to wear of the stylus or the object surface and furthermore is limited by the radius of stylus tip. Even though the laser-based ones do not suffer from this disadvantage, they are still limited to line-based assessment.

The ASTM E 2157 Circular Track Meter [10] represents a more sophisticated device compared to DSRM and covers the surface in a circular way. It is laser-based as well and thus also provides non-contact measurements. However, it also suffers from some shortcomings. The surface is covered only by a circular profile and hence it provides (like DSRM) only profile-based assessments. Moreover, even though it is portable and relatively small with a size of 40 cm × 40 cm × 27 cm, it weighs about 13 kg, and thus does not provide the applicability on highly slanted surfaces or ceilings (unless it is held by an extraordinarily strong athlete). In addition, the system is like most laser-based systems comparatively expensive.

In terms of building survey, Terrestrial Laser Scanners (TLS) and Mobile Laser Scanners (MLS) are gaining importance in the construction industry. However, although TLS and MLS are area-based and contactless measurement systems they typically provide geometric accuracies in the range of a few millimeters only [11–13] and are therefore not suitable for roughness determination.

Recently, laser-based triangulation methods have been added to international ISO standards as an alternative to the sand patch method [14]. The basic principle of laser triangulation methods is that either a laser point or laser line is projected by a laser diode onto the surface of an object. That point or line is detected by a position sensitive detector (PSD), which is placed at a fixed distance and angle to the laser diode. In this way, a change in the distance between the laser diode and object surface results in a change of the signal position on the PSD. Finally, the depth of the object can be determined by trigonometry.

Three laser-based triangulation systems for determining the roughness of concrete surfaces were introduced by Schulz [15–18]. Two of the custom-built systems, Profilometer and ELAtextur, use a point-based laser sensor, which is either mounted on a linear actuator and moved linearly over the surface (Profilometer) or rotated on a vertical axis (ELAtextur). The third measurement system is a laser sensor projecting a line instead of a single point. The length of the line is 100 mm and consists of 1280 single points. The results of the laser-based triangulation systems are strongly correlated with results obtained using the sand patch method.

Werner et al. [19] compared two off-the-shelf laser-based triangulation systems in terms of determining the surface parameters of concrete fractures. One is the cost-efficient system DAVID 3D and the other one is the high-end system LEICA T-Scan. They divide the surface of fractured concrete into different scales—micro, meso, and macro level. In the study, it is found that the DAVID 3D system is suitable for measuring at the meso level, which is specified as consisting of features measuring between 1 and 100 mm across. However, for smaller features they recommend the LEICA system. For the scale-independent determination of surface parameters they consider the fractal dimension.

Laser-based triangulation techniques are more sophisticated and permit contactless, user-independent, and reproducible measurements, unlike the sand patch method, but they provide, in general, only a point-based or line-based measurement. In addition, professional laser-based systems are comparatively expensive.

Image-based measurement methods such as digital photogrammetry, on the other hand, represent a competitive alternative to laser-based methods since only a camera is required. While close-range photogrammetry is generally used for 3D measurements of objects ranging in size from several millimeter to several meter, this method is suitable for measurements in micrometer range as well. In traditional digital photogrammetry, before the measurement process, objects to be measured usually have to be prepared by targets, which are later used to measure these in the images. This is often done manually or semi-automated, as, for example, in the software PHIDIAS [20].

For instance, digital photogrammetry was used to monitor cracks in concrete elements [21,22]. In the experiments, the surfaces of the concrete elements were covered by a grid of targets. When continuously increasing stress is applied to the concrete elements, the resulting cracks move the targets. Through these movements, the origin and the evolution of the cracks were observed. However, target-based measurements necessitate some preparation effort. In addition, just a sparse set of points can be measured.

However, in recent decades, the processing power of computers has increased immensely, allowing the development of powerful algorithms for fully automatic feature point detection. By using these feature detectors [23–26], there is no need for targets to be installed. Targets can be used in this case for georeferencing purposes or if measurements with particularly high accuracy are required. Moreover, if a monocular camera is used, units used will be dimensionless. In this case, targets can also be used to determine scale. Furthermore, algorithms for dense image matching allow measurements for every pixel thus achieving dense surface measurements.

A study that compares laser triangulation, photogrammetry, and the sand patch method with each other has been published by Wienecke et al. [27]. These three methods result in three different roughness coefficients. Although a comparison of the measured values indicates some correlation in the results of these three methods, there are still deviations evident. As possible reasons for these deviations, the authors mention, among other reasons, the lack of reproducibility of the sand patch method and the difficulty associated with measuring certain types of surface due to relatively large grain size used in the sand patch method.

In the literature, there are many other examples concerning optical profilometry for the characterization of surface texture. These methods are often based on white-light interferometry, confocal laser scanning microscopy, or focus variation microscopy.

For instance, fringe-based laser interferometry was used in [28] to investigate the concrete substrate roughness in patch repairs. To be precise, the authors examined two different methods for removing defective concrete, i.e., electric chipping hammers and Remote Robotic Hydro-erosion (RRH). A total of 60 slab specimens were analyzed and, for each of them, four different roughness parameters were calculated based on profile lines. The results reveal that RRHs are able to create rougher surfaces than chipping hammers. The main advantage of the proposed system is that it captures the surface in an area-based way and creates a detailed and accurate 3D topography of the surface.

Confocal microscopy was applied in [29] to characterize the fracture surface of six specimens of Portland cement pastes and mortar. The investigated specimens contain two types of sand, fine sand with an average diameter of 0.15 mm and coarse-grained sand with an average diameter of 0.75 mm. They estimated the surfaces of the specimens with a depth resolution of 10 μm and a total range for the depth of about 200 μm for the paste and 800 μm for the mortar specimens. The roughness value is finally determined by the ratio of the actual surface area and the nominal surface area. The results reveal that the actual surface area of cement pastes are about 1.8 times greater than the nominal surface area, while for mortars it is between 2.4–2.8 times the nominal surface. Further, it is found that the roughness is not simply a function of the largest aggregate size.

A range of other non-destructive methods (e.g., impulse response (IR) and impact-echo (IE)) are presented and applied for testing surface morphology of concrete substrates [30]. The equipment used by both IR and IE methods are portable due to their low weights of around 1 kg. However, the main drawbacks are the contact- and line-based assessment of the surface. Furthermore, the results are sensitive to mechanical noise created by equipment impacting.

In our prior research, we have conducted feasibility studies for the determination of the surface roughness of concrete using photogrammetry [31]. In the measurement setup, a digital single-lens reflex camera (DSLR) with a macro-objective lens was used in order to capture the surface of a particular concrete specimen. The reconstructed point cloud is visualized as a height map, showing the peaks and valleys of the concrete surface.

Building on our previous research, we propose that photogrammetry can also be used for analyzing the surface of building materials such as concrete elements, since it allows for a large-scale investigation of object surfaces with high precision. In particular, image-based methods provide an area-based measurement of the surface in contrast to some of the aforementioned methods. Additional benefits are the contactless and therefore non-destructive testing and the repeatability of results. Furthermore, a camera-based approach has the advantage of being portable and can be used on arbitrarily oriented surfaces.

3. Theoretical Background

In the following sections of this chapter, we provide some fundamentals that contribute to the understanding of the succeeding content of the paper. First, a brief definition of roughness is introduced by classifying it into orders of shape deviations and providing common roughness parameters. Subsequently, basics of digital photogrammetry are covered.

3.1. Defining Roughness

3.1.1. Shape Deviations

According to the German standard DIN 4760 [32], the deviation between the actual surface of an object, which is captured by a measurement instrument, and the geometric ideal surface is defined as shape deviation. Shape deviations are further classified into a total of six orders, as shown in Table 1.

While the first two order types specify the form deviation and waviness of surfaces, types 3–5 describe the properties of a surface that are relevant for our purposes, specifically the roughness. However, the actual surface of a technical material consists of all orders of shape deviations. Figure 1 shows an example of a surface profile composed of multiple types of shape deviation orders.

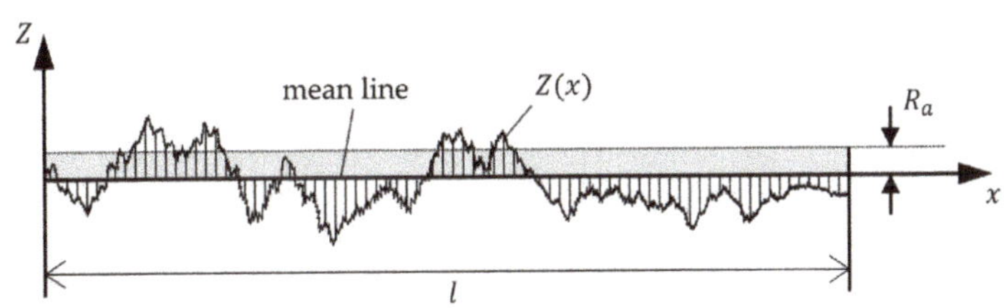

Figure 1. Profile line of a technical surface which is composed of multiple orders of shape deviations.

Table 1. Shape deviations of technical surfaces (according to the German standard DIN 4760).

	Shape Deviations	
1. Order Form deviation	Curvature, Unevenness	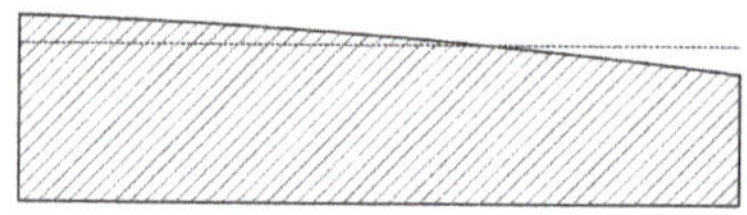
2. Order Waviness	Waves	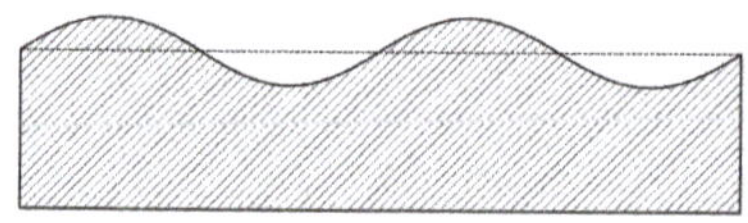
3. Order Roughness	Grooves	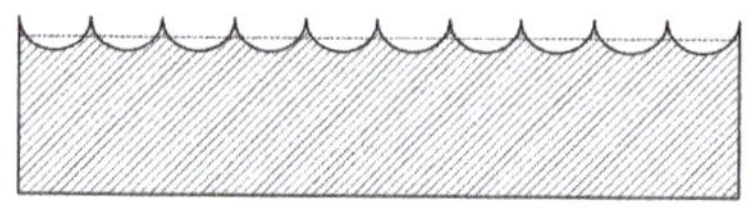
4. Order Roughness	Ridges, Scales, Crests	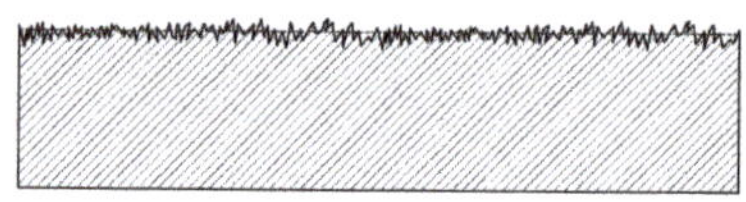
5. Order Roughness	Microstructure of the material	not easily presentable in image form
6. Order	Lattice structure of the material	not easily presentable in image form

3.1.2. Parameters

A range of parameters are available for the description of the roughness of technical surfaces and are specified in the standard DIN EN ISO 4287 [33]. Typically, most of these parameters are designed for single profile lines. In the following, a small selection of the most widely used parameters will be introduced and briefly discussed.

Arithmetical Mean Deviation of the Assessed Profile (R_a)

The arithmetical mean deviation R_a of an assessed profile is calculated by integration of the absolute values of the profile deviations along a reference line and dividing the sum by the length of the line. Thus, the parameter is calculated as follows:

$$R_a = \frac{1}{l}\int_0^l |Z(x)|dx. \tag{1}$$

Hence, R_a corresponds to the average distance of the profile line with respect to the mean line, as illustrated in Figure 1 using an exemplary profile line.

However, there are some drawbacks associated with using R_a to describe surface roughness with regard to adhesive properties. To demonstrate the problem, two profile lines with plainly different surface textures are shown in Figure 2. For simplicity, both lines are sketched as rectangular functions with the same height of rectangles. While the left profile line consists of less rectangles with a larger width, the right one is composed of more narrow rectangles. Due to the higher area of contact, basically the length of the profile line, the surface represented by the right line would lead to a greater adhesive bond. However, the arithmetical mean deviation is the same for both these cases.

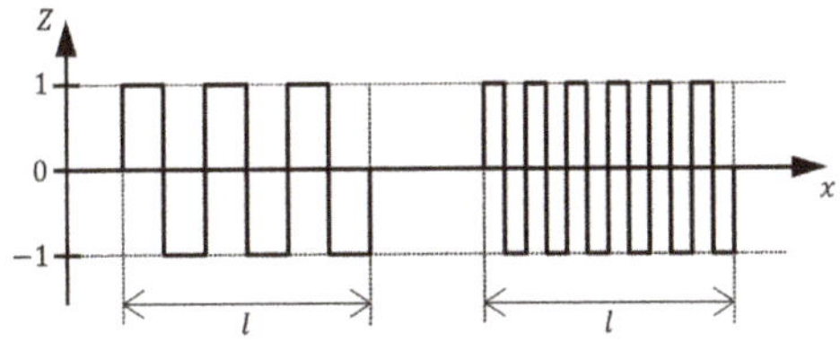

Figure 2. Two lines with different surface characteristics but same result for R_a.

Accordingly, arithmetical mean deviation is an inappropriate parameter for determining adhesive behavior but is suitable for the estimation of the necessary amount of coating material to completely cover the surface.

Mean Texture Depth (MTD)

Principally, the occurring sand patch after performing the sand patch method can be regarded as an approximation for a cylinder with an irregular depth (or height respectively). With the knowledge of the volume V of the applied amount of sand and the diameter d of the sand patch, the mean height h—in terms of the sand patch method also referred as mean texture depth (MTD)—can be calculated using the rearranged formula for the volume of a cylinder. The mean texture depth can thus be determined in the following way:

$$MTD = \frac{4 \cdot V}{\pi \cdot d^2}. \quad (2)$$

To simplify, MTD corresponds to the average distance of the object surface to a plane deliminated by the highest peaks. An illustration of mean texture depth is shown in Figure 3.

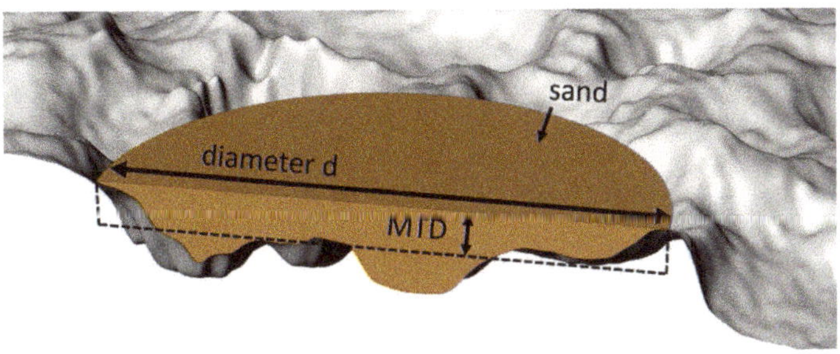

Figure 3. Illustration of the mean texture depth (MTD).

3.2. Digital Photogrammetry

3.2.1. Camera Model

The principle of photogrammetric imaging is based on a simplified model: The pinhole camera. When an object point is recorded by a camera, the optical ray from the object point runs straight through the optical center of the camera and is projected onto the image sensor as an image point. Thus, the optical center of a camera is the mathematical point in space through which all the optical rays of the captured object points pass. Accordingly, this imaging procedure is also known as central projection and can be described mathematically with the collinearity equations (see [34,35]).

The pinhole camera model, however, is just an idealized mathematical model of photogrammetric imaging. The physical model also plays an important role. The photographic objective lens of a camera usually consists of several different single lens elements through which the incoming optical rays are refracted several times before they come up onto the image sensor. The refraction through the lenses or an asymmetrical structure inside the photographic objective causes the straight lines of a viewed object to appear curved when projected onto the image sensor. These aberrations, also called lens distortion, consist mainly of radial and tangential components, and can be modeled, for example, with polynomial correction functions. While the influence of radial distortion increases depending on the distance of a pixel to the principal point, tangential distortion increases asymmetrically. To model photogrammetric imaging while considering lens distortion, additional modeling terms are applied to the collinearity equations. At this point, we do not go into further detail and refer to the relevant literature (see, inter alia, [34,35]).

The interior orientation of a camera describes the relationship between the image plane and the optical center, specified by the position of the principal point, the focal length, and the lens distortion coefficients. In contrast, exterior orientation refers to the pose (i.e., position and orientation) of a camera with respect to a world reference frame.

3.2.2. Epipolar Geometry

Given just one image with known pose and interior orientation of the camera, only the direction of a 3D world point corresponding to a certain image point can be determined. Schematically, for each image point, a ray can be generated that starts from the optical center of the camera and goes through the respective point in the positive image plane. The corresponding world point is then located somewhere on the ray.

However, to determine the actual spatial position of a world point, a second image, in which the same world point is captured from a different point of view, is required. Figuratively speaking, the optical ray of the first image is then depicted as a line—also called epipolar line—in the second image. Thus, given an image point in the first image, the location of its correspondence in the second image is restricted on its epipolar line. After identifying the corresponding image point, another optical ray for the same world point can be generated from the second image. The sought world point is then ideally located at the intersection of both rays.

The geometry between two images of the same scene is called epipolar geometry and can be encoded by the Fundamental Matrix F. In general, F maps an image point of the first image to its epipolar line in the second image. While F works with pixel coordinates, the Essential Matrix E, which is a specialization of F, deals with calibrated cameras and uses normalized image coordinates.

In general, the images are either divergent or convergent, which results in the epipolar lines running oblique. However, after rectifying both images to the same virtual plane and thus transforming to the canonical stereo configuration, the epipolar lines become parallel and corresponding image points lie in the same image row (or in the same column in the case of vertical stereo images). Figure 4 shows an image pair before and after a stereo rectification.

The shift—also known as the disparity—of two corresponding pixels in a rectified stereo pair can be encoded in a disparity map. The shorter the distance of a scene point to the camera, the greater the disparity of the pixel pair. For very distant objects, there is basically no displacement and thus the disparity approaches 0. Consequently, disparity is inversely proportional to depth.

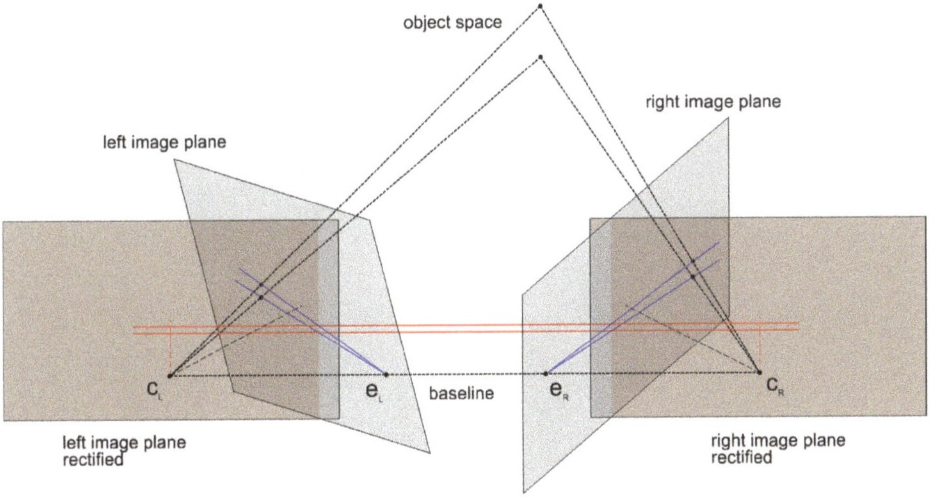

Figure 4. Epipolar geometry—initial image planes (gray) and stereo rectified images planes (orange).

4. Measurement System

4.1. Concept for Image Capture

Our 3D reconstruction procedure is based on image matching and requires photogrammetric images of the surface to be taken. Essentially, the interior orientation of the camera has to remain constant during the whole recording time and the scene should be captured from different points of view with a sufficiently high overlap.

As recording geometry for capturing images of object surfaces, we adapted the concept of traditional aerial photogrammetry since it allows a simpler hardware construction. For a more accurate measurement of the surface (especially in depth), the camera would have to capture the surface from different directions (not only with viewing direction perpendicular to the surface). However, to accomplish this, the hardware of the system would have to be designed much more complex (e.g., in order to tilt the camera to different directions during the capture). Since our aim is to develop a simpler, user-friendly, maintenance-low, lightweight, and cost-efficient measurement system, such a complex hardware construction would be a disadvantage in our case.

Since each object point has to be captured by at least 2 images to determine the 3D coordinate, at least 50% of overlap is required in the images to reconstruct the surface without gaps. A higher overlap, though, leads to over-determination (since in this case the surface will be captured by more than 2 images). This in turn leads to better estimates of the 3D object point coordinates. We use an overlap of 60–80%, which is common in aerial photogrammetry.

Accordingly, the measurement camera is moved in a meandering path parallel to the object surface. Further, the orientation of the camera remains constant with perpendicular viewing direction to the surface. Strongly overlapping images are taken in evenly spaced intervals. An illustration of our concept of the recording geometry using a monocular camera is depicted in Figure 5.

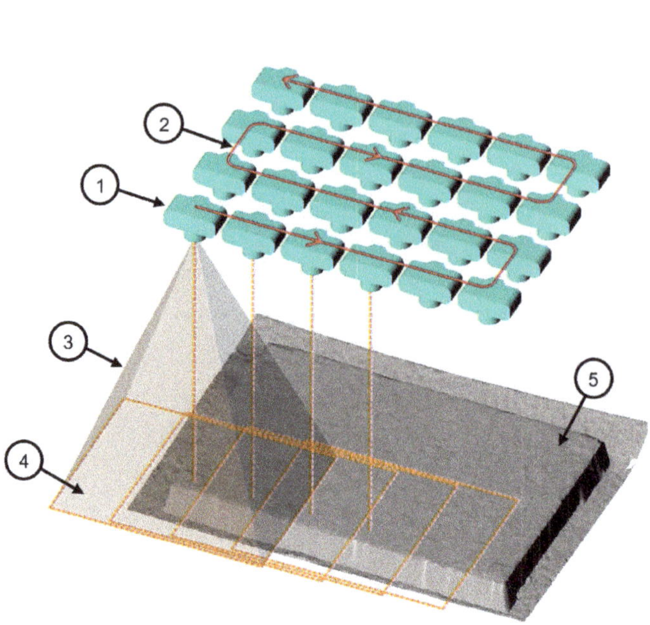

Figure 5. Concept of the recording geometry. 1—camera position; 2—camera trajectory; 3—field of view pyramid; 4—imaging area; 5—concrete surface.

4.2. Apparatus

The prototype of the measurement system consists of a cross slide with a controlling unit and propulsion, which guides an industrial camera in a meandering path automatically over the surface to be measured. The camera is moved on two axes and images are taken in stop and go mode. Subsequently, the captured images are transferred to a measurement computer via USB. Recording time depends on the number of images and the configured degree of overlap. It currently takes 30 images in around 5 min (5 images in x-direction and 6 images in y-direction). We use a Basler acA3800-14um monocular industrial camera (Ahrensburg, Schleswig-Holstein, Germany). Some main camera specifications are shown in Table 2.

Table 2. Specifications of the used industrial camera Basler acA3800-14um.

Specification	Value
Resolution (H × V)	3840 pixel × 2748 pixel
Pixel size (H × V)	1.67 µm × 1.67 µm
Bit depth	12 bits
Signal-to-noise ratio	32.9 dB
Mono/Colour	Mono
Shutter technology	Rolling shutter

Provided that the surface is not too inhomogeneous, evaluating an area of 10 cm × 10 cm should be sufficient to derive representative results for the whole surface of a concrete element. Since for an area to be reconstructed and subsequently evaluated, it has to have been captured by at least 2 images and with a sufficiently large baseline, an area of almost 20 cm × 20 cm has to be captured.

The measurement system can be powered either by an external power supply or by a built-in rechargeable battery. In addition, to ensure uniform, diffuse illumination of the area to be captured, LED strips are attached all around the perimeter of the apparatus. The intensity of the light can be dimmed smoothly. Consequently, the quality of measurement is independent of ambient illumination. Figure 6 shows the measurement system prototype.

Figure 6. Camera-based measurement system. 1—Industrial camera; 2—concrete specimen; 3—one of the four LED strips; 4—both moving axes; 5—power switch; 6—rotary switch for illumination adjustment; 7—rechargeable battery.

4.3. Custom-Built 3D Calibration Test-Field

For a sufficiently accurate calibration of the industrial camera, we designed a 3D calibration test-field with proper spatial distribution of the ground control points (GCPs). The custom-built calibration field consists of a 17 cm × 17 cm plate with a total of 144 columns on top. The columns have 3 different heights and are arranged with alternating heights in order to guarantee a proper spatial distribution of the GCPs. The appropriate spatial distribution ensures that in each of the images, GCPs from different depths are captured. The GCPs are distributed on the calibration field so that on each column there is one uncoded target and on the ground of the plate there are a total of 121 coded targets. In this way, the maximum height difference of the points is 2.25 cm. Due to the comparatively small size and the required fineness, the calibration field was manufactured by 3D printing. The required model for this was constructed using CAD software.

Additionally, four calibration rods with targets on each end of the rods are attached around the calibration field. These are used to determine the scale. The distances between the pairs of targets were measured by an interference comparator. Calculating the mean of the standard deviations of all 4 pairs leads to around 7 µm. The printed 3D calibration test-field, including the calibration rods, is shown in Figure 7.

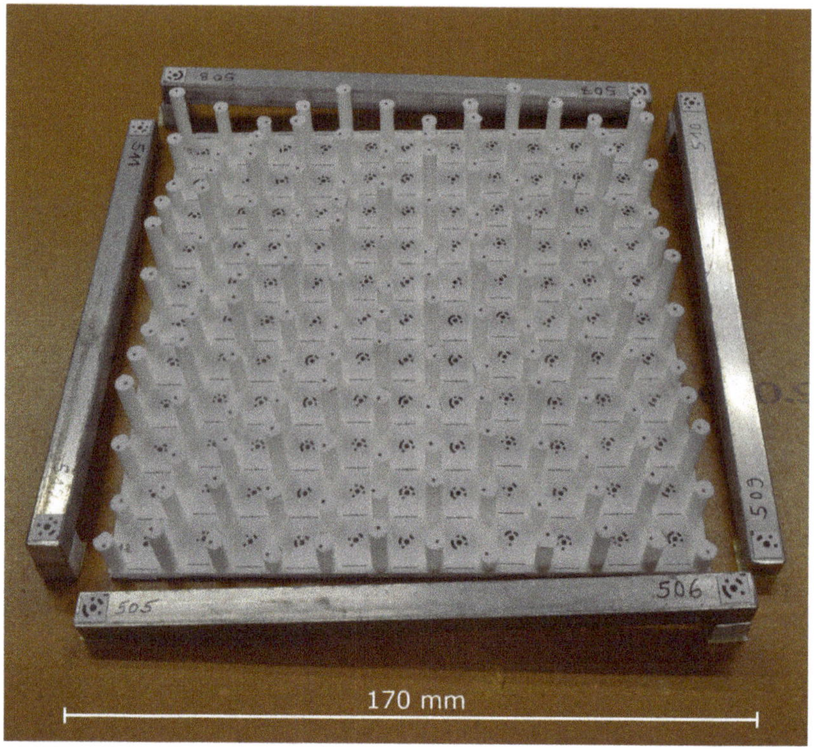

Figure 7. Custom-built 3D calibration test-field.

5. Methodology

5.1. 3D Reconstruction Pipeline

Although there are already software applications, which perform photogrammetric pipelines for 3D reconstruction (such as Agisoft Metashape [36]), we developed our own workflow, which fulfills our specific needs and furthermore is independent of commercial products. Hence, to perform 3D reconstruction of object surfaces using images taken by the calibrated industrial camera, we implemented an image matching pipeline, which includes both Structure from Motion (SfM) and Dense Image Matching (DIM) procedures. We use the SfM algorithm to refine the pose of the camera, which is roughly pre-determined due to the configured recording geometry. After determining the exact poses of the camera, we use DIM to obtain disparity maps of all image pairs and further, using these, we generate a point cloud of the scene as dense and gapless as possible. Our complete pipeline for incremental 3D reconstruction is presented in the following diagram (Figure 8).

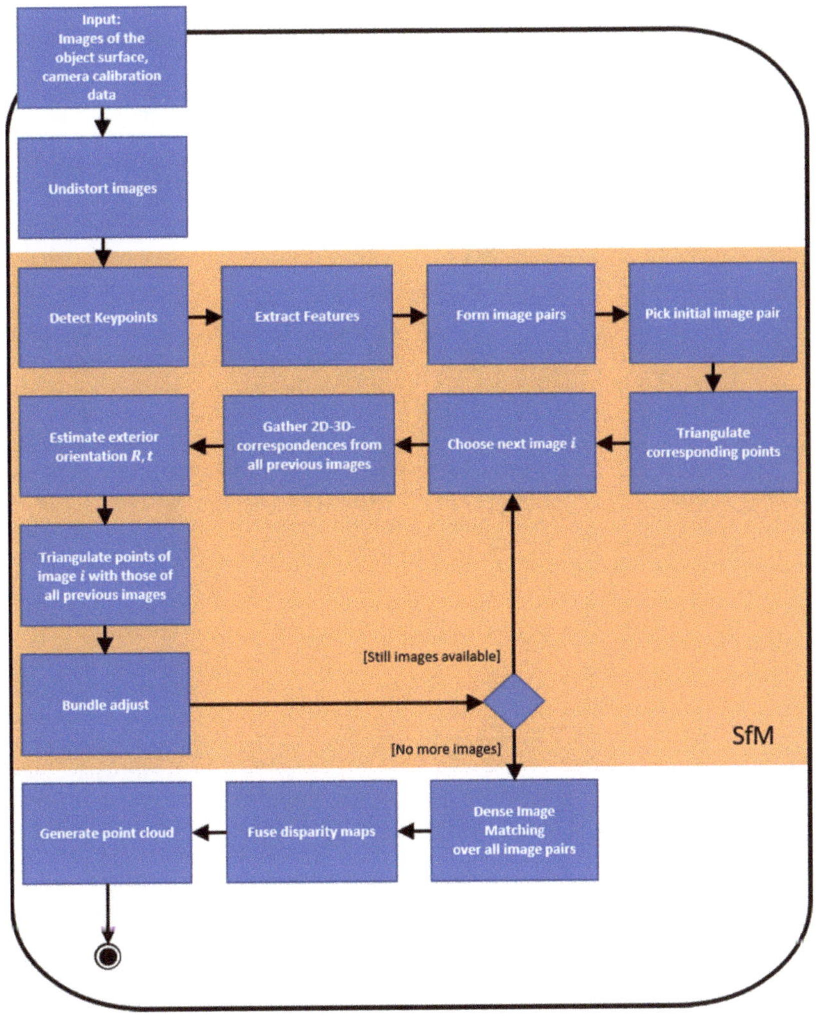

Figure 8. The incremental 3D reconstruction pipeline.

5.1.1. Preprocessing

Initially, the pipeline requires as input overlapping images of the object surface and camera calibration parameters, such as the interior orientation of the camera and lens distortion coefficients. Afterwards, all images are preprocessed to remove lens distortion and thus allow for the use of the simplified pinhole camera model.

5.1.2. Structure from Motion

An illustration of the single steps of the SfM procedure is provided in Figure 9.

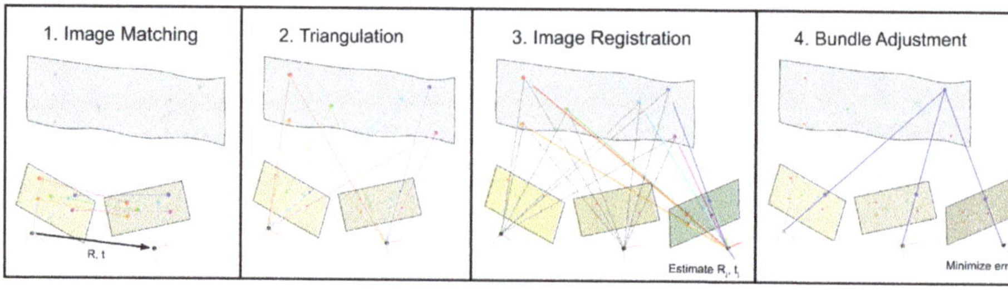

Figure 9. Visualization of the structure from motion procedure.

Interest operators, as, for instance, the widespread SIFT [25] and SURF [26], identify distinctive points in the images. These points—also called feature points—are basically corners, edges, or blobs, which lead to significant changes in the intensities of pixels and thus differ strongly from their neighborhood. Once the location of a feature point is detected, an image patch around the point is extracted, transformed in a manner so that it becomes scale- and rotation-invariant and stored in a vector called feature descriptor. In this way, every feature point in an image gets its own fingerprint, which can further be used to distinguish one point from the others. We use SURF as interest operator for feature detection and subsequent extraction since it provides around 3 times higher speed with similar accuracy when compared with SIFT [37].

Subsequently, matching of extracted feature descriptors in distinct images is utilized for identifying pairs of images looking at the same part of the scene as well as estimating the relative orientation between the image pairs. The relative orientation of an image pair is defined by the orientation, represented by a rotation matrix R, and the position, represented by a translation vector t, of the second image with respect to the first one. Since the exterior orientation of the camera is approximately pre-determined and basically corresponds to the case of aerial photogrammetry, we can exploit some key conditions. For example, the regularly spaced and strip-wise arranged images allow simplifications for establishing image pairs. Hence, we simply pick the first two images as an initial pair and consider only consecutive images for setting up further pairs. Provided that the system is configured with a sufficiently small movement of the camera between images, a high overlap is guaranteed. Thus, it can be assured that all images share common feature points, which prevents the occurrence of multiple models. Another benefit of the pre-known image sequence is that we neither have to deal with a possibly bad initialization of the model, which in turn could lead to a bad reconstruction due to the accumulation of errors.

The formation procedure of image pairs further runs through a multi-stage filtering step in order to achieve as robust matches as possible. Given an image I_i, we compare it with all previous images I_{i-j} and search for the two nearest matches of each feature point in the image I_i to that in the image I_{i-j} and vice versa. The first filtering step involves a distance ratio test as proposed in [25]. Accordingly, a match is considered poor if the distance ratio between the first-best and second-best match falls under a particular threshold. Following, we also eliminate matches based on a symmetry criterion and hence reject those that have not matched each other. The last filtering step involves a geometry verification process in which we use the RANSAC test for estimating the fundamental matrix based on the remaining point matches and identify possible outliers. Finally, an image pair is considered as valid with enough overlap if the number of filtered point matches exceeds a pre-defined threshold, e.g., in our case, 100. The diagram for establishing image pairs including the filtering procedure is shown in a box in Figure 10. The box basically represents a zoom-in of the fifth box (Form image pairs) in the diagram of the 3D reconstruction pipeline (Figure 8).

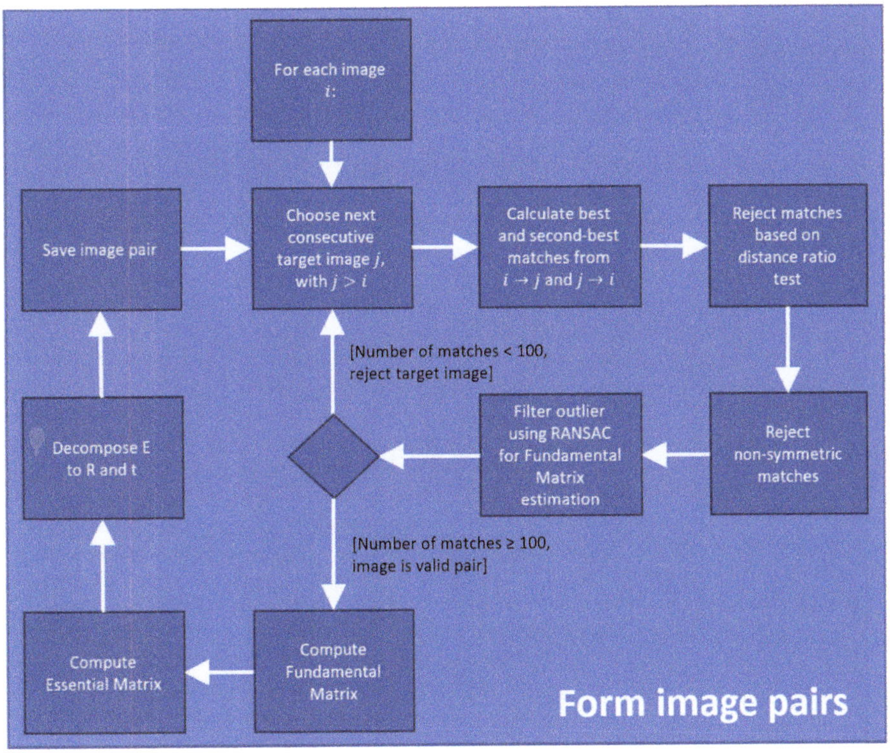

Figure 10. Procedure for image pair formation.

Based on the filtered point matches, the fundamental matrix F is computed as proposed in [38], and further, with a multiplication by the camera intrinsic matrix K, we obtain the essential matrix E. Subsequently, E is decomposed using singular value decomposition (SVD) into a rotation matrix R and a translation vector t, which represents the relative orientation between the image pair. Due to the unknown scale between both images, t is normalized.

Given the relative orientation of an initial image pair, we first triangulate corresponding points in space in order to determine the 3D coordinates of the associated world points. However, due to noise, the rays will never truly intersect. To solve that problem, we apply linear triangulation [39] for minimizing the algebraic error and thus obtain an optimal solution in the least squares sense.

Consecutive images are successively registered in the local model coordinate system utilizing 2D- to 3D-point correspondences as described in [40]. Therefore, with the correspondences of 2D feature points of a considered image and 3D world points triangulated by previous image pairs, we solve the perspective-n-point (PnP) problem [41]. As a result, we obtain the exterior orientation of the considered image with rotation matrix R and translation vector t in the given coordinate system. Finally, the new feature points of the registered image are also triangulated with those of the previous images.

Obtained results for the exterior orientation, world points, and feature points are considered to be approximations and subsequently optimized in the course of least-squares bundle adjustment. Thus, after each image registration, we apply Levenberg–Marquardt optimization [42,43] in order to simultaneously refine the exterior orientation of the images, the coordinates of the triangulated world points and the feature point coordinates. However, the interior orientation of the camera is considered fixed since it has been calibrated with high accuracy beforehand.

5.1.3. Dense Image Matching

Contrary to SfM that estimates 3D coordinates only from distinctive image points, the objective of DIM is to find the correspondence preferably for *each* pixel. These correspondences are then further used for triangulation in order to determine the 3D world coordinates.

Generally, algorithms for stereo matching can be divided into local and global matching methods. Local methods, as the name indicates, consider only local windows for the correspondence search. Specifically, for a certain pixel in the reference image, the window is sliding over a range of possible correspondences in the target image, restricted by a pre-defined disparity search range d. Matching costs, based on some similarity measure, are calculated for each possible correspondence. The correspondence is finally chosen as the pixel providing the minimum cost.

Local matching methods are in general comparatively easy to implement and very computational and memory efficient. However, disparity selection using a local Winner-takes-all (WTA) strategy struggles with point ambiguities like poor or repetitive textures and thus often leads to many mismatches.

Global matching methods, for example, based on Belief Propagation [44] or Graph Cuts [45], on the other hand, set up an energy function that takes all image points into account. An optimal disparity map is achieved by minimizing the energy function. Due to the inclusion of all image points in non-local constraints, global algorithms usually achieve a higher accuracy than local methods, but at the cost of a higher computational complexity.

In contrast, Semi-Global Matching (SGM), introduced by Hirschmuller [46], represents a compromise solution since it offers a reasonable trade-off between accuracy and computation complexity. SGM also sets up a global energy function E, including a data term E_{data} and a smoothness term E_{smooth}. The former measures the pixel-wise matching cost while the latter is for consistency, accomplished by penalizing discontinuities in the disparity map. The disparity map generation is performed by minimization of E over several one-dimensional paths using dynamic programming.

$$E(D) = \sum_{p} \left(\underbrace{C(p, D_p)}_{E_{data}} + \underbrace{\sum_{q \epsilon N_p} P_1 \cdot T[|D_p - D_q| = 1] + \sum_{q \epsilon N_p} P_2 \cdot T[|D_p - D_q| > 1]}_{E_{smooth}} \right). \quad (3)$$

In order to meet real-time requirements, we implemented the compute-intensive DIM, specifically SGM, for Graphics Processing Units (GPU) using Nvidia's programming model CUDA [47]. We parallelized each sub-algorithm of SGM for massively parallel processing and implemented parallel executable functions—also called kernel—for each of the algorithms. In the following, we recall the main steps of the SGM algorithm and, building on that, we introduce our parallelization scheme for GPU implementation.

A Cost Initialization:

An initial cost volume $C(p, D_p)$ of size $w \times h \times d$ has to be built up, where w and h are the width and height of the image and d is the disparity search range. For that purpose, pixel-wise matching costs based on some similarity measure have to be calculated for each pixel p in the reference image to its d potential correspondences in the target image.

The most widely used similarity measures are sum of absolute differences (SAD), sum of squared differences (SSD) and normalized cross correlation (NCC) [48]. Even though SAD and SSD are easy to implement and real-time capable, they assume that pixel intensities of the same object are almost equal in the images, which might not be always true. In contrast, NCC is less sensitive to changes of intensities as well as to gaussian noise, however it is computationally intensive.

In contrast, Census Transform (*CT*) [49] followed by a Hamming distance calculation represents an outstanding approach. *CT* compares the intensity of a considered pixel with that of its neighbors to generate a binary string—commonly referred as *CT* feature—as a representation for that pixel. The advantages of *CT* are illumination invariance, efficiency, and ease of implementation due to simple operations as XOR. We use a slightly modified version of *CT*—the Center-Symmetric Census Transform (*CS-CT*)—as proposed in [50], which compares neighboring pixels only to each other that are mirrored at the considered pixel. Therefore, the total number of operations decrease by around 50% and a more compact representation using just half of the memory is provided. The calculation of a *CS-CT* feature for a pixel at the location (x, y) is given as:

$$\begin{aligned} CS - CT_{m,n}(x,y) &= \bigotimes_{(i,j)\in L} s(I(x-i,y-j), I(x+i,y+j)), \\ \text{with } s(u,v) &= \begin{cases} 0, & if\ u \leq v \\ 1, & otherwise \end{cases} . \end{aligned} \tag{4}$$

Subsequently, the matching cost of two pixels is given by the Hamming distance of their *CT* features and is represented by:

$$C(p,q) = Hamming(p,q) = \|CS - CT_L(p) \oplus CS - CT_R(q)\|_1. \tag{5}$$

We implemented a kernel for *CS-CT*, which operates per image. The kernel is executed by a two-dimensional layout of threads in order to assign one pixel to each thread in the simplest possible way. The window size is chosen with 7 × 9 pixel and has the advantage that the resulting bit string of a *CS-CT* feature has a size of 31 bit and thus fits into a single 32-bit integer data type. Accordingly, an image with a size of 3840 × 2748 pixel leads to a total memory consumption of 40 MB.

A second kernel works on two *CS-CT* transformed images and calculates the matching costs. The kernel is again based on a two-dimensional thread layout with each thread calculating the Hamming distance for a specific *CS-CT* feature of the reference image to all d potential candidates in the target image. The XOR operation for two features and the subsequent population count in the resulting bit string—basically the number of ones—lead to values ranging from 0 to 31. Therefore, the use of a 1-byte data type for the Hamming distance and a disparity range of 80 values results in a total memory consumption of 805 MB for the initial cost volume $C(p, D_p)$.

B Cost Aggregation:

The costs of the initial cost volume have to be aggregated over several paths considering two penalties $P1$ and $P2$, with $P2 > P1$. The recursive aggregation on a path r at a specific pixel p and disparity value d is defined as:

$$\begin{aligned} L_r(p,d) &= C(p,d) \\ + min\Big(L_r(p-r,d), L_r(p-r,d-1) + P1, &L_r(p-r,d+1) + P1, \min_i L_r(p-r,i) + P2\Big) \\ &- \min_k L_r(p-r,k). \end{aligned} \tag{6}$$

Summing the aggregated costs over all paths delivers the aggregated cost volume S with:

$$S(p,d) = \sum_r L_r(p,d). \tag{7}$$

For the aggregation of the cost volume C, we implemented only the horizontal and vertical paths L_r in order to keep the computing time low. Thus, the aggregation is performed in the 4 directions left-to-right, right-to-left, top-to-bottom, and bottom-to-top. However, the recursive calculation prevents parallelization in each path. Though, we exploit the independency of different paths in the same direction (in the case of the horizontal paths) and different columns (in the case of vertical paths).

For instance, a kernel was implemented for the aggregation in the direction left-to-right and can be called with a specific column number. In that column, the parallelization occurs in the rows and disparity space of the cost volume. In this way, each thread is responsible for the aggregation of the cost in a particular row and disparity. The kernel is called for every column starting from the left to the right, with the first column being initialized by the initial matching cost. The procedure is analogous for the three other kernels of the remaining path directions.

Subtracting the minimum path cost $\underset{k}{min}\ L_r(p-r,k)$ in each step of the cost aggregation limits the maximum cost in a path with $C_{max} + P2$. Thus, the aggregated costs never exceed the range of 16 bits. Using images with a size of 3840 × 2748 pixel, a disparity range of 80 values and a data type of 2 bytes for the aggregated costs, the total memory consumption can be estimated at around 1.6 GB for each path.

C Disparity Selection:

For a simple case, the disparity d for each pixel p can finally be chosen based on the minimum cost in the aggregated cost volume. However, the subsequently generated point cloud based on that disparity map shows banding artifacts, since there are only 80 possible disparity values, which in turn leads to only 80 different coordinates for the depth.

Hence, we use sub-pixel interpolation, as proposed in [51], for choosing the disparity values in order to obtain a steady and continuous point cloud. Accordingly, the disparity with the minimum aggregated cost and both of its neighbors are considered for the estimation of a parabola. The disparity is finally chosen as the abscissa on which the parabola provides its minimum. Thus, the interpolated disparity for a pixel p corresponds to the vertex of the parabola and can be expressed by:

$$d_{min\_sub} = d_{min} - \frac{(S(p, d_{min+1}) - S(p, d_{min-1}))}{(2 \cdot S(p, d_{min-1}) + 2 \cdot S(p, d_{min+1}) - 4 \cdot S(p, d_{min}))}, \quad (8)$$

where d_{min} represents the integer disparity with the minimum aggregated cost.

Disparity Map Fusion and Point Cloud Generation

The actual depth z of a point can be derived in the following way:

$$z = \frac{b \cdot f}{d}, \quad (9)$$

with d the disparity of the point, f the focal length of the camera, and b the baseline of the camera between two images.

However, the obtained disparity maps were generated individually and usually have overlapping areas. Thus, in the case that the disparity maps are triangulated independently, multiple 3D points would be generated for some of the same physical object point. To avoid this, a three-dimensional grid of voxels could be built up for filtering multiple points. However, since our case corresponds to a 2.5D digital elevation model, we build up a two-dimensional regular grid with evenly sized cells parallel to the image planes. The points of the disparity maps are triangulated into these cells. In the case of multiple points falling into a particular cell, we simply pick the median of the depth of the points. Accordingly, for each cell we obtain at most one value for the depth.

5.2. Adapting Roughness Parameter to 3D Point Clouds

Reconstructed dense point clouds allow for the analysis of the surface topography. This can be used, for example, to estimate the roughness. Estimation of roughness can be performed (e.g., as in the conventional case) using single extracted profile lines as well as using the entire point cloud for an area-based determination. For this purpose, we adapted existing parameters (see Section 3.1.2) for 3D point clouds.

Arithmetical Mean Deviation R_a

For the calculation of the parameter R_a, first, the reference plane to which the mean deviation of the points is subsequently calculated has to be determined. In general, the plane needs to be estimated by minimizing the squared Euclidean distance of the measured surface points to the plane. That basically corresponds to the orthogonal distance regression (ODR), which also accounts for errors in the independent variables x and y in addition to the dependent variable z unlike the ordinary least squares (OLS) regression model. However, the main axes of the point cloud are basically parallel to the XY-plane of the coordinate system since the surface is captured parallel to the image plane and hence the difference between OLS and ODR becomes negligible. Consequently, we use OLS to estimate a plane so that the sum of squared vertical distances of the points to the plane becomes minimal.

With the points $p_i = (x_i, y_i, z_i)$ of the point cloud, we first set up an over-determined equation system for the plane π:

$$\underbrace{\begin{bmatrix} x_1 & y_1 & 1 \\ x_2 & y_2 & 1 \\ & \dots & \\ x_n & y_n & 1 \end{bmatrix}}_{A} \underbrace{\begin{bmatrix} a \\ b \\ c \end{bmatrix}}_{x} = \underbrace{\begin{bmatrix} z_1 \\ z_2 \\ \dots \\ z_n \end{bmatrix}}_{b}. \tag{10}$$

where a and b are the slopes of the plane in the x- and y-direction and c is the intersection of the plane with the z axis. Subsequently, a closed-form solution of the equation system is determined by multiplying both sides of the equation by the left pseudo inverse of the design matrix A. Thus, we estimate the parameters of the regression plane $\hat{\pi}$ with

$$\begin{bmatrix} \hat{a} \\ \hat{b} \\ \hat{c} \end{bmatrix} = \left(A^T A\right)^{-1} A^T b. \tag{11}$$

Finally, the calculation of the parameter R_a is done by numerical integration of the points. Accordingly, we sum the vertical distances of the points p_i to the regression plane $\hat{\pi}$, which are basically the residuals $\hat{e}_i$, and divide it by the total number of points n:

$$R_{a\,pointcloud} = \frac{1}{n}\sum_{i=1}^{n} d_{vertical}(p_i, \hat{\pi}) = \frac{1}{n}\sum_{i=1}^{n} \hat{e}_i. \tag{12}$$

6. Experiments

In the following, we describe our conducted experiments. Initially, the industrial camera being used was calibrated. Subsequently, the measurement system was assessed using a single concrete specimen. For evaluation of roughness estimation, we further conducted experiments using 18 concrete specimens. An overview of the performed experiments is shown in Figure 11.

6.1. Camera Calibration

In order to reconstruct the object surface as accurately as possible, the interior orientation of the camera has to be determined. This consists mainly of the focal length, the position of the principal point in the x- and y-directions, the radial distortion parameters $k1$–$k3$, and the tangential distortion parameters $p1$, $p2$. In the following subsections, the conducted experiments regarding camera calibration are presented. This involves two different approaches: A preliminary self-calibration and a more elaborate calibration using a custom-built 3D calibration test-field.

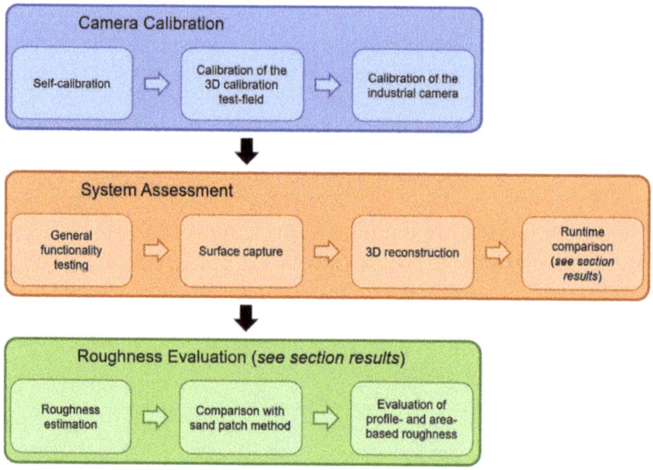

Figure 11. Overview of the conducted experiments.

6.1.1. Self-Calibration

Performing self-calibration, in which the camera is calibrated without ground control points (GCP), but only with measured image point coordinates of a (flat) concrete surface, leads to unsatisfactory results. The reason for this lies primarily in the lack of spatial distribution of the points and causes the parameters of the interior orientation of the camera to correlate with each other. Table 3 shows the dependencies of the parameters in the form of the correlation coefficients after the self-calibration procedure.

Table 3. Dependencies of the parameters of interior orientation after self-calibration.

	f	cx	cy	$k1$	$k2$	$k3$	$p1$	$p2$
f	1.00	−0.06	0.15	−1.00	0.99	−0.97	0.22	0.13
cx		1.00	0.05	0.06	−0.06	0.06	0.04	0.01
cy			1.00	−0.15	0.14	−0.13	0.04	−0.02
$k1$				1.00	−0.99	0.97	−0.22	−0.13
$k2$					1.00	−0.99	0.22	0.13
$k3$						1.00	−0.21	−0.13
$p1$							1.00	0.03
$p2$								1.00

The table shows that the radial distortion parameters ($k1$–$k3$) especially correlate with focal length (f). The subsequently reconstructed point cloud using these parameters shows a conspicuous curvature (Figure 12). To visualize the curvature, a single profile line was extracted from the reconstructed point cloud. This profile line, 10-fold scaled in the height direction, is shown at the bottom of Figure 12. The histogram of the height distribution of the point cloud also indicates a smeared normal distribution (Figure 12, right).

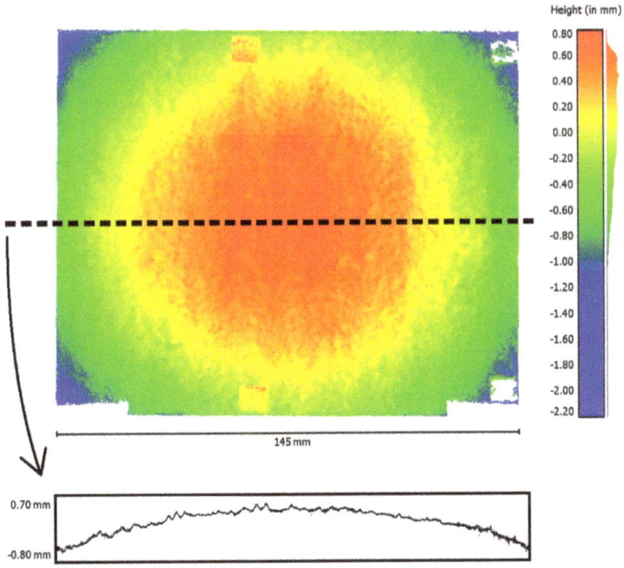

Figure 12. Reconstructed point cloud with conspicuous curvature after self-calibration (**top**, **left**), extracted profile line (**bottom**, 10-fold scaled in height), and histogram of the height values (**right**).

6.1.2. Calibration of the Test-Field

In order to determine the GCP coordinates of the custom-built 3D calibration test-field, we attached it to a tripod with a rotatable plate and took images using a digital single-lens reflex camera (DSLR). The DSLR was mounted on a ball joint, which can be moved on two axes parallel to the calibration field. The measurement setup is shown in Figure 13. A total of 69 images of the calibration field were taken systematically from different directions and subsequently processed using the photogrammetric software PHIDIAS (Version 2.21, Langerwehe, North Rhine-Westphalia, Germany) [20]. The targets were measured in the images and afterwards, the 3D world coordinates were determined using bundle adjustment. Bundle adjustment also provided information about image measurement accuracy, which is 0.57 µm or 0.12 pixels in the x-direction and 0.59 µm or 0.12 pixels in the y-direction.

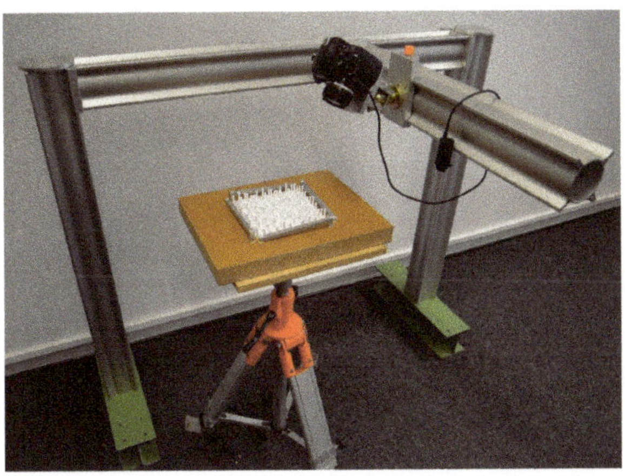

Figure 13. Measurement setup for the calibration of the 3D calibration test-field.

6.1.3. Calibration of the Industrial Camera

For the calibration of the interior orientation of the camera, we took a total of 30 images evenly distributed over the object. Subsequently, bundle adjustment with image measurements of the targets was performed in PHIDIAS, with only the parameters of the interior orientation included as unknowns. A cross validation of the adjusted point coordinates and subsequent calculation of the root mean square (RMS) of the point deviations leads to an object measurement accuracy of 3.40 µm for the x-coordinate, 4.19 µm for the y-coordinate, and 18.42 µm in depth. The conspicuously larger deviation for the z-component is related to the recording configuration and the associated poor intersection geometry of the optical rays. For the parameters of the interior orientation, we obtained the values and standard deviations given in Table 4. All parameters successfully passed a significance test with an error probability of 5%.

Table 4. Parameters of the interior orientation after calibration using the 3D calibration test-field.

Parameter		Value	Std. Dev.
f		8.2545 mm	0.0007 mm
cx		0.0737 mm	0.0012 mm
cy		0.0051 mm	0.0007 mm
$k1$	$(\cdot 10^{-4})$	−43.8231	0.2449
$k2$	$(\cdot 10^{-7})$	565.8091	36.7678
$k3$	$(\cdot 10^{-10})$	−10,555.4693	1665.3598
$p1$	$(\cdot 10^{-5})$	5.1708	0.2624
$p2$	$(\cdot 10^{-5})$	10.2455	0.2549

Parameter estimation using least squares adjustment based on the Gauss-Markov-Model also provided the correlations of the parameters in the form of the correlation coefficients. These are listed in Table 5. The previously occurring correlations of focal length and radial distortion parameters have been significantly reduced. The remaining correlations between the k-parameters or between $p1$ and cx as well as $p2$ and cy are justified in the mathematical-functional models and can therefore not be completely avoided (see e.g., [34]).

Table 5. Correlation coefficients of the parameters of interior orientation after calibration using the 3D calibration test-field.

	f	cx	cy	$k1$	$k2$	$k3$	$p1$	$p2$
f	1.00	0.00	0.01	0.03	−0.01	0.00	0.00	0.00
cx		1.00	0.00	−0.01	0.01	−0.01	0.72	0.00
cy			1.00	0.00	0.00	0.00	0.00	0.38
$k1$				1.00	−0.98	0.93	0.01	0.00
$k2$					1.00	−0.99	−0.01	0.00
$k3$						1.00	0.01	0.00
$p1$							1.00	0.00
$p2$								1.00

The visual impression of the point cloud reconstructed with the new set of parameters for the interior orientation confirms the significantly improved result as well. A curvature of the point cloud can no longer be noticed after reconstruction. For illustration, the point cloud and a profile line extracted from it and 10-fold scaled in height are shown in Figure 14. The histogram of the height distribution of the point cloud now approximates a normal distribution (Figure 14, right).

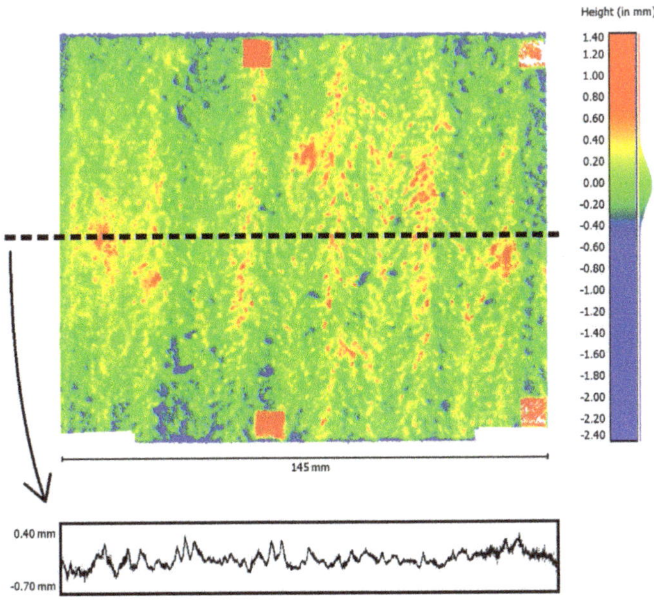

Figure 14. Point cloud (**top, left**), extracted profile line (**bottom**, 10-fold scaled in height), and histogram of the height values (**right**) after reconstruction with the new parameter set.

6.2. System Assessment

6.2.1. Test Objects

We used a set of 18 concrete specimens with plainly different surface textures as test objects. The specimens have a size of 40 cm × 40 cm × 10 cm. For each of them, there are reference values for the roughness, determined by the sand patch method, the laser triangulation method, and the paste method [8]. A selection of three specimens with different roughness is shown in Figure 15.

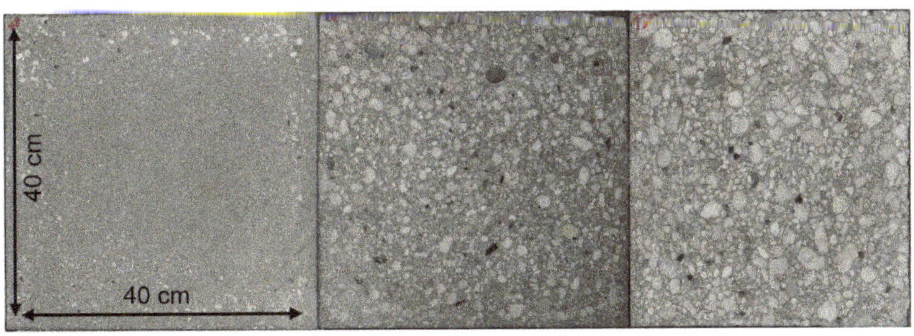

Figure 15. Selection of three concrete specimens with different roughness of the investigated 18.

6.2.2. Measurement Procedure

After calibration of the industrial camera using the custom-built 3D calibration testfield, we examined all 18 specimens with our proposed measurement system. In the following, the measurement procedure is described based on a single specimen.

The measurement system is centrally placed onto the concrete specimen and initially a general functionality testing of the measurement system is performed. After switching on the system, the lighting is adjusted so that the surface to be captured is well illuminated.

Following, the procedure for capturing images of the concrete surface is started by a controlling software. The images are captured with the following configuration: Five images in the x-direction and six images in the y-direction resulting in the images having an overlap of about 78% in x-direction and 69% in y-direction. The image capture of the surface with that particular configuration is finished in about 5 min. During capture, the images are transferred continuously to the mobile measurement computer (laptop).

After finishing the capture, the evaluation software is started and reads in the image dataset as well as the camera calibration data. In the first main step, the camera pose is refined using SfM. As a by-product, a sparse point cloud of the object surface is provided as well. The total time for this step is about 1 min. Figure 16 (top, left) shows the estimated camera poses with the sparse point cloud. Subsequently, a dense point cloud of the object surface is generated. Since this is the most compute-intensive step of the reconstruction pipeline, it takes about 6 min to complete. The generated dense point cloud and a zoom-in are presented at the top right corner and at the bottom in Figure 16. Finally, the reconstructed dense point cloud of the concrete surface is used to derive a roughness parameter. For this exemplary specimen, we obtain the value $R_a = 1.065$ mm.

(a) (b) (c)

Figure 16. 3D point cloud reconstruction of a particular concrete specimen: (**a**) Reconstructed camera poses including the sparse point cloud after SfM procedure, (**b**) generated dense point cloud after Dense Image Matching (DIM), and (**c**) zoom-in of the dense point cloud.

7. Results and Discussion

7.1. GPU Acceleration of SGM

The GPU-accelerated implementation of SGM was evaluated regarding the runtime. In order to determine the speed-up of the GPU acceleration, we also developed in C++ programming language a pure CPU-based implementation. In both cases, the same algorithms were used.

As a processing platform, we used Nvidia's graphics card GeForce GTX 1080 Ti. This high-end GPU contains 28 Streaming Multiprocessors (SM) with 128 CUDA Cores per SM and therefore provides a total of 3584 CUDA cores for parallel processing. The graphic card's throughput is around 11.70 TFLOPS and it has a memory space of around 11 GB.

As test data, we used stereo images of a concrete specimen previously captured by the measurement system. The area for dense image matching of the stereo images is restricted to the part of the images that is visible in both. Hence, we investigated our pipeline of SGM with grayscale images with a size of 2887 × 2652 pixels. The disparity search range was fixed to 80 pixels.

The comparison of the total runtime between the pure CPU implementation and the GPU-accelerated implementation is shown in Figure 17. We ran both implementations 10 times and charted the average. The runtime of the CPU implementation is therefore around 209 s, whereas the GPU-implementation achieves 4.4 s for the execution. Hence, with GPU acceleration, SGM is completed in only 2.1% of the time needed by the pure CPU implementation.

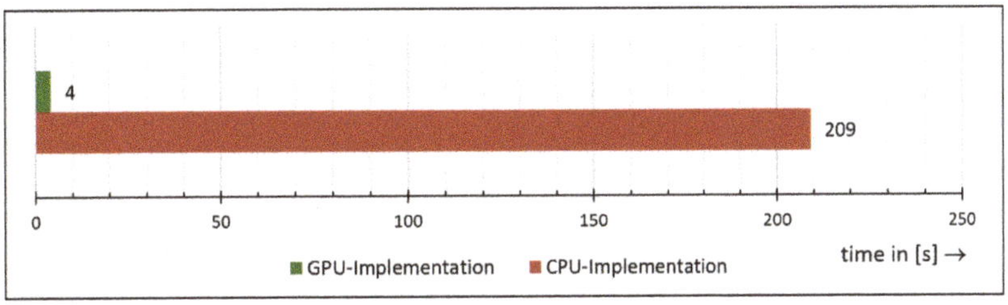

Figure 17. Comparison of the total runtime between the pure CPU implementation and the Graphics Processing Unit (GPU)-accelerated implementation.

For a more detailed comparison, we further plotted the runtime of every main algorithm of the SGM pipeline in Figure 18. The chart shows that basically each algorithm benefits from GPU acceleration. This is because the algorithms used by SGM are suitable for parallelization.

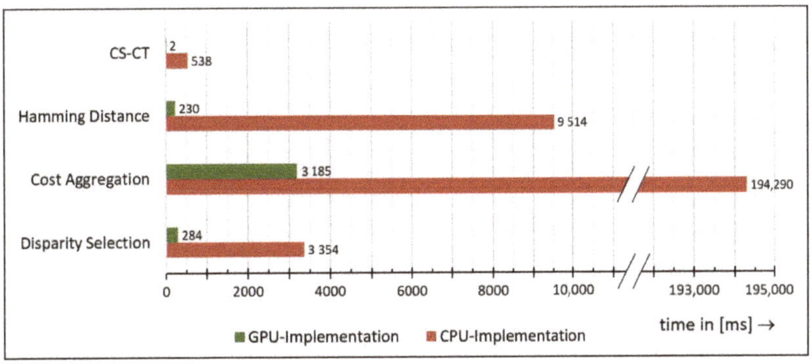

Figure 18. Runtime comparison of the single algorithms of the Semi-Global Matching (SGM) pipeline.

7.2. Comparison of the Results of Our Measurement System with the Sand Patch Method

The 18 concrete specimens were used to compare the results of our proposed measurement system with those obtained using sand patch method. For this purpose, we plotted our estimated results for the parameter R_a and the reference values for the MTD parameter determined by the sand patch method in Figure 19.

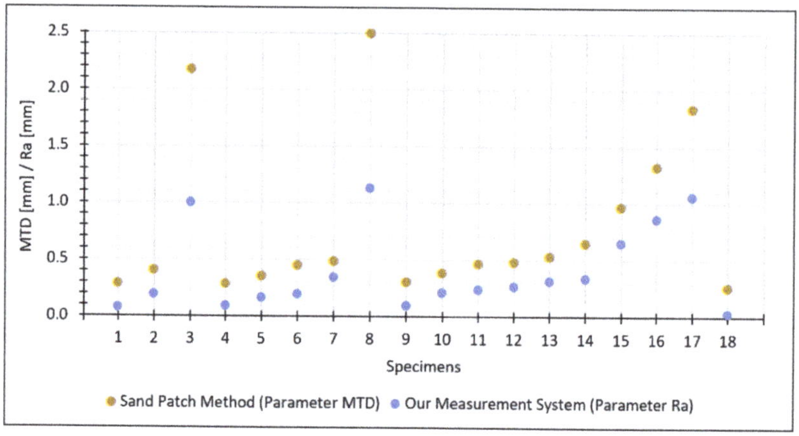

Figure 19. Comparison between R_a estimated by our measurement system and the references values for MTD determined by the sand patch method.

The values for the parameter R_a estimated by our measurement system are lower than the reference values for the MTD parameter. However, that was to be expected: Although both parameters represent the mean distance of the actual surface to a reference plane, the two reference planes used by the two methods are different. To be specific, R_a refers to the mean plane estimated through all points of the surface, while MTD refers to the plane placed onto the uppermost peaks (which leads in general to higher distances between the surface and the reference plane).

Furthermore, a closer look at the measurement values exposes some correlation between both methods. To show this more clearly, we plotted the values in a scatterplot with MTD in the x-axis and R_a in the y-axis (Figure 20). The Pearson correlation coefficient of both measurement series leads to 0.9681, which indicates a strong linear relationship.

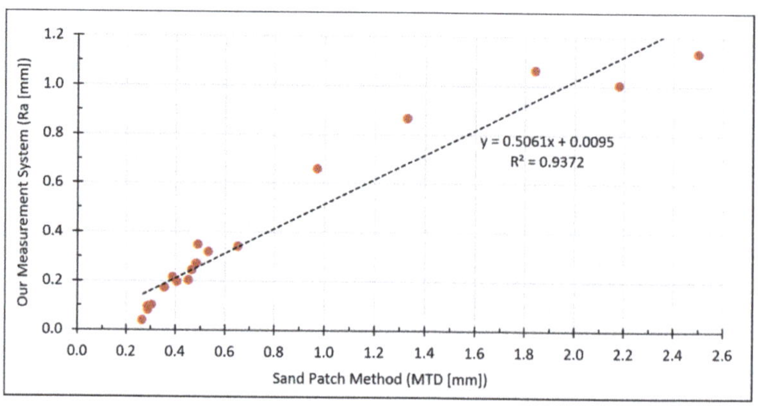

Figure 20. Correlations between R_a estimated by our measurement system and the reference values for MTD determined by the sand patch method.

7.3. Area- vs. Line-Based Estimation of the Roughness

To demonstrate the necessity of an area-based roughness determination, we conducted further examinations of the concrete specimens. In particular, we chose a specimen appearing to be very irregular and coarse to carry out further analyses.

Calculating, e.g., the roughness parameter R_a for that specimen in the area-based way leads to 0.905 mm. For comparison with the line-wise calculation, we extracted 11 equally spaced lines both in the horizontal and vertical directions from the same point cloud. The extracted lines have a spacing of 4 cm. Calculating the line-based roughness parameter leads to the following measurement series:

The individual values vary substantially as shown in Figure 21. For the horizontal lines, the arithmetic mean is $\overline{x}_h = 0.889$ mm and the standard deviation is $s_h = 0.249$ mm. For the vertical lines, the mean is $\overline{x}_v = 0.891$ mm and the standard deviation is $s_v = 0.199$ mm. Setting up a 95% confidence interval for the horizonal lines results in

$$P\{0.722 \leq \mu_h \leq 1.056\} = 95\% \tag{13}$$

and for the lines in the vertical direction, it results in

$$P\{0.757 \leq \mu_v \leq 1.025\} = 95\%. \tag{14}$$

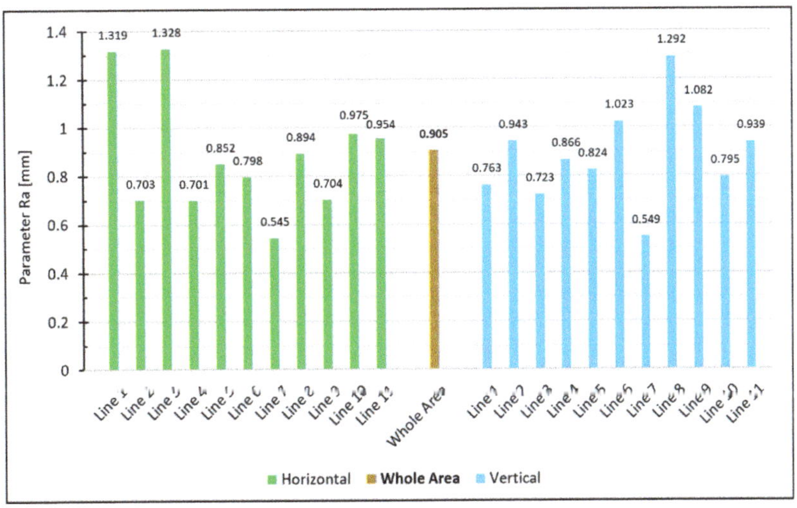

Figure 21. Calculated R_a for the extracted lines and for the whole area.

A normally distributed measurement series of the lines would lead to 5% of the values falling outside the 95% confidence interval (meaning either no line or just 1 line should fall outside the interval). However, in our case, six of the horizontal lines and four of the vertical lines do not fall into the calculated interval for the expected mean of the measurement series. Accordingly, the assumption that the individual profile lines are only subjected to normally distributed noise does not apply. Thus, the deviations of the lines can no longer be justified just stochastically, and this brings us to the conclusion that a single line is unsuitable to specify the roughness of an entire surface. An area-based determination, on the other hand, enables a reliable estimation of the roughness since all measuring points of the entire surface are concerned in the calculation.

8. Conclusions

8.1. Summary

This paper introduces a novel camera-based measurement system, which enables high-resolution analysis of technical surfaces to be performed. As a use-case, we demonstrated the roughness estimation using concrete specimens. However, basically every surface that is in the depth of field of the camera and hence can be captured sharp in the images can be measured with the proposed system. The image matching procedure, though, requires the surface to be non-reflective and to have an irregular and non-repetitive pattern in order to guarantee a unique matching of the image points in the images. However, even these constraints can be addressed by preparation of the object surface. For example, the surface can be sprayed with a very thin layer of paint in order to provide a unique pattern on the surface. Hence, generally any material (e.g., concrete elements, metals) could be measured by the system.

Before using the system for measurement purposes, the interior orientation of the industrial camera had to be calibrated. Self-calibration of the camera based on an object without control points—in our case a (flat) concrete surface—led to inadequate results. or this reason, we designed and manufactured a specific 3D calibration test-field with appropriate three-dimensional point distribution. The re-calibration of the camera using the new test-field significantly reduced the correlations of the interior orientation parameters.

For 3D reconstruction of object surfaces, we developed a two-step image matching pipeline, including SfM and DIM. The SfM algorithm is used for the estimation of the exterior orientation of the camera and is implemented using the open-source library OpenCV. For DIM, we utilize the SGM algorithm. Using the programming model of CUDA, we implemented SGM for GPUs in order to minimize the runtime and meet real-time requirements. As a result, the GPU implementation is 47.5 times faster than the pure CPU implementation.

To obtain initial results for the estimation of roughness, we adapted the roughness parameter R_a to 3D point clouds and investigated a total of 18 concrete specimens with different surface textures. Comparing the values for R_a estimated by our measurement system and the reference values for MTD determined by sand patch method shows a strong linear correlation. To show the necessity of an area-based measurement of the object surfaces, we carried out more detailed investigations with a particular specimen. It turned out that single lines are not significant for the representation of an entire surface, which in return confirms the importance of an area-based determination of the roughness.

8.2. Outlook

The determination of the parameter MTD using 3D point clouds could be performed in a similar way like the parameter R_a, with the difference that the reference plane is placed on the uppermost points of the point cloud. In addition, in this case, the distances between the points and the plane have to be summed and divided by the total number of points. This calculation procedure principally corresponds to the determination of MTD as derived by the sand patch method. In practice, though, the implementation of a calculation procedure for MTD holds some difficulties. The point clouds reconstructed by our measurement system usually consist of several million points, some of which, inevitably, are outliers. Therefore, choosing which points to use when defining the MTD reference plane would be difficult.

In the future, however, our measurement system enables novel opportunities for the investigation of technical surfaces. Many roughness parameters are designed for lines and are not suitable for the description of entire surfaces. Hence, besides the adaptation of further roughness parameters (e.g., MTD, R_v, R_p, R_t, Abbott-Firestone curve) to 3D point clouds, we think about designing new roughness parameters, which represent the surface properties in a better way. In particular, for different use-cases, different parameters should be considered. For example, a distinction has to be made between roughness parameters used to estimate the amount of coating material required to cover the surface and roughness in terms of adhesive bond of a surface. In this context, for example, the investigation of the gradients in the point clouds would be feasible.

The roughness of a 3D reconstructed surface essentially depends on the measurement resolution. For a resolution-independent estimation, the fractal dimension should be considered, as, for example, presented in [19].

As a further outlook, additional empirical studies, including comparisons with different methods (e.g., laser triangulation, sand patch method) have to be done in order to validate the camera-based system and present the method applicability. Therefore, as a first step, the measurement system could prove to be a supplementary measurement system to the sand patch method and after getting approval, it could perhaps become an alternative method.

Current high-end smartphones are equipped with an integrated graphics processor (IGP), which also enables parallel computing. In particular, smartphones with the Android operating-system provide through RenderScript [52] a powerful API for implementing parallel algorithms which can further be executed by the IGP. Therefore, it is conceivable that our image-based measurement method, as presented in this article, can be adapted onto mobile systems as smartphones since they provide all necessary components, such as, for example, a high-resolution camera and powerful processing unit. In addition, since nowadays almost everybody owns a—more or less powerful—smartphone no additional hardware would be necessary.

Author Contributions: B.Ö., R.S., and J.B. worked out the idea and designed the concept; B.Ö. designed and implemented the software and performed the experiments; R.S. and J.B. administrated and supervised the research project; B.Ö. wrote the paper, R.S. and J.B. reviewed the text and offered valuable suggestions for improving the manuscript. All authors have read and agreed to the published version of the manuscript.

Funding: This research was funded by Bundesministerium für Wirtschaft und Energie, grant number 16KN062126.

Informed Consent Statement: "Not applicable" for studies not involving humans or animals.

Data Availability Statement: "Not applicable" for studies not involving humans.

Acknowledgments: The authors acknowledge the company of Nedo GmbH & Co. KG for the excellent collaboration in the research project. The authors would further like to thank the Institute of Building Materials Research (ibac) of the RWTH Aachen University for providing the concrete specimens and corresponding reference values. Additionally, the authors would like to thank Louis Makiello for proofreading.

Conflicts of Interest: The authors declare no conflict of interest.

References

1. Julio, E.N.B.S.; Branco, F.A.B.; Silva, V.D. Concrete-to-concrete bond strength. Influence of the roughness of the substrate surface. *Constr. Build. Mater.* **2004**, *18*, 675–681. [CrossRef]
2. Santos, P.M.D.; Julio, E.N.B.S.; Silva, V.D. Correlation between concrete-to-concrete bond strength and the roughness of the substrate surface. *Constr. Build. Mater.* **2007**, *21*, 1688–1695. [CrossRef]
3. Santos, D.S.; Santos, P.M.D.; Dias-da-Costa, D. Effect of surface preparation and bonding agent on the concrete-to-concrete interface strength. *Constr. Build. Mater.* **2012**, *37*, 102–110. [CrossRef]
4. Santos, P.M.D.; Julio, E.N.B.S. A state-of-the-art review on roughness quantification methods for concrete surfaces. *Constr. Build. Mater.* **2013**, *38*, 912–923. [CrossRef]
5. Bikerman, J.J. *The Science of Adhesive Joints*; Academic Press: Cambridge, MA, USA, 1968.
6. Kaufmann, N. Das Sandflächenverfahren. *Straßenbautechnik* **1971**, *24*, 131–135.
7. Mellmann, G.; Oppat, K. Maß für Maß. Rautiefen-Bestimmung von Betonoberflächen mittels Laserverfahren. *Bautenschutz Bausanier B B* **2008**, *31*, 30–32.
8. Steinhoff, A.; Holthausen, R.S.; Raupach, M.; Schulz, R.-R. Entwicklung eines Pastenverfahrens zur Bestimmung der Rautiefe an vertikalen Betonoberflächen. Entwicklungsschwerpunkte und Ergebnisse einer Studie. *Beton* **2020**, *70*, 182–186.
9. China, S.; James, D.E. Comparison of Laser-Based and Sand Patch Measurements of Pavement Surface Macrotexture. *J. Transp. Eng.* **2012**, *138*, 176–181. [CrossRef]

10. ASTM E2157-15. *Standard Test Method for Measuring Pavement Macrotexture Properties Using the Circular Track Meter*; ASTM: West Conshohocken, PA, USA, 2019.
11. Ma, L.F.; Li, Y.; Li, J.; Wang, C.; Wang, R.S.; Chapman, M.A. Mobile Laser Scanned Point-Clouds for Road Object Detection and Extraction: A Review. *Remote Sens.* **2018**, *10*, 1531. [CrossRef]
12. Gonzalez-Jorge, H.; Rodriguez-Gonzalvez, P.; Shen, Y.Q.; Laguela, S.; Diaz-Vilarino, L.; Lindenbergh, R.; Gonzalez-Aguilera, D.; Arias, P. Metrological intercomparison of six terrestrial laser scanning systems. *IET Sci. Meas. Technol.* **2018**, *12*, 218–222. [CrossRef]
13. Stal, C.; Verbeurgt, J.; De Sloover, L.; De Wulf, A. Assessment of handheld mobile terrestrial laser scanning for estimating tree parameters. *J. For. Res.* **2020**. [CrossRef]
14. DIN EN ISO 13473-1:2017-08. *Characterization of Pavement Texture by Use of Surface Profiles—Part 1: Determination of Mean Profile Depth*; ISO: Geneva, Switzerland, 2017.
15. Schulz, R.-R.; Schmidt, T.; Hardt, R.; Riedl, R. Baustellengerechte Laser-Profilmessverfahren für die Steuerung und Eigenüberwachung der Oberflächentexturierung von Verkehrsflächen aus Beton. *Straße Autob.* **2013**, *64*, 911–920.
16. Schulz, R.-R. Fortschritte bei der Rauheitsbewertung von Betonoberflächen. Alternativen zum Sandflächenverfahren. *Beton* **2016**, *66*, 502–510.
17. Schulz, R.-R. Laser schlägt Sand—Rautiefenmessung an Betonoberflächen. *Bau. Im Bestand B B* **2017**, *40*, 44–48.
18. Schulz, R.-R. Roughness and anti-slip properties of concrete surfaces—Electro-optical measuring systems to determine roughness parameters. *Bft Int.* **2008**, *74*, 4–15.
19. Werner, S.; Neumann, I.; Thienel, K.C.; Heunecke, O. A fractal-based approach for the determination of concrete surfaces using laser scanning techniques: A comparison of two different measuring systems. *Mater. Struct.* **2013**, *46*, 245–254. [CrossRef]
20. PHIDIAS. The Complete Solution for Photogrammetric Close Range Applications. Available online: http://www.phocad.com/en/en.html (accessed on 19 October 2020).
21. Benning, W.; Lange, J.; Schwermann, R.; Effkemann, C.; Görtz, S. Monitoring crack origin and evolution at concrete elements using photogrammetry. *Int. Arch. Photogramm. Remote Sens. Spat. Inf. Sci.* **2004**, *35*, 678–683.
22. Benning, W.; Lange, J.; Schwermann, R.; Effkemann, C.; Gortz, S. Photogrammetric measurement system for two-dimensional deformation and crack analysis of concrete constructions. *Sens. Meas. Syst.* **2004**, *1829*, 813–817.
23. Calonder, M.; Lepetit, V.; Strecha, C.; Fua, P. BRIEF: Binary Robust Independent Elementary Features. In Proceedings of the Computer Vision ECCV, Heraklion, Greece, 5–11 September 2010; Volume 6314, pp. 778–792. [CrossRef]
24. Rosten, E.; Porter, R.; Drummond, T. Faster and Better: A Machine Learning Approach to Corner Detection. *IEEE Trans. Pattern Anal. Mach. Intell.* **2010**, *32*, 105–119. [CrossRef]
25. Lowe, D.G. Distinctive image features from scale-invariant keypoints. *Int. J. Comput. Vis.* **2004**, *60*, 91–110. [CrossRef]
26. Bay, H.; Ess, A.; Tuytelaars, T.; Van Gool, L. Speeded-Up Robust Features (SURF). *Comput. Vis. Image Underst.* **2008**, *110*, 346–359. [CrossRef]
27. Wieneke, K.; Herbrand, M.; Vogler, N.; Schwermann, R.; Blankenbach, J. Measurement methods for determining the roughness of concrete surfaces. *Bauingenieur* **2018**, *93*, 365–373.
28. Grigoriadis, K. Use of laser interferometry for measuring concrete substrate roughness in patch repairs. *Autom. Constr.* **2016**, *64*, 27–35. [CrossRef]
29. Lange, D.A.; Jennings, H.M.; Shah, S.P. Analysis of Surface-Roughness Using Confocal Microscopy. *J. Mater. Sci.* **1993**, *28*, 3879–3884. [CrossRef]
30. Sadowski, L. Methodology of the assessment of the interlayer bond in concrete composites using NDT methods. *J. Adhes. Sci. Technol.* **2018**, *32*, 139–157. [CrossRef]
31. Özcan, B.; Schwermann, R.; Blankenbach, J. Kamerabasiertes Messsystem zur Bestimmung der Rauigkeit von Bauteiloberflächen—Kalibrierung und erste Ergebnisse. In Proceedings of the 19. Internationaler Ingenieurvermessungskurs, München, Germany, 3–6 March 2020.
32. DIN 4760:1982-06. *Form Deviations; Concepts; Classification System*; Beuth: Berlin, Germany, 1982.
33. DIN EN ISO 4287:2010-07. *Geometrical Product Specifications (GPS)—Surface Texture: Profile Method—Terms, Definitions and Surface Texture Parameters*; ISO: Geneva, Switzerland, 2010.
34. Luhmann, T.; Robson, S.; Kyle, S.; Boehm, J. *Close-Range Photogrammetry and 3d Imaging*, 3rd ed.; Walter de Gruyter GmbH: Berlin, Germany, 2020.
35. Hartley, R.; Zisserman, A. *Multiple View Geometry in Computer Vision*, 2nd ed.; Cambridge University Press: Cambridge, UK; New York, NY, USA, 2003.
36. Agisoft Metashape. Available online: https://www.agisoft.com/features/professional-edition/ (accessed on 19 October 2020).
37. Panchal, P.M.; Panchal, S.R.; Shah, S.K. A Comparison of SIFT and SURF. *Int. J. Innov. Res. Comput. Commun. Eng.* **2013**, *1*, 323–327.
38. Hartley, R.I. In defense of the eight-point algorithm. *IEEE Trans. Pattern Anal. Mach. Intell.* **1997**, *19*, 580–593. [CrossRef]
39. Hartley, R.I.; Sturm, P. Triangulation. *Comput. Vis. Image Underst.* **1997**, *68*, 146–157. [CrossRef]
40. Blut, C.; Blankenbach, J. Three-dimensional CityGML building models in mobile augmented reality: A smartphone-based pose tracking system. *Int. J. Digit. Earth* **2020**. [CrossRef]

41. Fischler, M.A.; Bolles, R.C. Random Sample Consensus—A Paradigm for Model-Fitting with Applications to Image-Analysis and Automated Cartography. *Commun. ACM* **1981**, *24*, 381–395. [CrossRef]
42. Levenberg, K. A method for the solution of certain non-linear problems in least squares. *Q. Appl. Math.* **1944**, *2*, 164–168. [CrossRef]
43. Marquardt, D.W. An Algorithm for Least-Squares Estimation of Nonlinear Parameters. *J. Soc. Ind. Appl. Math.* **1963**, *11*, 431–441. [CrossRef]
44. Sun, J.; Zheng, N.N.; Shum, H.Y. Stereo matching using belief propagation. *IEEE Trans. Pattern Anal. Mach. Intell.* **2003**, *25*, 787–800. [CrossRef]
45. Kolmogorov, V.; Zabih, R. Computing visual correspondence with occlusions using graph cuts. In Proceedings of the Eighth Ieee International Conference on Computer Vision, Vol Ii, Proceedings, Vancouver, BC, Canada, 7–14 July 2001; pp. 508–515.
46. Hirschmuller, H. Accurate and efficient stereo processing by semi-global matching and mutual information. In Proceedings of the IEEE Computer Society Conference on Computer Vision and Pattern Recognition, San Diego, CA, USA, 20–25 June 2005; Volume 2, pp. 807–814. [CrossRef]
47. CUDA Zone | NVIDIA Developer. Available online: https://developer.nvidia.com/cuda-zone (accessed on 19 October 2020).
48. Zhao, F.; Huang, Q.M.; Gao, W. Image matching by normalized cross-correlation. In Proceedings of the IEEE International Conference on Acoustics, Speech and Signal Processing, Toulouse, France, 14–19 May 2006; Volume 1–13, pp. 1977–1980.
49. Zabih, R.; Woodfill, J. *Non-Parametric Local Transforms for Computing Visual Correspondence*; Springer: Berlin/Heidelberg, Germany, 1994; pp. 151–158.
50. Spangenberg, R.; Langner, T.; Rojas, R. *Weighted Semi-Global Matching and Center-Symmetric Census Transform for Robust Driver Assistance*; Springer: Berlin/Heidelberg, Germany, 2013; pp. 34–41.
51. Tian, Q.; Huhns, M.N. Algorithms for Subpixel Registration. *Comput. Vis. Graph. Image Process.* **1986**, *35*, 220–233. [CrossRef]
52. RenderScript. Available online: https://developer.android.com/guide/topics/renderscript/compute (accessed on 19 October 2020).

materials

MDPI

Article

Contactless Measurements of Carrier Concentrations in InGaAs Layers for Utilizing in InP-Based Quantum Cascade Lasers by Employing Optical Spectroscopy

Marcin Kurka [1,*], Michał Rygała [1], Grzegorz Sęk [1], Piotr Gutowski [2], Kamil Pierściński [2] and Marcin Motyka [1]

1 Laboratory for Optical Spectroscopy of Nanostructures, Department of Experimental Physics, Wrocław University of Science and Technology, Wybrzeże Wyspiańskiego 27, 50-370 Wrocław, Poland; michal.daniel.rygala@gmail.com (M.R.); grzegorz.sek@pwr.edu.pl (G.S.); marcin.motyka@pwr.edu.pl (M.M.)

2 Łukasiewicz Research Network–Institute of Electron Technology, Al. Lotników 32/48, 02-668 Warszawa, Poland; gutowski@ite.waw.pl (P.G.); pierscin@ite.waw.pl (K.P.)

* Correspondence: marcin.kurka@pwr.edu.pl

Received: 11 June 2020; Accepted: 9 July 2020; Published: 12 July 2020

Abstract: The precise determination of carrier concentration in doped semiconductor materials and nanostructures is of high importance. Many parameters of an operational device are dependent on the proper carrier concentration or its distribution in both the active area as well as in the passive parts as the waveguide claddings. Determining those in a nondestructive manner is, on the one hand, demanded for the fabrication process efficiency, but on the other, challenging experimentally, especially for complex multilayer systems. Here, we present the results of carrier concentration determination in $In_{0.53}Ga_{0.47}As$ layers, designed to be a material forming quantum cascade laser active areas, using a direct and contactless method utilizing the Berreman effect, and employing Fourier-transform infrared (FTIR) spectroscopy. The results allowed us to precisely determine the free carrier concentration versus changes in the nominal doping level and provide feedback regarding the technological process by indicating the temperature adjustment of the dopant source.

Keywords: Berreman effect; quantum cascade lasers; gas sensing; carrier concentration; mid-infrared

1. Introduction

Mid-infrared is a spectral region of ever increasing significance. Numerous applications include free-space communication, imaging, and gas sensing [1]. The latter itself covers a vast area, ranging from medical diagnosis, e.g., detecting cancer markers [2] by breath analysis, localizing toxic or explosive leaks at factories and waste disposal sites, in-situ industrial process control, up to remotely checking alcohol content in exhaled air inside vehicles [3,4]. Absorption spectra with strong characteristic lines for different gases in that range make it possible to unambiguously identify gas mixture composition, and the state-of-the-art optical sensing systems are able to detect them at ppb concentrations in sub-second temporal resolution using quantum cascade lasers (QCLs) [5–8] or interband cascade lasers (ICLs) [9,10]. Many properties of those lasers are determined by properly chosen carrier concentration in respective areas thereof. Especially, the active regions of these lasers contain n-type doping, the concentration of which has to be precisely controlled to ensure the operation of the device and its performance. Moreover, they must be cladded by a layer of semiconductors with proper refraction index, in order to make a waveguide for the generated radiation. This can be achieved by changing the charge carrier concentration via precisely controlling the amount of doping during growth. Verifying the concentration levels during post-processing is normally destructive, by using such techniques as Hall effect measurements or capacity-voltage measurements [11]. There are, however, several optical experiments which not only are contactless, but in general, do not require affecting the sample, which can be then further processed with already known characteristics. With no preprocessing requirements, and the possibility of measuring grown wafer during the process, it seems a very promising upgrade. In this paper, we show results of carrier concentration measurements using so the called "Berreman effect" [12]. Our purpose was to establish the carrier concentration of $In_{0.53}Ga0_{.47}As$ layers and compare it with "nominal", i.e., obtained by interpolation of Hall measurements of reference samples' concentration, and to make improvements to the doping process. The material of the samples was purposed for active areas of QCLs for Long Wavelength Infrared (LWIR) range. The method itself allows to measure various kinds of layers with thicknesses below 10 nm [13], with reports that the effect takes places in ~1 nm thick region [14,15]. The concentrations measured can be as low as 10^{17} cm^{-3} while performing measurements in transmission mode [16], given sufficient sensitivity of the setup, which can be improved using modulation techniques such as fast differential reflectance [17,18], or photoreflectance in step-scan mode of the FTIR spectrometer [19,20]. The phenomenon can be observed as an enhancement of absorption of p-polarized radiation where the dielectric function reaches zero at the plasmon frequency. In heavily doped semiconductors, this can be ascribed to free carriers in surface plasmon polaritons, and its frequency is directly dependent on the carrier concentration. Therefore, it is possible to determine the concentration of free carrier in a sample by measuring the changes in the absorption of polarized light.

2. Materials and Methods

In this paper, we show the results of reflectance measurements of the Berreman effect of $In_{0,53}Ga_{0,47}As$ layers samples with different nominal carrier concentrations. These are 1-µm thick Si-doped $In_{0.53}Ga_{0.47}As$ layers lattice-matched to the InP substrate, grown by solid source molecular beam epitaxy (MBE) with a Riber Compact 21T reactor [21]. Different temperatures of the Si-source were used in order to vary the doping concentrations. The nominal carrier concentrations in the investigated structures were calculated by interpolation of Hall measurement curves measured with a Bio-Rad HL5500 system (Bio-rad, Hercules, CA, USA), and are shown in Table 1.

The optical measurements were performed in Bruker Vertex 80 FTIR spectrometer (Bruker, Billerica, MA, USA) with additional custom-designed evacuated external chamber to provide an oblique angle of 45 degree for the incident light [17,18] (which is necessary to observe the Berreman effect). As the light source, a glowbar was used, whereas the detector was a liquid-Nitrogen-cooled Mercury Cadmium Telluride photodiode.

Table 1. The samples' description.

Sample	Hall-Measured Carrier Concentration (cm^{-3})	Si-Source Nominal Temperature (°C)
(A) C766	1.9×10^{19}	1300
(B) C764	9.9×10^{18}	1275
(C) C763	5.2×10^{18}	1250
(D) C759	1.1×10^{18}	1190

3. Results

Figure 1 shows reflectance spectra for all samples measured at 300 K for two orthogonal linear polarizations of the incident light. A distinct absorption dip can clearly be seen for TM polarization (red curves), which corresponds to a case where there is an electric field component perpendicular to the sample surface. Black curves denote spectra obtained for s polarized light (TE) where there is no corresponding minimum observed, which is consistent with theory. Also visible in the red curves of the TM polarization is how the Berreman minima shift towards lower wavenumber (longer wavelength) with decreasing doping and finally even disappear for sample D due to the cut-off wavelength of our setup (detector limit at 600 cm^{-1}).

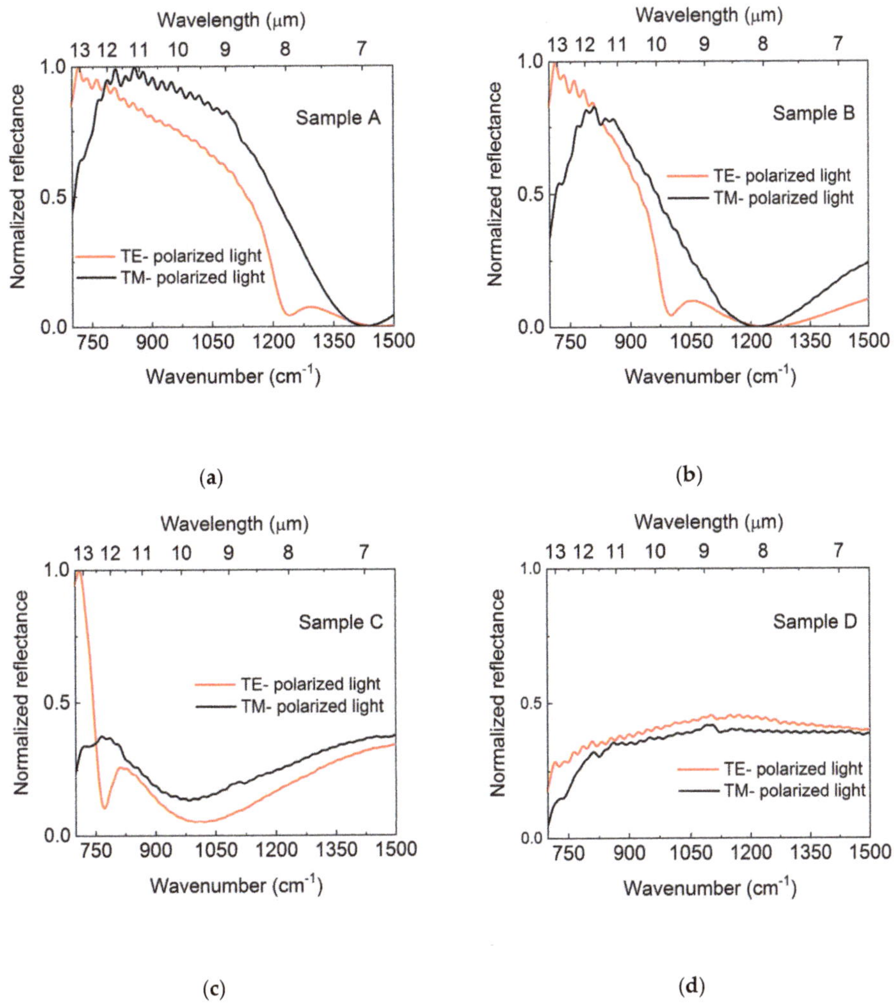

Figure 1. The reflectance spectra for two orthogonal polarizations of the probing light denoted TE (red curves) and TM (black curves): (**a**) 1.9×10^{19} cm^{-3}; (**b**) 9.9×10^{18} cm^{-3}; (**c**) 5.2×10^{18} cm^{-3}; (**d**) 1.1×10^{18} cm^{-3}.

Figure 2 shows intensity normalized reflectance spectra for the p-polarized light of three samples A, B and C. The absorption minima can be ascribed to plasma frequency ω_p, of free electrons in the layers. We can see that ω_p shifts to lower energies with carrier concentration decreasing according to Equation (1).

$$\omega_p^2 = \frac{ne^2}{\varepsilon_0 \varepsilon_\infty m^*} \quad (1)$$

where ε_∞ is the material dielectric constant equal to 11.64 and ε_0 is the permittivity of free space. The value of m^* is 0.0453 of electron rest mass.

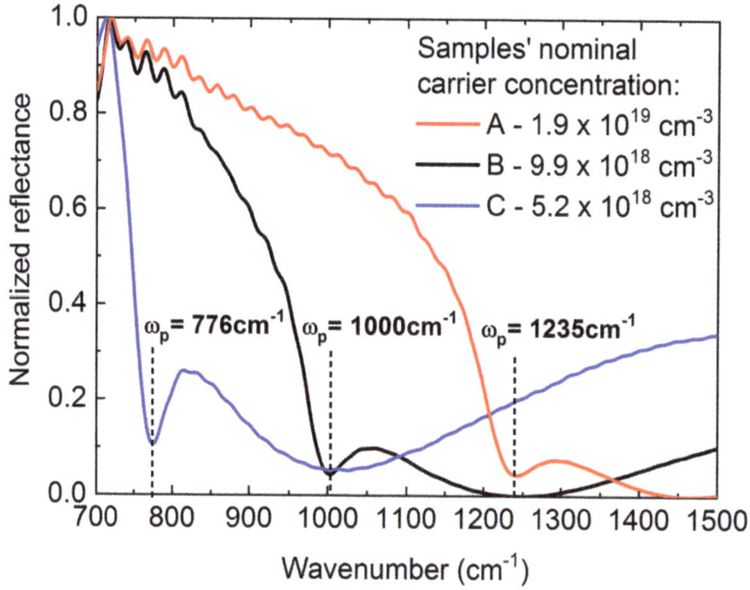

Figure 2. Normalized p-polarized reflectance spectra for Sample A (red curve), Sample B (black curve), and Sample C (blue curve) together with given characteristic reflectance minima at plasma frequencies and indication of carrier nominal concentration.

Figure 3 shows a comparison of the dependence of the obtained nominal carrier concentration versus plasma frequencies (black squares) obtained for InGaAs alloy with similar composition by Charache et al. [22]. This allowed us to get a useful function after exponential fitting (red line). A similar procedure has been proven to be successful when applied for the analysis of Berreman minima in the case of carrier concentration determination in InAs layers [23], using data from Hinkey et al. [24]. By open squares we marked nominal concentration values on the fitted line at the respective wavenumbers. Open solid points denote the plasmon frequencies measured by reflectivity measurements (given in wavenumbers). By placing these frequencies on the fitted curve, we were able to determine the actual electron concentrations of 1.3×10^{19} cm^{-3}, 7.0×10^{18} cm^{-3}, and 4.0×10^{18} cm^{-3} for samples A, B, and C, respectively. Furthermore, we can see that the revealed differences between the nominal and the measured concentrations, despite not being large, show for all cases the measured values to be smaller than the nominal ones. This is just the first approximation approach, which can be used as an attempt for growth parameters verification in order to establish a better match between nominal and achieved concentrations in the layers.

A more sophisticated and precise method requires including the effective mass dependence on the carrier concentration in the calculation of the plasma frequencies as a function of concentration, since it cannot be neglected for concentrations above 10^{18} cm^{-3}, as in ref. [25]. Due to non-parabolic behavior of energy dispersion far from Γ point of the Brillouin zone, a correction of carrier effective mass must be taken into account, which can be expressed by Equation (2) taken from ref. [26].

$$m^*(n) = m_e\left[1+\frac{4P^2}{3E_g}\left(1+\frac{8P^2\hbar^2\left(3\pi^2 n\right)^{\frac{2}{3}}}{3m_e E_g^2}\right)^{\frac{-1}{2}}\right]^{-1}, \tag{2}$$

where m_e is the electron rest mass and P^2 is the momentum matrix element of coupling between valence and conduction bands, calculated to be 15.4 eV.

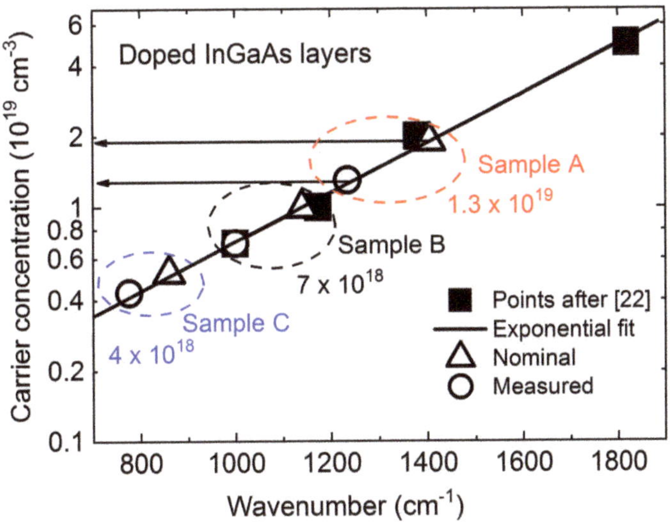

Figure 3. Carrier concentration in a function of Plasma frequency. Black squares denotes points after ref. [22] together with respective fit (red line). Triangular points represents nominal concentrations and open circles those established.

Figure 4a shows the calculated m^*/m_e ratio as a function of electron concentration. As we can see, the role of correction is more important for larger concentrations in the range of nominal concentrations considered within this paper (indicated by the red dotted square), where the relative changes of the effective mass up to around 30% can occur. Figure 4b shows calculated plasma frequencies assuming the mass correction (black curve) together with the nominal values (black squares) and those determined here (red circles). The obtained values are still slightly lower than nominal, however closer to the nominal values when compared to those obtained in the first, simpler approach. The respective summary is shown in Table 2. The resolution of our experimental setup was 2 cm^{-1}, while the full widths at half maximum (FWHM) of measured spectra were ~50 cm^{-1}. This allowed us to estimate uncertainty of the carrier concentration at ±2%, which is not enough to explain the difference between nominal concentration.

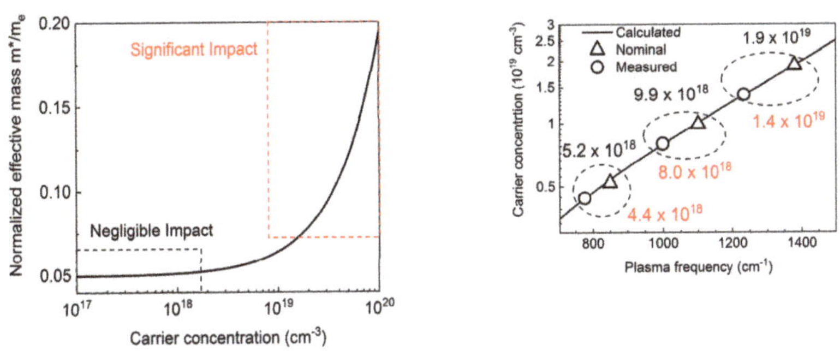

(a) (b)

Figure 4. (**a**) Electron mass changes as a function of their concentration after Equation (2); (**b**) Plasma frequencies as a function of electron concentration. Black curve represents calculated Plasma frequencies assuming mass correction shown in panel (**a**). Open triangles depicts nominal values and open circles those measured.

Table 2. The samples' description and growth protocol.

Sample	Nominal Carrier Concentration	First Approach	Second Approach
(A) C766	1.9×10^{19}	1.3×10^{19}	1.38×10^{19}
(B) C764	9.9×10^{18}	7.0×10^{18}	8.0×10^{18}
(C) C763	5.2×10^{18}	4.0×10^{18}	4.4×10^{18}
(D) C759	1.1×10^{18}	N/A	N/A

4. Discussion

Figure 5 shows the nominal electron concentrations for all investigated samples versus the applied Si source temperature together with a fit (black curve) which shows this dependence in a broader range of concentrations and temperatures. Red open circles denote determined (by second more accurate approach) electron concentrations for samples A–C together with respective fit (red curve). The comparison of the nominal dependency and the corrected experimental results shows that the respective adjustments become important for larger concentrations (higher source temperatures).

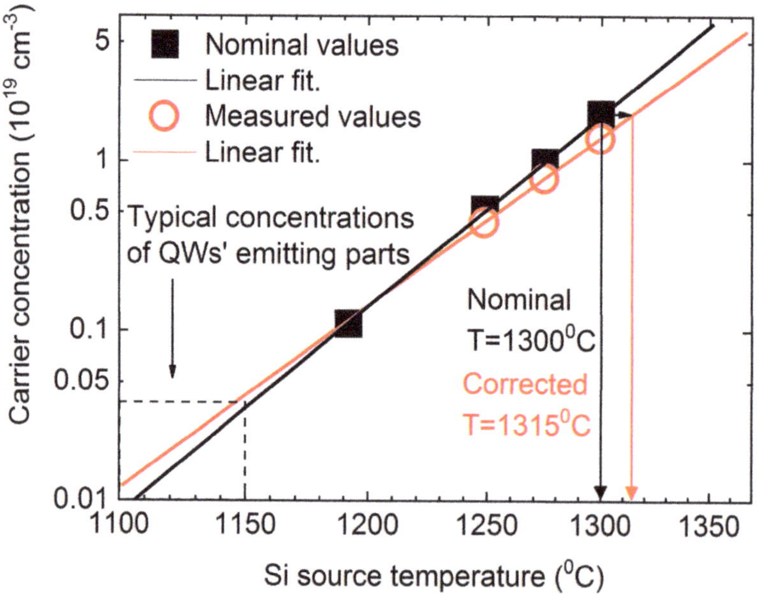

Figure 5. Nominal carrier concentrations (black squares) as a function of applied Si source temperature together with fit (black curve). Open red circles denotes corrected (in the way of second approach) values of the carrier concentration in a function of the applied Si source temperature together with respective fit (red curve).

Therefore, in a range of concentrations presented in this paper, a slight temperature increase seems to be appropriate. For instance, to obtain a nominal concentration for sample A, Si source temperature should be increased by ~15 K from 1300 °C to 1315 °C to offset the observed differences. On the other hand, in the range of typical doping (~10^{16}–10^{17} cm^{-3}) of the lasing active areas of quantum cascade lasers, little temperature compensation would be required.

5. Conclusions

In this paper, we have demonstrated an optical method used for electron concentration determination in the calibration of InGaAs layers designed for waveguides of InP-based quantum cascade lasers. This method requires neither contacts nor an external magnetic field as are typically required in direct-contact methods such as Hall and C-V measurements. Moreover, the method can be applied to various types of structures, made of different kinds of materials and grown in different techniques such as metal-organic chemical vapor deposition, as well as different techniques of doping. It is important to note that 15 K is a significant difference from the point of view of manufacturing technology. There are several possible causes that would explain reported discrepancies between Hall effect method and the Berreman effect method, such as the degree of ionization of the dopant atoms in the sample, or gauging the magnetic field of the magnet, while using Hall setup. Further establishing causes of shown differences is crucial to optimize the manufacturing process and therefore the quality of ready devices due to the feasibility of the optical method, which makes it a perfect candidate to supersede other methods. By determination of the so called "Berreman effect minima" in the reflectance spectra and the derived carrier concentration versus plasma frequency, we were able to verify the nominal concentration, and finally establish Si source temperature versus concentration dependence.

Author Contributions: Conceptualization, M.M.; methodology, M.K. and M.M.; software, M.K.; validation, M.K. and M.M; formal analysis, M.K. and M.M.; investigation, M.K. and M.R.; resources, K.P. and P.G.; data curation, M.K. and M.R.; writing—original draft preparation, M.K.; writing—review and editing, M.M. and G.S.; visualization, M.K., M.R., and M.M.; supervision, M.M.; project administration, M.M.; funding acquisition, M.M. and G.S. All authors have read and agreed to the published version of the manuscript.

Funding: This work was supported by the project SENSE, founded by National Centre for Research and Development TECHMAT-STRATEG1/347510/15/NCBR/2018.

Acknowledgments: The authors would like to thank K. Ryczko for his comments and discussions regarding the performed calculations.

Conflicts of Interest: The authors declare no conflict of interest. The funders had no role in the design of the study; in the collection, analyses, or interpretation of data; in the writing of the manuscript, or in the decision to publish the results.

References

1. Schiff, H.I.; Mackay, G.I.; Bechara, J. *Air Monitoring by Spectroscopy Techniques*; Wiley: New York, NY, USA, 1994.
2. Wang, C.; Sahay, P. Breath Analysis Using Laser Spectroscopic Techniques: Breath Biomarkers, Spectral Fingerprints, and Detection Limits. *Sensors* **2009**, *9*, 8230–8262. [CrossRef] [PubMed]
3. Airoptic. Available online: https://www.airoptic.pl (accessed on 30 January 2020).
4. Nanoplus GmbH. Available online: https://nanoplus.com (accessed on 30 January 2020).
5. Kosterev, A.; Wysocki, G.; Bakhirkin, Y.; So, S.; Lewicki, R.; Fraser, M.; Tittel, F.; Curl, R.F. Application of quantum cascade lasers to trace gas analysis. *Appl. Phys. B* **2008**, *90*, 165–176. [CrossRef]
6. Nikodem, M.; Gomółka, G.; Klimczak, M.; Pysz, D.; Buczyński, R. Demonstration of mid-infrared gas sensing using an anti-resonant hollow core fiber and a quantum cascade laser. *Opt. Express* **2019**, *27*, 36350–36357. [CrossRef] [PubMed]
7. Kluczyński, P.; Lundqvist, S.; Westberg, J.; Axner, O. Faraday rotation spectrometer with sub-second response time for detection of nitric oxide using a cw DFB quantum cascade laser at 5.33 um. *Appl. Phys. B* **2011**, *103*, 451–459. [CrossRef]
8. Lu, Q.Y.; Manna, S.; Wu, D.H.; Slivken, S.; Razeghi, M. Shortwave quantum cascade laser frequency comb for multi-heterodynespectroscopy. *Appl. Phys. Lett.* **2018**, *112*, 141104. [CrossRef]
9. Lundqvist, S.; Kluczynski, P.; Weih, R.; von Edlinger, M.; Naehle, L.; Fischer, M.; Bauer, A.; Hoefling, S.; Koeth, J. Sensing of formaldehyde using a distributed feedback interband cascade laser emitting around 3493 nm. *Appl. Opt.* **2012**, *51*, 6009–6013. [CrossRef]

10. Tutuncu, E.; Nageke, M.; Becker, S.; Fischer, M.; Koeth, J.; Wolf, C.; Kostler, S.; Ribitsch, V.; Teuber, A.; Groeger, M.; et al. Advanced Photonic Sensors Based on Interband Cascade lasers for Real-Time Mouse Breath Analysis. *ACS Sens.* **2018**, *3*, 1743–1749. [CrossRef]
11. Tomokage, H.; Yanahira, T.; Yoshida, M. On the Basic Assumption to Obtain Carrier Concentration Profile by Differential Hall Coefficient Measurement. *Jpn. J. Appl. Phys.* **1996**, *35*, 1824–1825. [CrossRef]
12. Berreman, D.W. Infrared absorption at longitunal optic frequency in cubic crystal films. *Phys. Rev.* **1966**, *130*, 2193–2198. [CrossRef]
13. Harbecke, B.; Heinz, B.; Grosse, P. Optical Properties of Thin Films and the Berreman Effect. *Appl. Phys. A* **1985**, *38*, 263–267. [CrossRef]
14. Shaykhutdinov, T.; Furchner, A.; Rappich, J.; Hinrichs, K. Mid-infrared nanospectroscopy of Berreman mode and epsilon-near-zero local field confinement in thin films. *Opt. Mater. Express* **2017**, *7*, 3706–3714. [CrossRef]
15. Calvani, P.; Kalaboukhov, A.; Shibayev, P.P.; Salehi, M.; Moon, J.; Oh, S.; Falsetti, E.; Ortolani, M.; Granozio, F.M.; Brubach, J.-B.; et al. Infrared spectroscopy of two-dimensional systems. *Eur. Phys. J. Special Topics* **2019**, *228*, 669–673. [CrossRef]
16. Ibáñez, J.; Tarhan, E.; Ramdas, A.K.; Hernández, S.; Cuscó, R.; Artús, L.; Melloch, M.R.; Hopkinson, M. Direct observation of LO phonon-plasmon coupled modes in the infrared transmission spectra of n-GaAs and n-$In_xGa_{1-x}As$ epilayers. *Phys. Rev. B* **2004**, *69*, 075314. [CrossRef]
17. Motyka, M.; Misiewicz, J. Fast Differential Reflectance Spectroscopy of Semiconductor Structures for Infrared Applications by Using Fourier Transform Spectrometer. *Appl. Phys. Express* **2010**, *3*, 112401. [CrossRef]
18. Motyka, M.; Sek, G.; Janiak, F.; Misiewicz, J.; Klos, K.; Piotrowski, J. Fourier-transformed photoreflectance and fast differential reflectance of HgCdTe layers. The issues of spectral resoultion and Fabry-Perot oscillations. *Meas. Sci. Technol.* **2011**, *22*, 125601. [CrossRef]
19. Motyka, M.; Sęk, G.; Misiewicz, J.; Bauer, A.; Dallner, M.; Höfling, S.; Forchel, A. Fourier Transformed Photoreflectance and Photoluminescence of Mid infrared GaSb-Based Type II Quantum Wells. *Appl. Phys. Express* **2009**, *2*, 126505. [CrossRef]
20. Dyksik, M.; Motyka, M.; Kurka, M.; Ryczko, K.; Schade, A.; Kamp, M.; Höfling, S.; Sęk, G. Electrical Tuning of the oscillator strength in type II InAs/GaInSb quantum wells for active regions of passively mode-locked interband cascade lasers. *Jpn. J. Appl. Phys* **2017**, *56*, 110301. [CrossRef]
21. Gutowski, P.; Sankowska, I.; Karbownik, P.; Pierscinska, D.; Serebrennikova, O.; Morawiec, M.; Pruszynska-Karbownik, E.; Golaszewska-Malec, K.; Pierscinski, K.; Muszalski, J.; et al. MBE growth of strain-compensated InGaAs/InAlAs/InP quantum cascade lasers. *J. Cryst. Growth* **2017**, *466*, 22–29. [CrossRef]
22. Charache, G.; DePoy, D.; Raynolds, J.; Baldasaro, P.; Miyano, K.; Holden, T.; Pollak, F.; Sharps, P.; Timmons, M.; Geller, C.; et al. Moss-Burstein and plasma reflection characteristics of heavily doped n-type InxGa1-xAs and InPyAs1-y. *J. Appl. Phys.* **1999**, *86*, 242452. [CrossRef]
23. Kozub, M.; Motyka, M.; Dyksik, M.; Sek, G.; Misiewicz, J.; Nishisaka, K.; Maemoto, T.; Sasa, S. Non-destructive carrier concentration determination in InAs thin films for THz radiation generating devices using fast differential reflectance spectroscopy. *Opt. Quant. Electron.* **2016**, *48*, 384. [CrossRef]
24. Hinkey, R.; Tian, Z.; Yang, R.; Mishima, T.; Santos, M. Reflectance spectrum of plasmon waveguide interband cascade lasers and observation of the Berreman effect. *J. Appl. Phys.* **2011**, *110*, 043113. [CrossRef]
25. Dyksik, M.; Motyka, M.; Sek, G.; Misiewicz, J.; Dallner, M.; Höfling, S.; Kamp, M. Influence of carrier concentration on properties of InAs waveguide layers in interband cascade laser structures. *J. Appl. Phys.* **2016**, *120*, 043104. [CrossRef]
26. Li, Y.; Stradling, R.; Knight, T.; Birch, J.; Thomas, R.; Phillips, C.; Ferguson, I. Infrared reflection and transmission of undoped and Si-doped InAs grown on GaAs by molecular beam epitaxy. *Semicond. Sci. Technol.* **1993**, *8*, 101. [CrossRef]

materials

MDPI

Article

Detection and Quantification of Cracking in Concrete Aggregate through Virtual Data Fusion of X-ray Computed Tomography Images

Tyler Oesch [1,*], Frank Weise [1] and Giovanni Bruno [1,2]

[1] Bundesanstalt für Materialforschung und–prüfung, BAM (Federal Institute for Materials Research and Testing), 12205 Berlin, Germany; frank.weise@bam.de (F.W.); giovanni.bruno@bam.de (G.B.)

[2] Institute of Physics and Astronomy, University of Potsdam, Karl-Liebknecht-Str.24-25, 14476 Potsdam, Germany

* Correspondence: tyler.oesch@bam.de

Received: 24 July 2020; Accepted: 3 September 2020; Published: 4 September 2020

Abstract: In this work, which is part of a larger research program, a framework called "virtual data fusion" was developed to provide an automated and consistent crack detection method that allows for the cross-comparison of results from large quantities of X-ray computed tomography (CT) data. A partial implementation of this method in a custom program was developed for use in research focused on crack quantification in alkali-silica reaction (ASR)-sensitive concrete aggregates. During the CT image processing, a series of image analyses tailored for detecting specific, individual crack-like characteristics were completed. The results of these analyses were then "fused" in order to identify crack-like objects within the images with much higher accuracy than that yielded by any individual image analysis procedure. The results of this strategy demonstrated the success of the program in effectively identifying crack-like structures and quantifying characteristics, such as surface area and volume. The results demonstrated that the source of aggregate has a very significant impact on the amount of internal cracking, even when the mineralogical characteristics remain very similar. River gravels, for instance, were found to contain significantly higher levels of internal cracking than quarried stone aggregates of the same mineralogical type.

Keywords: X-ray computed tomography (CT); concrete; alkali-silica reaction (ASR); ASR-sensitive aggregate; solubility test; specific surface area; crack detection; automated image processing; damage quantification

1. Introduction

1.1. Alkali-Silica Reaction (ASR)

Despite decades of research, the problem of harmful alkali-silica reaction (ASR) in the field of concrete construction has not yet been satisfactorily solved. For the first time in 1940, Stanton [1] reported damaging strains within concrete due to chemical reactions of cement and aggregate. In the 1950s, Powers and Steinour [2,3] developed initial models of ASR's damage mechanism. In the 1970s, Locher and Sprung [4] identified opal and porous flint as alkali-sensitive aggregates and developed theories on their reaction mechanisms. In the 1980s, various researchers conducted in-depth studies on the influence of alkali metal salts on the swelling pressures of the ASR gel [5,6]. The current state of knowledge in the field of ASR has also been extensively described in number of recent publications [7,8].

During the ASR process, the reactive SiO_2 within aggregates reacts with alkalis (supplied from the cement paste or from an external source) in the presence of water to form expansive alkali silicate hydrates (ASR gels). Because the tensile strength of road surface concrete is often significantly lower than the swelling pressures caused by ASR gels, cracking can be induced [9]. The progression and the extent of the resulting cracking processes are determined to a large extent by the type of aggregate. For example, fast-reacting aggregates (among others, flint, opaline sandstone and mudstone) are characterized by gel and crack formation emanating from the transition zone between the grain and the mortar matrix [7]. On the other hand, in the case of the slow-reacting aggregates (for instance greywacke or quartz porphyry), which were of primary of interest in this project, the gel formation takes place above all inside the aggregate itself, which results in the formation of internal aggregate cracks [7].

The severity of the ASR degradation process is thought to be partially dependent on the amount of porosity within a given aggregate that is accessible to liquid penetrating from the sample surface. Against this background and to evaluate the alkali sensitivity of the aggregate, the influence of the specific surface area on the solubility behaviour of four different aggregates in 0.1 M potassium hydroxide solution without and with defined addition of NaCl at a temperature of 80 °C was thoroughly investigated in a joint project [10]. To quantify this relationship, a non-destructive method is needed for measuring both the external surface area of aggregates and the internal surface area of aggregate cracks and pores, including a differentiation of internal voids connected to the sample surface from those isolated from the surface. The primary focus of this publication is devoted to the crack detection method. A detailed analysis of the implications for ASR damage, including a comparison of the CT results with those from other porosity measurement methods, such as mercury porosimetry and the Brunauer–Emmett–Teller (BET) method, can be found in Oesch et al. 2020 [11] and in Weise et al. 2019 [10].

1.2. X-Ray Computed Tomography (CT)

The development of X-ray computed tomography (CT) began in the 1960s and clinical X-ray CT investigations have been widely conducted since the 1970s [12,13]. Since that time, many different reconstruction algorithms have been developed for clinical use, including algorithms based on the algebraic reconstruction technique (ART), filtered back projection (FBP), and iterative reconstruction (IR) [14]. Medical X-ray CT scanning systems are, however, unsuitable for many applications in materials science given their lower X-ray energy characteristics and coarser resolution compared to specialized laboratory-based X-ray CT systems used for materials research [15]. These differences occur both because the size of the intended scanning objects tends to significantly differ between clinical and materials science applications and because the X-ray absorption characteristics of live tissues are much lower than those of materials such as concrete and steel.

X-ray CT has been used in non-destructive concrete research applications for more than 30 years [16,17]. In this scanning method, a sample is placed on a rotating table between an X-ray source and an X-ray detector [18]. By adjusting the distances between the X-ray source, the sample and the X-ray detector, it is possible to vary the voxel (i.e., 3D pixel) resolution in the resulting images. The penetration of the sample by the X-ray beam causes an X-ray attenuation image of the sample to be projected upon the detector. By recording these projected images during the 360° rotation of the sample, the projections can be inverted using volume reconstruction algorithms, which produce a 3D representation of X-ray attenuation within the sample [19]. X-ray attenuation is approximately proportional to local material density and can be used to identify single objects within a material (or structure) and to individually separate and analyze those objects.

Previous research has shown that X-ray CT scans can be taken during incremental testing. This includes mechanical testing (such as unconfined compression, split cylinder, triaxial, and reinforcing bar pull-out testing [20–24]), chemical testing (such as the measurement of progressive corrosion in reinforced concrete during repeated exposure to chloride [25] and the transport of water [26–28]) and thermal testing (such as water migration in heated concrete [29]).

1.3. Crack Detection and Quantification

Crack detection and quantification is important for understanding and modelling a series of material behaviours. Precise measurement of crack surface area is needed, for instance, in order to calculate the fracture energy expended during damage processes using basic fracture mechanics relationships [30]. The crack orientations have also been observed to exhibit a behaviour that is highly dependent on the anisotropy of the material structure, such as fibre orientation within fibre-reinforced concretes [31].

Crack detection and quantification within X-ray CT images have been the subject of extensive past research. Most of these crack detection methods have leveraged one or more unique characteristics of cracks, which differentiate them from the surrounding material. Possibly the most popular research approaches have focused on the use of template-matching methods in order to separate cracks from the surrounding materials [32,33]. This method of crack detection relies on the similarity of cracking structures to certain template shapes, such as small planes or discs. Although impressive results have been demonstrated using the template-matching method, the template parameters are not universal and must generally be tailored for each material and each imaging scenario.

Research was recently carried out by Paetsch (2019) [34] with the goal of partially overcoming these challenges related to using template matching approaches. This research indicated that the results of a series of analyses carried out using different template shapes can be combined in order to obtain a greater accuracy of the detected cracks. However, Paetsch (2019) [34] has underlined that further problems remain to be solved that are common to most template-matching methods, such as difficulties detecting cracks in areas where significant crack branching or widening occurs.

Another method that takes advantage of the narrow shape characteristics of cracks is a Hessian-based approach [35,36], which identifies regions that exhibit sharp changes in image intensity. Percolation methods have also shown significant promise in detecting cracks across a range of materials [37]. These methods leverage the fact that most cracks are continuous, narrow objects with relatively consistent (low) density. Percolation methods have proven insufficient, however, to accurately detect complex cracks of varying size in most materials. Impressive results have also been obtained through a combination of the Hessian and percolation-based methods into the Hessian-driven percolation approach, although the processing time required for such an analysis remains prohibitive for most high-resolution CT images [32].

Many methods have also been employed that leverage the unique characteristics of a specific material or damage scenario. One excellent example of this approach is the use of digital volume correlation (DVC) to detect cracks in samples subjected to in-situ loading [38–40]. DVC is used to measure strains within samples by calculating voxel movements between subsequent CT images (such as images calculated before and after a loading increment). Cracks can, thus, be identified as areas of either high strain or poor correlation within DVC images or through more complex analysis methods, such as phase-congruency analysis. Although these methods appear to exhibit a relatively high accuracy (even sub-pixel, see [41]), they are only useful for detecting cracks caused by progressive in-situ testing with simultaneous CT. They do not provide any benefit for detecting cracks that are already present in specimens prior to testing. Phase congruency can, however, also be used to detect edge features (including cracks) within the greyscale images without direct reference to DVC [42].

Another example of a material-specific crack detection approach is the leveraging of typical wood structure within logs to identify cracks. These cracks typically run perpendicular to the growth rings of the logs and can, thus, be easily identified by their orientation characteristics [43].

Despite the many promising crack detection methods outlined here, a series of obstacles remain that have prevented the implementation of consistent, accurate, and quantitative crack analyses as part of CT scanning. First, most of these methods require some amount of tailoring for specific material properties (as in the case of template matching and percolation) or specific crack conditions (such as measuring only those arising from in-situ testing using DVC methods). Second, none of these methods has been successfully validated for the quantitative determination of crack properties (such as surface area) based on other standard measurement techniques.

In this study, we will show, on the example of different aggregate types, how the obstacles mentioned above can be circumvented by a novel data fusion strategy.

2. Materials and Methods

2.1. Sample Selection

For this testing series, a group of aggregates from four different categories were selected and analysed. These categories were selected in order to include stones with a variety of different mineralogical compositions, deterioration conditions, and alkali-sensitivity characteristics (Table 1 and Figure 1) [44]. All stones included in this analysis were in the 8 mm to 16 mm size range: During the sieving process, the individual stones passed successfully through a sieve with a 16 mm mesh but were unable to pass through a sieve with an 8 mm mesh [45].

Table 1. Categories of stone selected for cracking analysis.

Category	Stone Type	Alkali Sensitivity [40 °C-BV]
GK1	River Gravel	EIII-S
GK2	Quarried Stone (Greywacke)	EIII-S
GK3	Quarried Stone (Rhyolite)	EI-S
GK4	River Gravel	EI-S

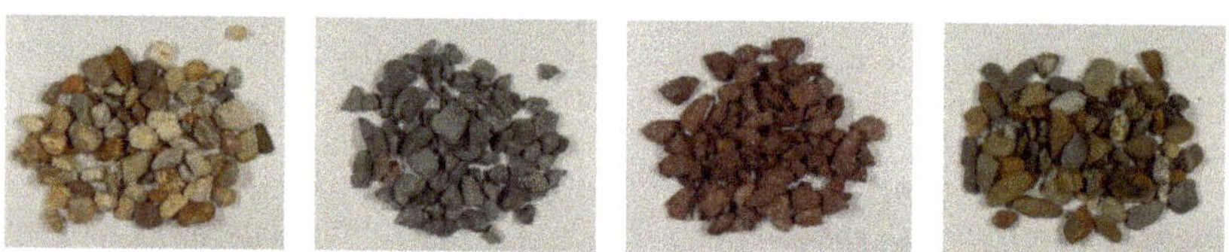

GK1 GK2 GK3 GK4

Figure 1. Photographs of typical individual grains from each of the stone categories.

For the categories GK1 and GK4, the selected stones were sieved from a natural river gravel. Such river gravels are typically characterized by significant mineralogical deterioration due to naturally occurring weathering processes. In order to characterize these river gravels, which have a heterogeneous mineralogical composition, the primary types of rock (7 types in total) occurring in GK1 and GK4 were determined and 10 individual grains of each type were selected for CT analysis.

For the categories GK2 and GK3, the selected stones were sieved from crushed stone chips that were quarried from solid rock deposits. Such quarried stones are typically characterized by greater mineralogical integrity than river gravels because they have not been exposed to significant weathering. Due to the homogeneity of the quarried stone, the selection of individual grains was limited to 10 each for GK2 and GK3. Thus, combining all the individual grains from each of the four stone categories included in this research study, a total of 90 different individual grains were investigated.

2.2. CT Scanning

During this research program, an acceleration voltage of 130 kV and current of 180 µA were used for the X-ray source. The X-ray beam was also filtered using a 0.5 mm thick Copper plate immediately upon leaving the source in order to remove (unwanted) photons of small energies from the X-ray beam, thereby increasing the contrast of the resulting images. The flat panel detector used for this scanning contained a 2048 × 2048 pixel field.

Individual aggregates were sorted based on mineralogical characteristics and placed within corresponding plastic tubes with small pieces of foam separating the aggregates from one another.

As a result of heating and deterioration of the target material within the X-ray tube as well as changes in detector sensitivity over time, significant variations in the measured X-ray beam intensity and distribution can occur. To compensate for these variations, "dark-field" and "bright-field" images, which correspond to blank images (i.e., containing no sample) acquired with no illumination and full illumination, respectively, were acquired prior to the scanning of each plastic tube of samples. These images were then used to calibrate the X-ray images of the samples.

The CT machine was pre-programmed to collect a complete scan of each aggregate before repositioning the plastic tube using a manipulator and beginning the scan of the next aggregate. Thus, the scanning conditions for all stones within any given plastic tube were identical. All scanning conditions other than resolution were also held constant for all plastic tubes.

It was important to maintain scanning conditions that were as consistent as possible to ensure that the results of the crack analysis would be comparable. In spite of this, the resolution was maximised for each stone type, if they significantly varied in size and shape. This was done by adjusting the distance between the plastic tube and the X-ray source. The corresponding voxel sizes were then calculated directly from the measured distances between the X-ray source, the sample holder and the X-ray detector for each individual set of scans.

Although these variations in image resolution are known to directly affect measurements of crack properties, such as surface area (increased surface area is generally detected with decreasing voxel sizes), some estimation of the magnitude of this effect can already be accounted for based on a recent study [46]. The results of this study indicate that even a doubling of voxel size does not appear to generally change the measured crack surface area by more than a factor around two. We will see below that the small voxel size variations had little influence on the much larger variations in measured crack surface area among the different aggregate types, so that comparisons could be made, and conclusions drawn.

2.3. Image Analysis

2.3.1. Data Fusion Approach

In such an environment, where many analysis methods are readily available, but no individual method is sufficiently accurate to provide the needed measurements, the use of a data-fusion inspired approach becomes very promising and attractive. Data fusion enables researchers to combine data from multiple sources in order to produce more consistent and accurate results than those provided by any single source [47]. In non-destructive testing, the different sources used in data fusion typically result from differing non-destructive measurement approaches. Such data-fusion based approaches have even been successfully applied to the application of crack detection [48].

It is, however, often the case that researchers only possess meaningful data from a single measurement technique, such as CT (often indeed used as a reference), or do not have access to further non-destructive testing equipment. Even in this case, it should be possible to use the theory behind the data fusion approach to improve the overall quality of the quantitative determinations resulting from image analysis. Rather than relying on a wide variety of physical measurement techniques, data fusion in this approach would be carried out by using the output from an array of different image analysis techniques. Since such an approach relies on the fusion of different computationally generated data sources resulting from the same original CT image rather than on a variety of different physical measurement techniques, it can be more appropriately described as virtual data fusion. This is similar to the process that is thought to occur when an expert identifies cracks using the human eye. The expert has only one measurement technique (a visual image), but by considering its many different aspects (such as coloration, shape, and relationship to surrounding objects or planes of stress), the expert is able to quite easily and accurately detect a crack on the surface of a specimen.

Figure 2 shows a diagram of this virtual data fusion approach. In the diagram, the characteristics of a crack are displayed in dark grey boxes and the image processing steps for identifying objects with those characteristics are depicted in light grey boxes. The list provided here is only meant to serve as an example and is by no means exhaustive. It is clear, however, that results obtained through a fusion of the results from twelve such independent analyses will be much more accurate and resilient to varying material conditions than the results from any single analysis technique. An example of how such a step-by-step approach can identify individual cracks within a generic concrete sample is shown in Figure 3.

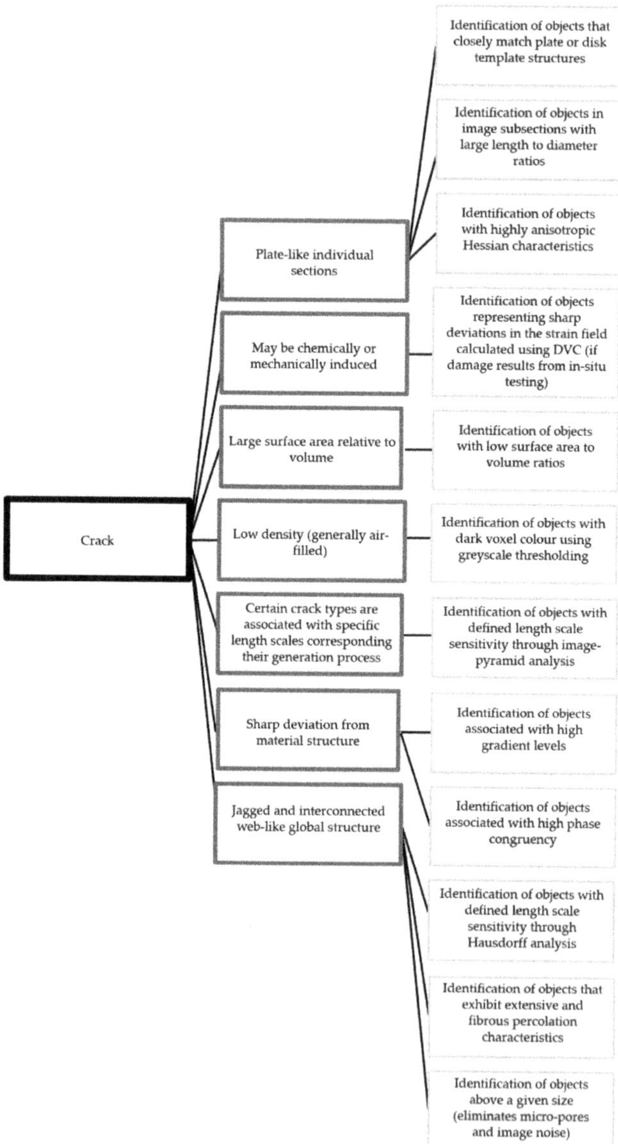

Figure 2. Virtual data fusion web linking cracks (black) with their characteristics (dark grey) and image analysis techniques for identifying objects with those characteristics (light grey).

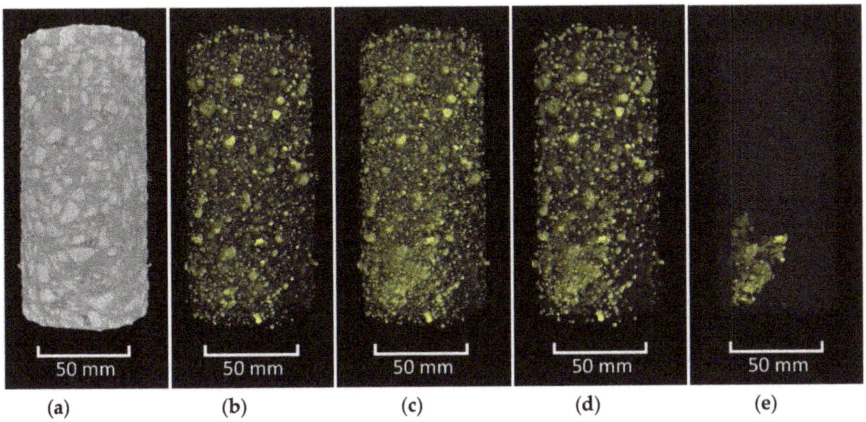

Figure 3. Step-by-step isolation of cracking by means of virtual data fusion. CT image of a concrete sample (**a**), voids identified using a greyscale threshold (**b**), voids identified in (**b**) added to interfacial zones identified using a gradient-based analysis (**c**), removal of small connected components (noise and isolated pores) from image (**d**), and elimination of components with small specific surface areas from image (**e**). Raw data courtesy of U.S. Army Engineer Research and Development Center (ERDC).

2.3.2. Implementation

Although, in theory, twelve or more different analysis methods could be used in the virtual data fusion implementation for analysing the ASR-sensitive aggregate dataset, in practice such a full implementation was impractical. We selected a partial implementation of the scheme including only three analysis approaches. The choice included the approaches that would produce cracking data with sufficient accuracy for this specific application while simultaneously testing the effectiveness of the virtual data fusion concept for obtaining accurate and consistent results. A successful demonstration of the virtual data fusion method with only three analysis components would then give justification for the further development of the algorithm to gradually include additional analysis modules, simultaneously growing in accuracy and resiliency. Moreover, the modular architecture of the strategy could more easily be adapted to different problems than a full but rigid analysis.

In fact, for the specific ASR-sensitive aggregate analysis described in this paper, it was not important to separate internal pores from internal cracks since the surface area of both pores and cracks was vulnerable to ASR degradation. Thus, modules related to this differentiation could be left out of the analysis. Furthermore, since scans were not available at varying levels of degradation, all methods relying on changes in sample state relative to time (such as DVC-based methods) were also left out of this analysis. All image analysis described in this paper was completed using custom algorithms developed and implemented in MATLAB [49].

Module 1: Identification of Objects with Low Density

When cracks or pores have widths exceeding two voxels, the voxels in their centres are completely filled with air. Thus, these central voxels are characterized by a particularly low greyscale value. In order to separate these voxels, a threshold has to be selected and subsequently used for image binarization. All voxels darker than the threshold would then be transformed to white and all voxels lighter than the threshold transformed to black. Given that this threshold must be consistently applied for a large range of datasets, an automated selection method was implemented.

Using the triangle selection algorithm [50,51], a virtual line is drawn from the origin of the image histogram to the top of the largest histogram peak (excluding the initial peak at zero, which represents voxels from the air around the specimen) (Figure 4). A calculation is then conducted to determine which point on the histogram is furthest from the virtual line along an intersecting, perpendicular line. The location of that point is identified as a potential threshold. It was found, however, that such a threshold is rather over-encompassing, leading to the introduction of considerable noise into the resulting binarized images. Thus, a slightly more conservative threshold value was also implemented for this module that is 25% lower than the original threshold. An example of the results from this analysis can be observed in Figure 5.

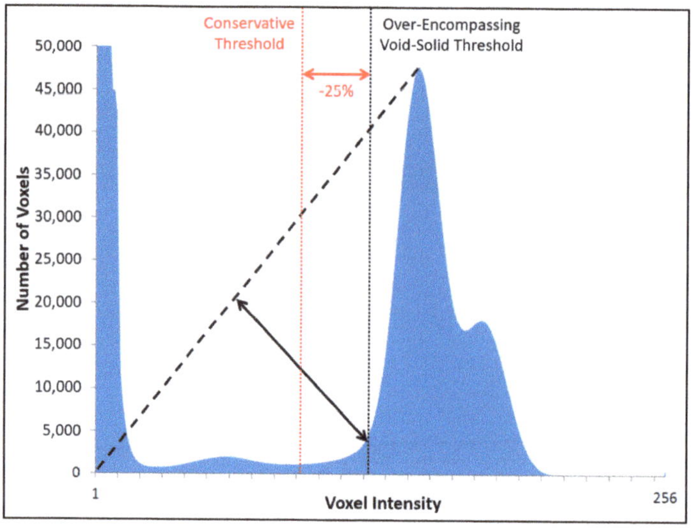

Figure 4. Automated threshold selection using the triangle approach (based on figure in [10]).

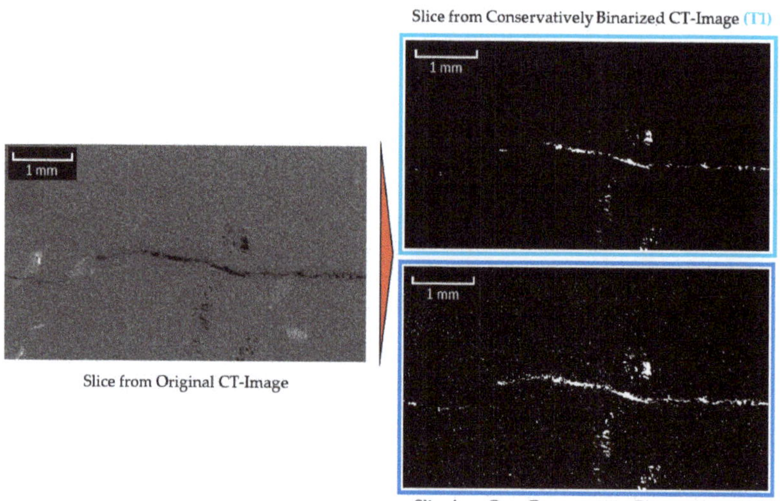

Figure 5. Binarization of greyscale image using the conservative and over-encompassing thresholds identified using the automated selection method (based on figure in [10]).

Module 2: Identification of Objects with High Gradient

When an object in a CT-image has either a much higher or a much lower greyscale value compared to the surrounding material, its edges can be identified as regions of high gradient. For this purpose, a gradient-magnitude image is calculated from the original CT-image through the use of the Sobel operator (Figure 6) [52]. Since the resulting gradient image consists of a wide range of grey values, it must also be binarized. For this purpose, a histogram of the gradient image grey values is computed and another automated triangular threshold selection is completed, this time from the right edge of the histogram (Figure 7). The resulting binary image contains the edges of both bright and dark objects in the original CT image (Figure 8). In order to remove the edges associated with the bright (high-density) objects in the CT-image as well as some of the noise, the binarized gradient image can be multiplied by a binarized CT-image (this time using the "over-encompassing void-solid threshold" identified in Figure 4) (Figure 9).

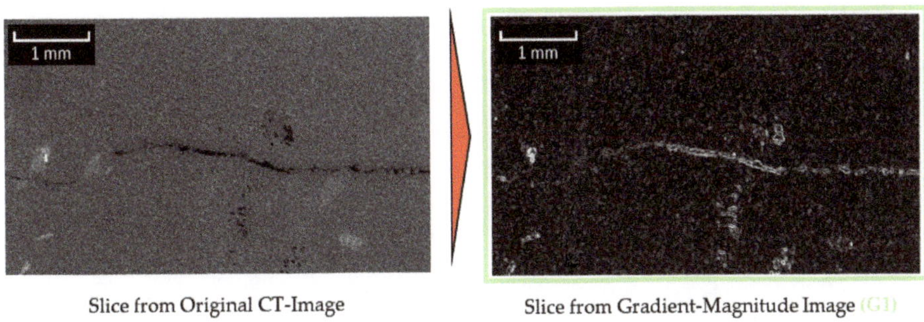

Figure 6. Generation of the gradient-magnitude image (based on figure in [10]).

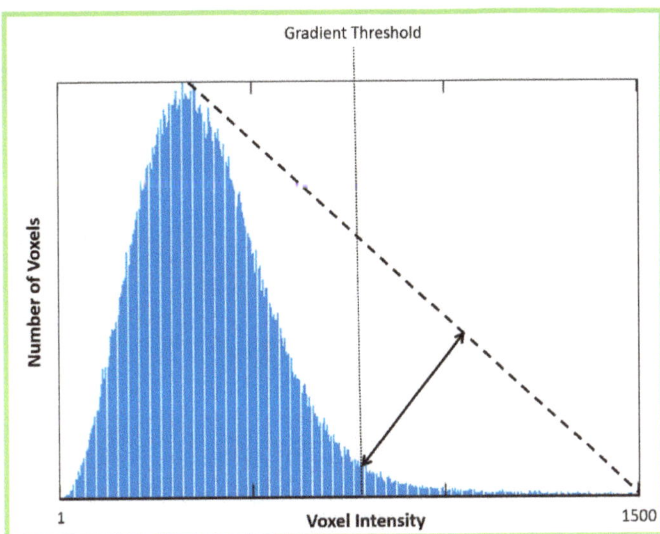

Figure 7. Automated gradient threshold selection using the triangle approach (based on figure in [10]).

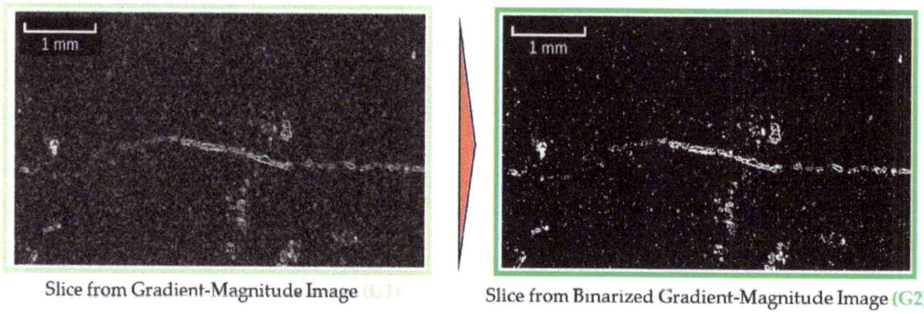

Figure 8. Binarization of the gradient image (based on figure in [10]).

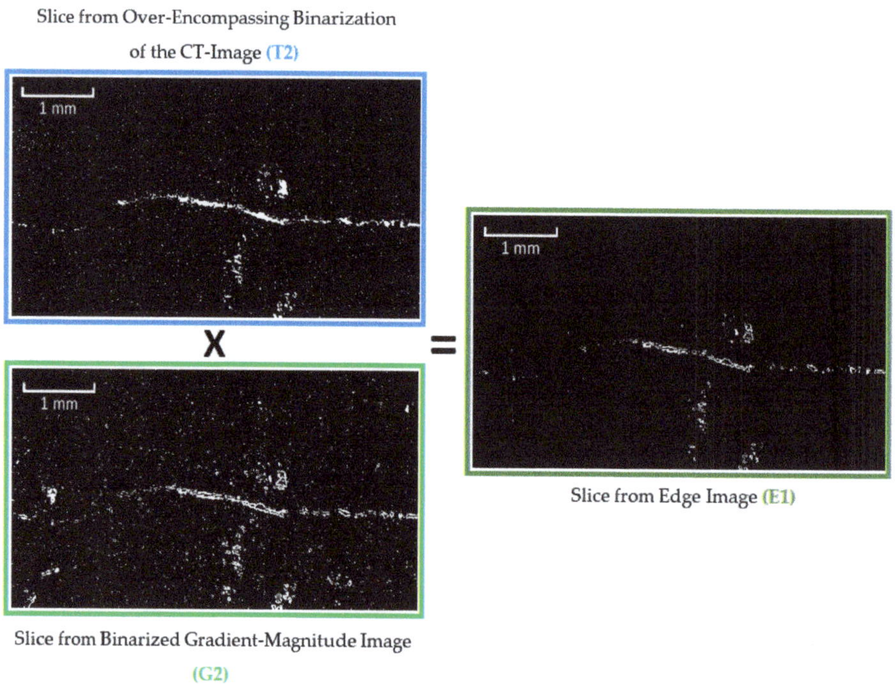

Figure 9. Generation of an image containing the outer edges of only low-density objects (based on figure in [10]).

Module 3: Identification of Objects above a Given Size

Prior to noise identification and removal, the results from Modules 1 and 2 were combined into a single image (Figure 10). This ensured that the centres and edges of the cracks/pores would both be present and accounted for during object size analysis. The object size analysis was completed using a connected components algorithm. During this analysis, individual voxels are assessed to determine whether they are part of a larger object in the binary image by analysing whether they adjoin other voxels of the same color. For this analysis, white voxels with touching faces, edges or corners (also known as "26-connected") were identified as connected. Objects with voxel volumes smaller than 125 voxels were subsequently identified and eliminated (Figure 11).

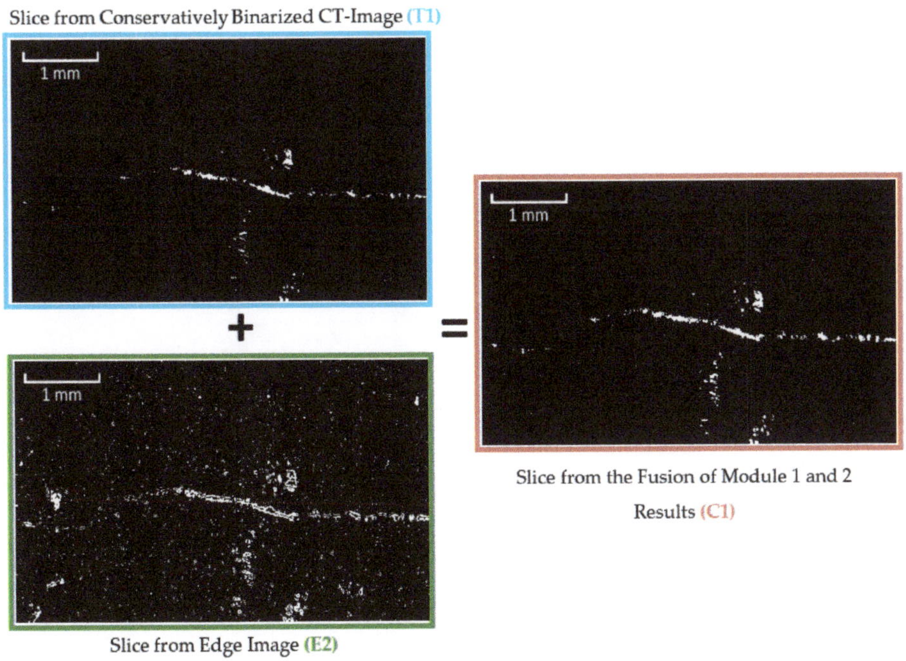

Figure 10. Fusion of images resulting from Modules 1 and 2 (based on figure in [10]).

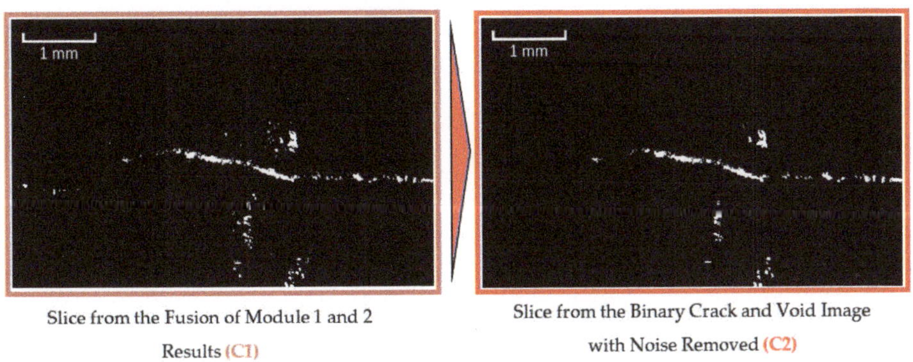

Figure 11. Removal of noise from binary crack and void image (based on figure in [10]).

It was found that the use of smaller cubes than 125 voxels (with five voxel long sides) tended to leave considerable noise within the image while the use of larger cubes resulted in the loss of a considerable number of voxels along the path of the cracks. Given that the full CT images each contained over seven billion voxels, such an object with a volume of 125 voxels represented less than one ten-millionth of the total image volume. A flowchart of the entire crack-detection process is provided in Figure 12.

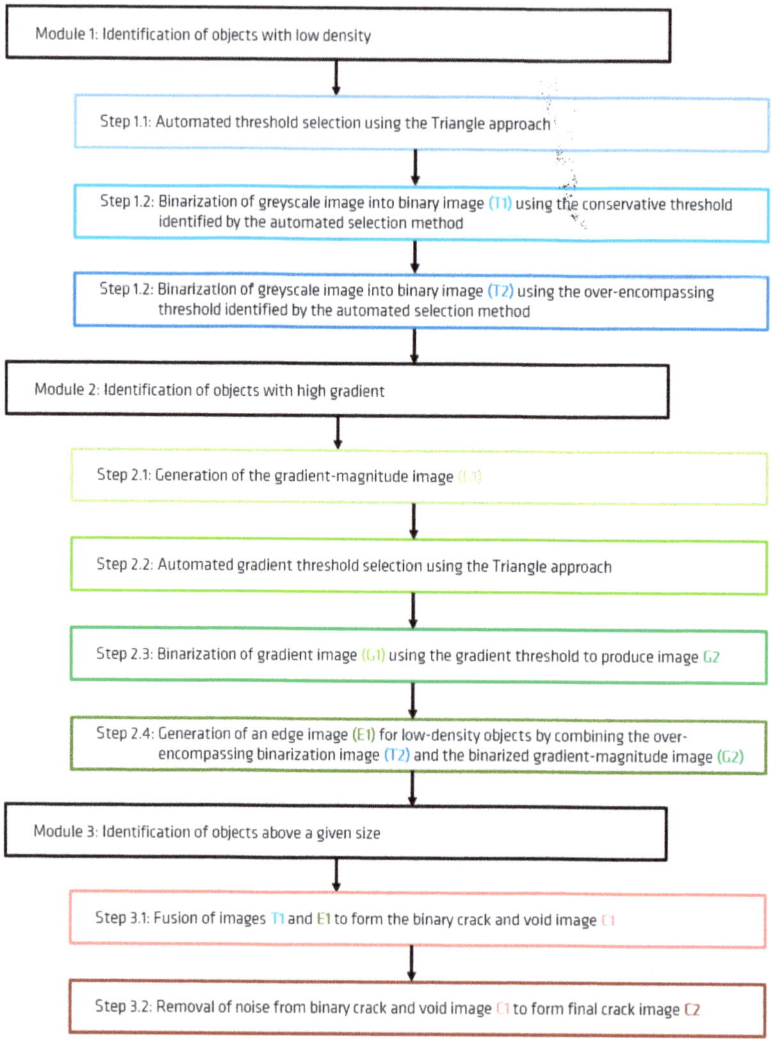

Figure 12. Flowchart of crack-detection process (based on figure in [10]).

3. Results

After the cracks and internal voids were identified within each of the aggregates, their characteristics could be quantified and compared. Of primary interest for this analysis was the determination of volume and surface area characteristics. In particular, in order to compare the results of the CT-analysis with those of other non-destructive measurement techniques, it was important to separate surface-connected cracks and voids (referred here to as "open voids") from internally isolated cracks and voids (referred to here as "closed voids"). This is because measurements of internal surface area using the Brunauer–Emmett–Teller (BET) method [53,54] only account for internal voids accessible from outside of the sample. Similarly, the measurement of void volume through mercury porosimetry is thought to depend primarily on the saturation of internal voids that are connected to the stone surface. Such a separation of surface-connected cracks could be completed using a connected components analysis in which only cracks and voids containing voxels that touched the stone surface were retained.

The crack/pore surface area measurements obtained using this approach are provided for all of the river-gravel type aggregates in Figure 13 and for both river-gravel and quarried-stone aggregates of the minerals rhyolite and greywacke in Figure 14. Note that the measurements for the greywacke (GK4) and rhyolite (GK1) river gravels appear in both figures for comparison purposes. Figure 15 also provides 3D images of cracking distributions for two selected individual grains of the same mineral (greywacke), where one grain has been extracted from a quarry and the other has been taken from river gravel. For images of all analysed aggregate types, see Appendix A. Tabulated values of the surface area measurements are also provided for each individual sample in Appendix B.

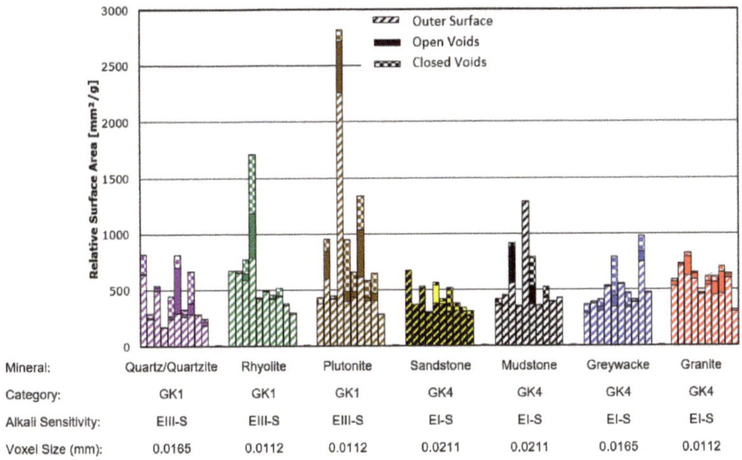

Figure 13. Measured surface area for river-gravel type aggregates (based on figure in [10]).

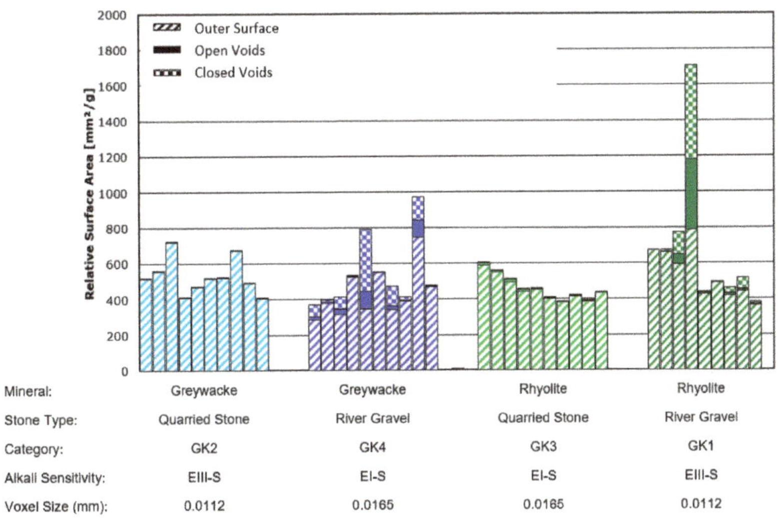

Figure 14. Surface area measurement comparison between quarried-stone (GK2 and GK3) and river-gravel (GK1 and GK4) aggregates of the same mineral types (based on figures in [10,11]).

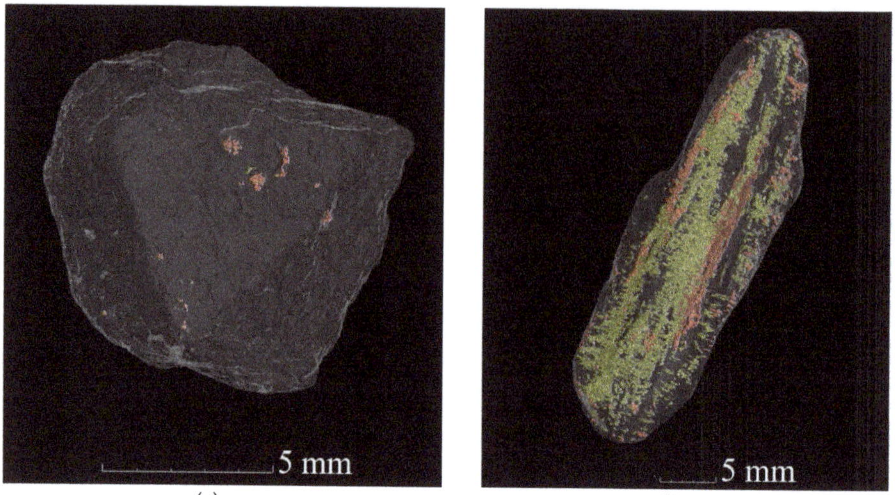

(a) (b)

Figure 15. Example images of cracking within a quarried greywacke aggregate (**a**) and a river-gravel type greywacke aggregate (**b**). Red pixels denote externally accessible cracks/pores and yellow pixels denote externally inaccessible (closed) cracks/pores (reproduction of figures from [10,11]).

Error bars in Figures 13 and 14 could not be calculated. This is because there is still no universally accepted method for estimating the combined error introduced by CT measurement systems and image processing algorithms. Although numerous approaches for estimating such error bars have been proposed [55,56], these tend to be rather computationally intensive and time consuming and remain an active area of research. For pure dimensional measurements, it is common to use either the voxel size or the focal spot size of the X-ray tube as an estimation of error. Given that the focal spot size of this scanning system was significantly smaller than the voxel sizes obtained during these investigations, the voxel sizes for each scan (also listed in Figures 13 and 14) can be taken as an estimation of possible error in the dimensional measurements.

From Figure 14 it is clear that the amount of internal cracking (including both surface-connected cracking and non-surface-connected cracking) in the river gravel aggregates was much higher than that in the quarried aggregates, even when their mineralogical characteristics were similar. The magnitude of this effect is also much too large to be attributed to variations in CT resolution. The underlying basis for these differences in quantitative crack measurements can also be estimated through visual observation of CT images, such as those displayed in Figure 15. Clear, layered cracking is visible within river gravel greywacke stones; this is not present within the greywacke stones extracted from quarries. This is thought to result from the aggressive weathering process that river gravel is subjected to during its lifecycle prior to construction use. This indicates that the selection of high-quality aggregate based on mineralogical characteristics alone may be insufficient.

It is also clear from Figure 13 that even for stones from a single source and with a single mineral composition, a significant amount of variability in internal porosity and cracking is present. The variation between individual stones of a single type (such as rhyolite (GK1)) is often much larger than the average difference between two entirely different stone types (such as between rhyolite (GK1) and granite (GK4)). Thus, we recommend the use of large statistical samples to properly characterize each stone type for ASR sensitivity.

4. Discussion and Conclusions

This research clearly demonstrates the need for universal, automated, and consistent crack detection methods that allow the cross comparison of results from large quantities of CT-scan data from different sample types. A framework, called "virtual data fusion", was developed that has the potential to successfully provide such a method. A partial implementation of this method in a custom program was developed for use in research focused on crack measurement in ASR-sensitive aggregates. Our results demonstrated the success of the program in effectively identifying crack-like structures and measuring their characteristics such as crack extension (relative surface area) and surface connectivity.

These results demonstrate the significant impact that the source of extraction can have on the characteristics of aggregates. Even for aggregates of the same mineral type, river gravels contain significantly higher levels of internal porosity and cracking than quarried stone. This is thought to result from the aggressive weathering process that river gravel is subjected to prior to its selection and use for construction. This indicates that the selection of high-quality aggregate based on mineralogical characteristics alone may be insufficient. It is also clear from these results that there is a significant amount of variability in internal porosity and cracking even for stones with the same mineralogical characteristics and extraction source. Thus, large statistical samples will be necessary to properly characterize each stone type for ASR sensitivity.

Author Contributions: Conceptualization, T.O.; methodology, T.O. and F.W.; software, T.O.; investigation, T.O. and F.W.; resources, T.O., F.W., and G.B.; writing—original draft preparation, T.O. and F.W.; writing—review and editing, T.O., F.W., and G.B.; visualization, T.O.; supervision, F.W. and G.B.; project administration, F.W.; funding acquisition, F.W. All authors have read and agreed to the published version of the manuscript.

Funding: This presentation is based on parts of the research project carried out at the request of the Federal Ministry of Transport and Digital Infrastructure, represented by the Federal Highway Research Institute, under research project No. 06.0108/2014/BRB. The author is solely responsible for the content. (Dieser Präsentation liegen Teile der im Auftrag des Bundesministeriums für Verkehr und digitale Infrastruktur, vertreten durch die Bundesanstalt für Straßenwesen, unter FE-Nr. 06.0108/2014/BRB durchgeführten Forschungsarbeit zugrunde. Die Verantwortung für den Inhalt liegt allein beim Autor).

Acknowledgments: The authors would like to thank Dietmar Meinel from the Bundesanstalt für Materialforschung und—prüfung (BAM, Federal Institute for Materials Research and Testing) for his guidance and support during the CT scanning and data analysis.

Conflicts of Interest: The authors declare no conflict of interest. The funders had no role in the design of the study; in the collection, analyses, or interpretation of data; in the writing of the manuscript, or in the decision to publish the results.

Appendix A. Visualization of Individual Grains

A visual impression of the internal microstructure of the individual grains is provided by selected CT-based visualizations (Figures A1–A4). In these images, the solid material of the individual grains is shown semi-transparently. This allows a better spatial visualization of the cracks and pores emanating from the outer surface (i.e., open voids—colored red) and the cracks and pores not accessible from the outside (i.e., closed voids—colored yellow).

	Selected CT Images of Individual Aggregate of Type GK1		Short Description
Quartz/Quartzite	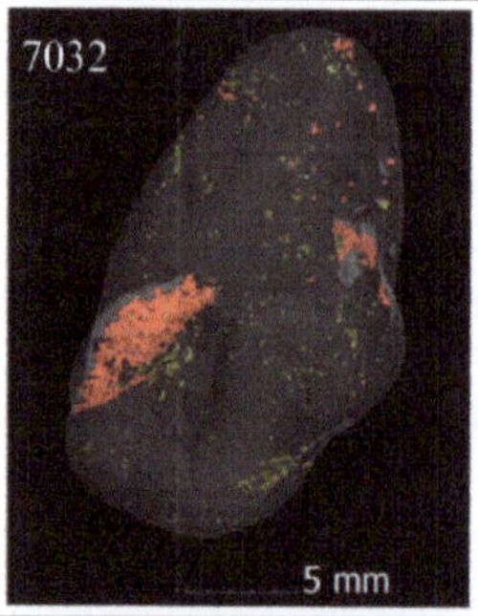	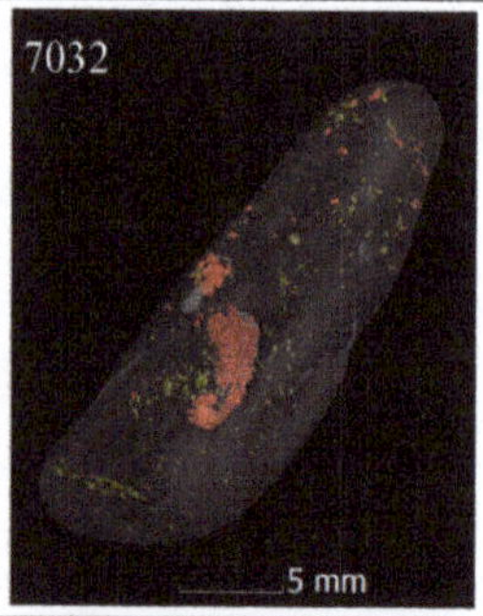	Relatively low content of both open and closed void space
			Relatively high content of both open and closed void space
Rhyolite	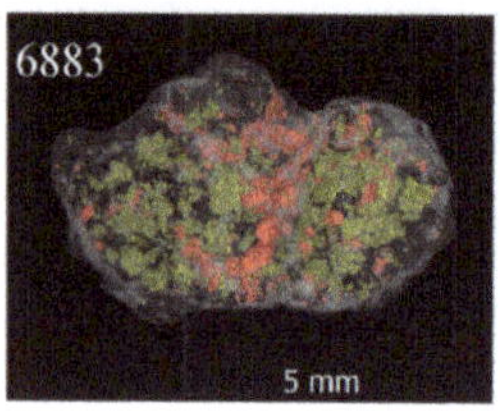	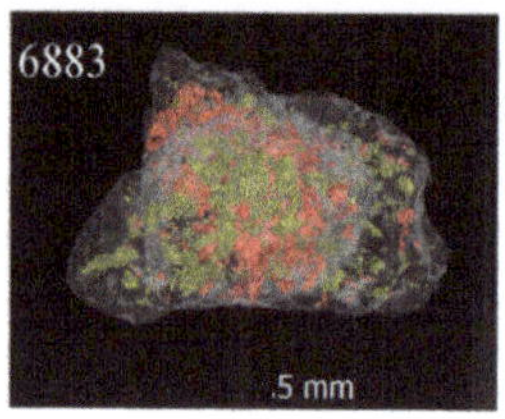	Relatively moderate content of both open and closed void space

Figure A1. *Cont.*

Plutonite			Relatively high content of both open and closed void space
			Relatively low content of open void space and relatively high content of closed void space
Legend:	▬ Open voids	▬ Closed voids	

Figure A1. CT-visualizations for individual river gravel aggregate of type GK1 (the left and right images for each aggregate show views from the 0 and 90 degrees, respectively) (based on figures in [10,11]).

	Selected CT Images of Individual Aggregate of Type GK4		**Short Description**
Sandstone	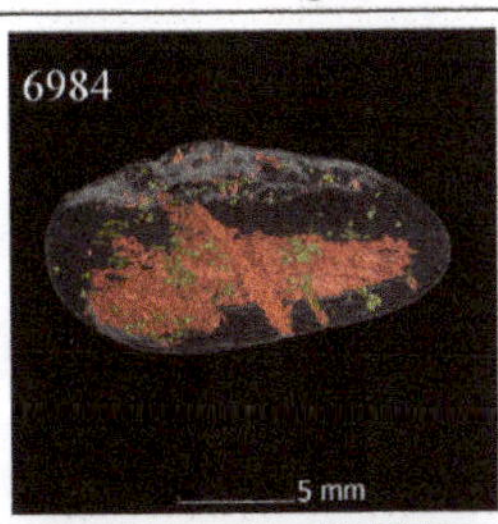	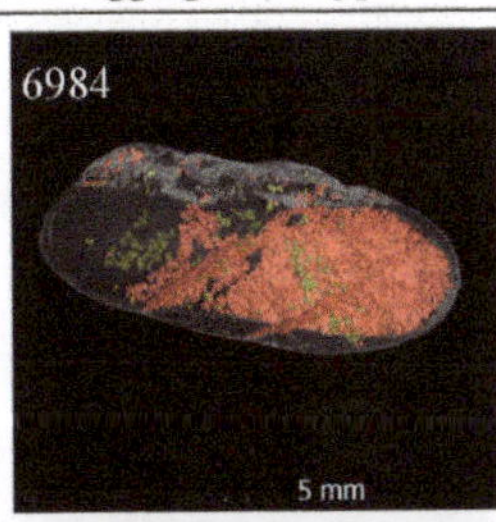	Relatively high content of open void space and relatively low content of closed void space (layering also visible)
Mudstone	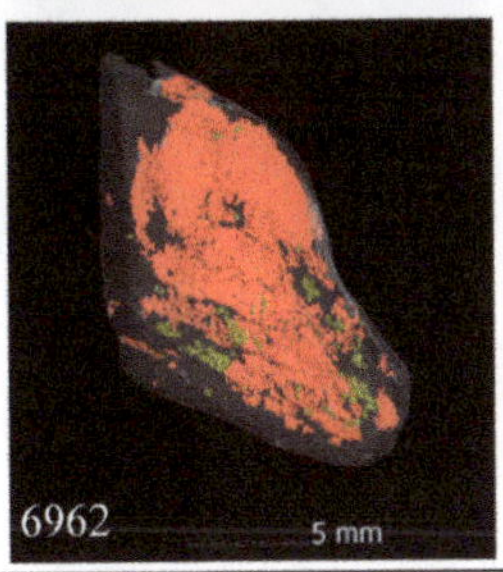		Relatively high content of open void space and relatively low content of closed void space (layering also visible)

Figure A2. *Cont.*

Greywacke			Relatively moderate content of open void space and relatively high content of closed void space (layering also visible)
Granite			Relatively high content of open void space and relatively low content of closed void space
Legend:	Open voids (orange)	Closed voids (green)	

Figure A2. CT-visualizations for individual river gravel aggregate of type GK4 (the left and right images for each aggregate show views from the 0 and 90 degrees, respectively) (based on figures in [10,11]).

	Greywacke Aggregate	
	Quarried Stone (GK2)	**River Gravel (GK4)**
Selected CT Images of Individual Aggregate		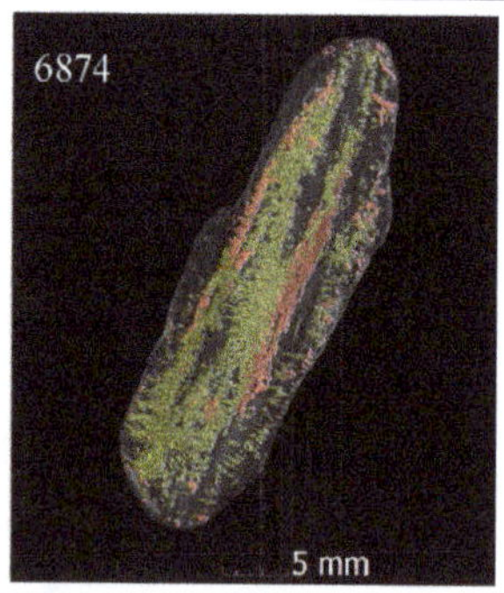
Short Description	Relatively low content of both open and closed void space	Relatively moderate content of open void space and relatively high content of closed void space (layering also visible)
Legend:	Open voids (orange)	Closed voids (green)

Figure A3. CT-visualizations for individual greywacke aggregate from quarried stone (GK2) and river gravel (GK4) (based on figures in [10,11]).

	Rhyolite Aggregate	
	Quarried Stone (GK3)	**River Gravel (GK1)**
Selected CT Images of Individual Aggregate		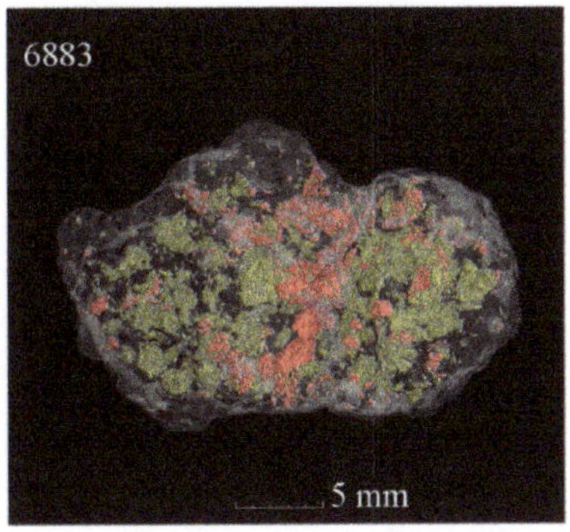
Short Description	Relatively low content of both open and closed void space	Relatively moderate content of both open and closed void space
Legend:	Open voids	Closed voids

Figure A4. CT-visualizations for individual rhyolite aggregate from quarried stone (GK3) and river gravel (GK1) (based on figures in [10,11]).

The strong fluctuation in the amount of open and closed voids in the individual quartz/quartzite grains of aggregate type GK1 is clearly visible in Figure A1. For the rhyolite of aggregate type GK1, Figure A1 includes a CT-based visualization of the single grain with the highest surface areas of open and closed voids. The broad spectrum of open and closed void surface characteristics for plutonite is demonstrated by the CT-based visualizations of the two individual grains shown in Figure A1.

In contrast to the GK1 aggregate, the surface characteristics of the individual grains of aggregate type GK4 tend to vary much less. In Figure A2, a CT-based visualization of the individual sandstone grain with the highest surface area of open voids of all examined sandstone grains is provided. As expected, this image shows a relatively high content of open voids and a low content of closed voids. The spatial arrangement of the open voids suggests a layered structure. The examined single grains of mudstone and greywacke also show a stratification due to the similar formation history. In the latter case, however, the closed voids predominate over open voids. The final CT-based visualization in Figure A2 is a single grain of granite and demonstrates that this type of rock can also have high contents of open voids.

As mentioned in the paper, the individual grains of the quarried stone have very small open and closed void surface areas compared to the river gravel grains. This is impressively documented by the CT-based visualizations for greywacke and rhyolite, which are shown in Figures A3 and A4 and which compare individual grains originating from quarried stone and river gravel. Aside from those shown in Figures A3 and A4, further visualizations of the quarried stone aggregate (GK2 and GK3) have not been included in the appendix. This is because these stone types have little to no visible porosity in the CT-based visualizations.

Appendix B. Tabulated Results of Cracking Analysis

Table A1. Individual grain results from the cracking analysis of stone category GK1 (river gravel; alkali sensitivity EIII-S).

Stone Type	Mass of the Individual Grain [g]	CT Scan Number	Voxel Size [µm]	Surface Areas Measured Using CT [m^2]				
				Closed Porosity	Outer Surface	Open Porosity	Outer Surface + Open Porosity	
							Absolute [m^2]	Relative [m^2/g]
Quartz/Quartzite	3.37	7030	16.5	5.69×10^{-4}	2.17×10^{-3}	2.20×10^{-5}	2.18×10^{-3}	6.47×10^{-4}
	4.49	7031		1.42×10^{-4}	1.10×10^{-3}	6.20×10^{-5}	1.12×10^{-3}	2.49×10^{-4}
	2.89	7032		3.85×10^{-5}	1.44×10^{-3}	9.65×10^{-5}	1.47×10^{-3}	5.09×10^{-4}
	5.63	7033		3.94×10^{-6}	9.50×10^{-4}	1.43×10^{-6}	9.50×10^{-4}	1.69×10^{-4}
	6.26	7034		1.10×10^{-3}	1.55×10^{-3}	1.46×10^{-4}	1.57×10^{-3}	2.51×10^{-4}
	8.04	7035		9.13×10^{-4}	2.39×10^{-3}	3.24×10^{-3}	2.79×10^{-3}	3.47×10^{-4}
	7.33	7036		2.62×10^{-4}	1.95×10^{-3}	1.88×10^{-4}	1.97×10^{-3}	2.69×10^{-4}
	8.32	7037		2.33×10^{-3}	2.38×10^{-3}	8.53×10^{-4}	2.48×10^{-3}	2.98×10^{-4}
	6.50	7038		3.64×10^{-5}	1.84×10^{-3}	2.89×10^{-6}	1.84×10^{-3}	2.83×10^{-4}
	11.17	7039		3.10×10^{-4}	2.09×10^{-3}	3.71×10^{-4}	2.13×10^{-3}	1.91×10^{-4}
	Mean			5.71×10^{-4}	1.79×10^{-3}	4.98×10^{-4}	1.85×10^{-3}	3.21×10^{-4}
	Standard Deviation			6.88×10^{-4}	4.82×10^{-4}	9.46×10^{-4}	5.52×10^{-4}	1.40×10^{-4}
Rhyolite	0.77	6880	11.2	0.00	5.18×10^{-4}	0.00	5.18×10^{-4}	6.73×10^{-4}
	0.74	6881		8.60×10^{-6}	4.87×10^{-4}	1.71×10^{-6}	4.88×10^{-4}	6.60×10^{-4}
	1.27	6882		1.59×10^{-4}	7.50×10^{-4}	7.07×10^{-5}	8.21×10^{-4}	6.47×10^{-4}
	1.24	6883		6.55×10^{-4}	9.78×10^{-4}	4.87×10^{-4}	1.46×10^{-3}	1.18×10^{-3}
	1.96	6884		9.57×10^{-6}	8.32×10^{-4}	1.49×10^{-5}	8.47×10^{-4}	4.32×10^{-4}
	1.16	6885		1.30×10^{-6}	5.65×10^{-4}	1.33×10^{-6}	5.67×10^{-4}	4.89×10^{-4}
	2.90	6886		8.63×10^{-5}	1.21×10^{-3}	3.28×10^{-5}	1.25×10^{-3}	4.30×10^{-4}
	2.74	6887		1.79×10^{-4}	1.20×10^{-3}	3.35×10^{-5}	1.24×10^{-3}	4.52×10^{-4}
	2.85	6888		3.67×10^{-5}	1.03×10^{-3}	1.80×10^{-5}	1.05×10^{-3}	3.68×10^{-4}
	5.70	6889		6.38×10^{-5}	1.62×10^{-3}	3.81×10^{-6}	1.62×10^{-3}	2.85×10^{-4}
	Mean			1.20×10^{-4}	9.20×10^{-4}	6.64×10^{-5}	9.86×10^{-4}	5.62×10^{-4}
	Standard Deviation			1.89×10^{-4}	3.44×10^{-4}	1.42×10^{-4}	3.82×10^{-4}	2.41×10^{-4}

Table A1. *Cont.*

Stone Type	Mass of the Individual Grain [g]	CT Scan Number	Voxel Size [μm]	Surface Areas Measured Using CT [m²]				
				Closed Porosity	Outer Surface	Open Porosity	Outer Surface + Open Porosity	
							Absolute [m²]	Relative [m²/g]
Plutonite	2.43	6970	11.2	1.21×10^{-4}	9.24×10^{-4}	1.17×10^{-5}	9.35×10^{-4}	3.85×10^{-4}
	1.50	6971		1.57×10^{-4}	9.03×10^{-4}	3.64×10^{-4}	1.27×10^{-3}	8.44×10^{-4}
	1.62	6972		4.53×10^{-5}	6.74×10^{-4}	9.49×10^{-6}	6.84×10^{-4}	4.22×10^{-4}
	2.09	6973		2.12×10^{-4}	4.72×10^{-3}	9.71×10^{-4}	5.69×10^{-3}	2.72×10^{-3}
	2.67	6974		1.23×10^{-3}	1.07×10^{-3}	2.26×10^{-4}	1.30×10^{-3}	4.87×10^{-4}
	1.96	6975		3.46×10^{-4}	8.53×10^{-4}	1.02×10^{-4}	9.54×10^{-4}	4.87×10^{-4}
	5.58	6976		1.66×10^{-3}	3.43×10^{-3}	2.35×10^{-3}	5.79×10^{-3}	1.04×10^{-3}
	2.47	6977		3.58×10^{-4}	1.00×10^{-3}	7.83×10^{-5}	1.08×10^{-3}	4.38×10^{-4}
	4.08	6978		7.40×10^{-4}	1.65×10^{-3}	2.67×10^{-4}	1.91×10^{-3}	4.69×10^{-4}
	4.44	6979		0.00	1.24×10^{-3}	9.91×10^{-8}	1.24×10^{-3}	2.79×10^{-4}
	Mean			4.86×10^{-4}	1.65×10^{-3}	4.38×10^{-4}	2.08×10^{-3}	7.57×10^{-4}
	Standard Deviation			5.27×10^{-4}	1.27×10^{-3}	6.95×10^{-4}	1.85×10^{-3}	6.89×10^{-4}

Table A2. Individual grain results from the cracking analysis of stone category GK2 (quarried stone; alkali sensitivity EIII-S).

Stone Type	Mass of the Individual Grain [g]	CT Scan Number	Voxel Size [μm]	Surface Areas Measured Using CT [m^2]				
				Closed Porosity	Outer Surface	Open Porosity	Outer Surface + Open Porosity	
							Absolute [m^2]	Relative [m^2/g]
Greywacke	1.03	6860	11.2	2.94×10^{-7}	5.31×10^{-4}	2.67×10^{-6}	5.34×10^{-4}	5.18×10^{-4}
	1.12	6861		5.87×10^{-8}	6.23×10^{-4}	0.00	6.23×10^{-4}	5.56×10^{-4}
	0.66	6862		7.67×10^{-7}	4.74×10^{-4}	8.48×10^{-7}	4.74×10^{-4}	7.19×10^{-4}
	2.23	6863		2.59×10^{-7}	9.14×10^{-4}	5.49×10^{-7}	9.15×10^{-4}	4.10×10^{-4}
	2.04	6864		0.00	9.55×10^{-4}	3.03×10^{-7}	9.55×10^{-4}	4.68×10^{-4}
	1.9	6865		0.00	9.78×10^{-4}	1.72×10^{-6}	9.80×10^{-4}	5.16×10^{-4}
	1.39	6866		1.14×10^{-7}	7.21×10^{-4}	0.00	7.21×10^{-4}	5.19×10^{-4}
	1.08	6867		0.00	7.26×10^{-4}	1.48×10^{-7}	7.26×10^{-4}	6.72×10^{-4}
	2.11	6868		9.99×10^{-8}	1.03×10^{-3}	6.28×10^{-7}	1.03×10^{-3}	4.87×10^{-4}
	3.98	6869		4.92×10^{-8}	1.61×10^{-3}	1.21×10^{-6}	1.61×10^{-3}	4.04×10^{-4}
	Mean			1.64×10^{-7}	8.56×10^{-4}	8.07×10^{-7}	8.57×10^{-4}	5.27×10^{-4}
	Standard Deviation			2.24×10^{-7}	3.11×10^{-4}	8.09×10^{-7}	3.11×10^{-4}	9.63×10^{-5}

Table A3. Individual grain results from the cracking analysis of stone category GK3 (quarried stone; alkali sensitivity EI-S).

Stone Type	Mass of the Individual Grain [g]	CT Scan Number	Voxel Size [μm]	Surface Areas Measured Using CT [m²]				
				Closed Porosity	Outer Surface	Open Porosity	Outer Surface + Open Porosity	
							Absolute [m²]	Relative [m²/g]
Rhyolite	1.53	6890	16.5	5.14×10^{-6}	9.00×10^{-4}	1.86×10^{-5}	9.18×10^{-4}	6.00×10^{-4}
	1.66	6891		4.52×10^{-7}	9.12×10^{-4}	8.54×10^{-6}	9.21×10^{-4}	5.55×10^{-4}
	2.68	6892		6.18×10^{-6}	1.32×10^{-3}	4.16×10^{-5}	1.36×10^{-3}	5.07×10^{-4}
	2.56	6893		6.92×10^{-6}	1.13×10^{-3}	2.93×10^{-5}	1.16×10^{-3}	4.52×10^{-4}
	2.89	6894		1.36×10^{-7}	1.30×10^{-3}	1.24×10^{-5}	1.32×10^{-3}	4.56×10^{-4}
	3.08	6895		4.42×10^{-6}	1.24×10^{-3}	6.88×10^{-6}	1.25×10^{-3}	4.05×10^{-4}
	5.18	6896		7.97×10^{-7}	1.97×10^{-3}	6.97×10^{-6}	1.98×10^{-3}	3.82×10^{-4}
	3.44	6897		9.38×10^{-6}	1.42×10^{-3}	2.21×10^{-5}	1.44×10^{-3}	4.19×10^{-4}
	5.07	6898		1.44×10^{-6}	1.96×10^{-3}	6.57×10^{-6}	1.96×10^{-3}	3.87×10^{-4}
	3.47	6899		8.69×10^{-7}	1.50×10^{-3}	3.59×10^{-6}	1.50×10^{-3}	4.32×10^{-4}
	Mean			3.57×10^{-6}	1.36×10^{-3}	1.57×10^{-5}	1.38×10^{-3}	4.59×10^{-4}
	Standard Deviation			3.10×10^{-6}	3.52×10^{-4}	1.16×10^{-5}	3.48×10^{-4}	6.92×10^{-5}

Table A4. Individual grain results from the cracking analysis of stone category GK4 (river gravel; alkali sensitivity EI-S).

Stone Type	Mass of the Individual Grain [g]	CT Scan Number	Voxel Size [µm]	Surface Areas Measured Using CT [m^2]				
				Closed Porosity	Outer Surface	Open Porosity	Outer Surface + Open Porosity	
							Absolute [m^2]	Relative [m^2/g]
Sandstone	0.84	6980	21.1	0.00	5.67×10^{-4}	0.00	5.67×10^{-4}	6.75×10^{-4}
	1.95	6981		0.00	7.24×10^{-4}	0.00	7.24×10^{-4}	3.71×10^{-4}
	1.35	6982		1.06×10^{-5}	6.95×10^{-4}	8.69×10^{-6}	7.03×10^{-4}	5.21×10^{-4}
	3.63	6983		5.04×10^{-6}	1.07×10^{-3}	2.26×10^{-7}	1.07×10^{-3}	2.96×10^{-4}
	3.35	6984		7.44×10^{-5}	1.28×10^{-3}	5.34×10^{-4}	1.82×10^{-3}	5.42×10^{-4}
	4.19	6985		7.09×10^{-5}	1.56×10^{-3}	1.09×10^{-4}	1.67×10^{-3}	4.00×10^{-4}
	4.78	6986		1.01×10^{-3}	1.42×10^{-3}	4.23×10^{-5}	1.47×10^{-3}	3.07×10^{-4}
	5.28	6987		3.25×10^{-4}	1.66×10^{-3}	2.01×10^{-5}	1.68×10^{-3}	3.18×10^{-4}
	6.26	6988		1.80×10^{-4}	1.91×10^{-3}	4.91×10^{-5}	1.95×10^{-3}	3.12×10^{-4}
	8.3	6989		3.80×10^{-4}	2.20×10^{-3}	1.59×10^{-5}	2.21×10^{-3}	2.67×10^{-4}
	Mean			2.05×10^{-4}	1.31×10^{-3}	7.79×10^{-5}	1.39×10^{-3}	4.01×10^{-4}
	Standard Deviation			2.97×10^{-4}	5.17×10^{-4}	1.55×10^{-4}	5.52×10^{-4}	1.28×10^{-4}

Table A4. *Cont.*

Stone Type	Mass of the Individual Grain [g]	CT Scan Number	Voxel Size [μm]	Surface Areas Measured Using CT [m²]				
				Closed Porosity	Outer Surface	Open Porosity	Outer Surface + Open Porosity Absolute [m²]	Outer Surface + Open Porosity Relative [m²/g]
Mudstone	3.6	69€0	21.1	1.72×10^{-4}	1.30×10^{-3}	2.56×10^{-5}	1.33×10^{-3}	3.68×10^{-4}
	1.59	69€1		4.28×10^{-7}	7.14×10^{-4}	9.34×10^{-7}	7.15×10^{-4}	4.49×10^{-4}
	1.61	69€2		5.36×10^{-5}	9.01×10^{-4}	5.17×10^{-4}	1.42×10^{-3}	8.81×10^{-4}
	3.27	69-3		0.00	1.15×10^{-3}	1.26×10^{-6}	1.15×10^{-3}	3.51×10^{-4}
	1.13	69-4		0.00	1.45×10^{-3}	0.00	1.45×10^{-3}	1.28×10^{-3}
	2.78	6955		7.04×10^{-4}	1.02×10^{-3}	4.66×10^{-4}	1.49×10^{-3}	5.36×10^{-4}
	2.39	6956		4.35×10^{-6}	8.54×10^{-4}	0.00	8.54×10^{-4}	3.57×10^{-4}
	2.63	6957		2.16×10^{-4}	1.12×10^{-3}	4.55×10^{-5}	1.16×10^{-3}	4.42×10^{-4}
	4.34	6958		0.00	1.70×10^{-3}	8.01×10^{-7}	1.70×10^{-3}	3.93×10^{-4}
	2.51	6969		1.13×10^{-5}	1.06×10^{-3}	1.56×10^{-6}	1.06×10^{-3}	4.23×10^{-4}
	Mean			1.16×10^{-4}	1.13×10^{-3}	1.06×10^{-4}	1.23×10^{-3}	5.48×10^{-4}
	Standard Deviation			2.10×10^{-4}	2.79×10^{-4}	1.94×10^{-4}	2.88×10^{-4}	2.86×10^{-4}
Greywacke	8.19	6870	16.5	5.61×10^{-4}	2.33×10^{-3}	1.25×10^{-4}	2.46×10^{-3}	3.00×10^{-4}
	5.28	6871		6.17×10^{-5}	1.99×10^{-3}	3.93×10^{-5}	2.03×10^{-3}	3.84×10^{-4}
	8.67	6872		5.56×10^{-4}	2.70×10^{-3}	3.04×10^{-4}	3.01×10^{-3}	3.47×10^{-4}
	5.99	6873		1.71×10^{-6}	3.13×10^{-3}	3.88×10^{-5}	3.17×10^{-3}	5.29×10^{-4}
	4.89	6874		1.68×10^{-3}	1.69×10^{-3}	4.92×10^{-4}	2.18×10^{-3}	4.46×10^{-4}
	2.61	6875		3.26×10^{-6}	1.43×10^{-3}	4.02×10^{-6}	1.44×10^{-3}	5.50×10^{-4}
	6.3	6876		7.10×10^{-4}	2.16×10^{-3}	1.11×10^{-4}	2.27×10^{-3}	3.61×10^{-4}
	3.86	6877		8.41×10^{-5}	1.51×10^{-3}	7.54×10^{-6}	1.51×10^{-3}	3.92×10^{-4}
	3.09	6878		4.12×10^{-4}	2.31×10^{-3}	2.89×10^{-4}	2.60×10^{-3}	8.41×10^{-4}
	2.63	6879		9.88×10^{-6}	1.24×10^{-3}	2.88×10^{-6}	1.24×10^{-3}	4.72×10^{-4}
	Mean			4.08×10^{-4}	2.05×10^{-3}	1.41×10^{-4}	2.19×10^{-3}	4.62×10^{-4}
	Standard Deviation			4.96×10^{-4}	5.68×10^{-4}	1.58×10^{-4}	6.19×10^{-4}	1.47×10^{-4}

Table A4. *Cont.*

Stone Type	Mass of the Individual Grain [g]	CT Scan Number	Voxel Size [μm]	Surface Areas Measured Using CT [m^2]				
				Closed Porosity	Outer Surface	Open Porosity	Outer Surface + Open Porosity	
							Absolute [m^2]	Relative [m^2/g]
Granite	1.53	6950	11.2	2.14×10^{-5}	8.15×10^{-4}	6.23×10^{-5}	8.77×10^{-4}	5.74×10^{-4}
	0.72	6951		1.58×10^{-6}	5.08×10^{-4}	1.41×10^{-5}	5.22×10^{-4}	7.26×10^{-4}
	1.09	6952		3.63×10^{-5}	6.80×10^{-4}	1.84×10^{-4}	8.64×10^{-4}	7.93×10^{-4}
	0.98	6953		1.17×10^{-5}	5.76×10^{-4}	5.53×10^{-5}	6.31×10^{-4}	6.44×10^{-4}
	1.92	6954		1.01×10^{-6}	8.71×10^{-4}	1.59×10^{-5}	8.87×10^{-4}	4.62×10^{-4}
	1.76	6955		7.23×10^{-5}	9.39×10^{-4}	6.90×10^{-5}	1.01×10^{-3}	5.73×10^{-4}
	3.17	6956		1.36×10^{-4}	1.43×10^{-3}	3.65×10^{-4}	1.80×10^{-3}	5.68×10^{-4}
	2.77	6957		1.37×10^{-4}	1.28×10^{-3}	5.26×10^{-4}	1.81×10^{-3}	6.53×10^{-4}
	2.69	6958		4.95×10^{-5}	1.62×10^{-3}	5.91×10^{-5}	1.67×10^{-3}	6.23×10^{-4}
	6.15	6959		1.09×10^{-4}	1.87×10^{-3}	9.91×10^{-8}	1.87×10^{-3}	3.04×10^{-4}
	Mean			5.76×10^{-5}	1.06×10^{-3}	1.35×10^{-4}	1.19×10^{-3}	5.92×10^{-4}
	Standard Deviation			5.06×10^{-5}	4.42×10^{-4}	1.66×10^{-4}	5.04×10^{-4}	1.29×10^{-4}

References

1. Stanton, T. Expansion of Concrete Through Reaction Between Cement and Aggregate. *Proc. Am. Soc. Civ. Eng.* **1940**, *66*, 1781–1812.
2. Powers, T.; Steinour, H. An Interpretation of Some Published Researches on the Alkali-Aggregate Reaction Part 1—The Chemical Reactions and Mechanism of Expansion. *J. Am. Concr. I* **1955**, *26*, 497–516.
3. Powers, T.; Steinour, H. An Interpretation of Some Published Researches on the Alkali-Aggregate Reaction Part 2—A Hypothesis Concerning Safe and Unsafe Reactions with Reactive Silica in Concrete. *J. Am. Concr. I* **1955**, *51*, 785–812.
4. Locher, F.; Sprung, S. *Ursache und Wirkungsweise der Alkalireaktion*; Betontechnische Berichte; Forschungsinstitut der Zementindustrie: Düsseldorf, Germany, 1973; pp. 101–123.
5. Chatterji, S.; Thaulow, N.; Jensen, A. Studies of alkali-silica-reaction. Part 5. Verification of a newly proposed reaction mechanism. *Cem. Concr. Res.* **1989**, *19*, 177–183. [CrossRef]
6. Chatterji, S.; Thaulow, N.; Jensen, A. Studies of alkali-silica-reaction. Part 6. Practical implications of a proposed reaction mechanism. *Cem. Concr. Res.* **1988**, *18*, 363–366. [CrossRef]
7. Stark, J.; Erfurt, D.; Freyburg, E.; Giebson, C.; Seyfarth, K.; Wicht, B. *Alkali-Kieselsäure-Reaktion*; F. A. Finger-Instituts für Baustoffkunde: Weimar, Germany, 2008.
8. Thomas, M.D.A.; Fournier, B.; Folliard, K.J. *Alkali-Aggregate Reactivity (AAR) Facts Book*; U.S. Department of Transportation, Federal Highway Administration (FHWA): Washington, DC, USA, 2013.
9. Qi, Y.; Ziyun, W. Study of expansion mechanism of A.S.R. using sol-gel expansion method. In Proceedings of the 12th ICAAR, Beijing, China, 15–19 October 2004; pp. 226–229.
10. Weise, F.; Kositz, M.; Oesch, T.; Huenger, K.-J.; Wilsch, G.; Sigmund, S. *Analyse des Gefügeabhängigen Löslichkeitsverhaltens Potenziell AKR-Empfindlicher Gesteinskörnungen*; Bundesanstalt für Straßenwesen (BASt), Schlussbericht FE 06.0108/2014/BRB: Bergisch Gladbach, Germany, 2019.
11. Oesch, T.; Weise, F.; Marx, H.; Kositz, M.; Huenger, K.-J. Analysis of the porosity of alkali-sensitive aggregates for the assessment of microstructure-dependent solubility in the context of ASR (ACCEPTED). In Proceedings of the 4th International RILEM Conference on Microstructure Related Durability of Cementitious Composites, Den Haag, The Netherlands, 28–30 April 2021.
12. Buzug, T.M. Computed Tomography. In *Springer Handbook of Medical Technology*; Kramme, R., Hoffmann, K.-P., Pozos, R.S., Eds.; Springer: Berlin/Heidelberg, Germany, 2011. [CrossRef]
13. Kalender, W.A. *Computed Tomography: Fundamentals, System Technology, Image Quality, Applications*, 3rd ed.; Publicis Publishing: Erlangen, Germany, 2011.
14. Willemink, M.J.; Noël, P.B. The evolution of image reconstruction for CT-from filtered back projection to artificial intelligence. *Eur. Radiol.* **2019**, *29*, 2185–2195. [CrossRef]
15. du Plessis, A.; le Roux, S.G.; Guelpa, A. Comparison of medical and industrial X-ray computed tomography for non-destructive testing. *Case Stud. Nondestruct. Test. Eval.* **2016**, *6*, 17–25. [CrossRef]
16. Martz, H.E.; Schneberk, D.J.; Roberson, G.P.; Monteiro, P.J.M. Computerized-Tomography Analysis of Reinforced-Concrete. *ACI Mater. J.* **1993**, *90*, 259–264.
17. Morgan, I.L.; Ellinger, H.; Klinksiek, R.; Thompson, J.N. Examination of Concrete by Computerized-Tomography. *J. Am. Concr. I* **1980**, *77*, 23–27.
18. Flannery, B.P.; Deckman, H.W.; Roberge, W.G.; D'Amico, K.L. Three-Dimensional X-ray Microtomography. *Science* **1987**, *237*, 1439–1444. [CrossRef]
19. Feldkamp, L.A.; Davis, L.C.; Kress, J.W. Practical cone-beam algorithm. *J. Opt. Soc. Am. A* **1984**, *1*, 612–619. [CrossRef]
20. Landis, E.N.; Nagy, E.N.; Keane, D.T. Microtomographic measurements of internal damage in portland-cement-based composites. *J. Aerosp. Eng.* **1997**, *10*, 2–6. [CrossRef]
21. Asahina, D.; Landis, E.N.; Bolander, J.E. Modeling of phase interfaces during pre-critical crack growth in concrete. *Cem. Concr. Compos.* **2011**, *33*, 966–977. [CrossRef]
22. Poinard, C.; Piotrowska, E.; Malecot, Y.; Daudeville, L.; Landis, E.N. Compression triaxial behavior of concrete: The role of the mesostructure by analysis of X-ray tomographic images. *Eur. J. Environ. Civ. Eng.* **2012**, *16*, S115–S136. [CrossRef]
23. Oesch, T.; Landis, E.; Kuchma, D. Conventional Concrete and UHPC Performance-Damage Relationships Identified Using Computed Tomography. *J. Eng. Mechan.* **2016**, *142*. [CrossRef]

24. Oesch, T. In-Situ CT Investigation of Pull-Out Failure for Reinforcing Bars Embedded in Conventional and High-Performance Concretes. In Proceedings of the 6th Conference on Industrial Computed Tomography (ICT), Wels, Austria, 9–12 February 2016.
25. Paetsch, O.; Baum, D.; Prohaska, S.; Ehrig, K.; Meinel, D.; Ebell, G. 3D Corrosion Detection in Time-dependent CT Images of Concrete. In Proceedings of the Digital Industrial Radiology and Computed Tomography Conference, Ghent, Belgium, 22–25 June 2015; p. 10.
26. Yang, L.; Zhang, Y.S.; Liu, Z.Y.; Zhao, P.; Liu, C. In-situ tracking of water transport in cement paste using X-ray computed tomography combined with CsCl enhancing. *Mater. Lett.* **2015**, *160*, 381–383. [CrossRef]
27. Boone, M.A.; De Kock, T.; Bultreys, T.; De Schutter, G.; Vontobel, P.; Van Hoorebeke, L.; Cnudde, V. 3D mapping of water in oolithic limestone at atmospheric and vacuum saturation using X-ray micro-CT differential imaging. *Mater. Charact.* **2014**, *97*, 150–160. [CrossRef]
28. Oesch, T.; Weise, F.; Meinel, D.; Gollwitzer, G. Quantitative In-situ Analysis of Water Transport in Concrete Completed Using X-ray Computed Tomography. *Trans. Porous Med.* **2019**, *127*, 371–389. [CrossRef]
29. Powierza, B.; Stelzner, L.; Oesch, T.; Gollwitzer, C.; Weise, F.; Bruno, G. Water Migration in One-Side Heated Concrete: 4D In-Situ CT Monitoring of the Moisture-Clog-Effect. *J. Nondestruct. Eval.* **2019**, *38*, 15. [CrossRef]
30. Bažant, Z.P.; Planas, J. *Fracture and Size Effect in Concrete and Other Quasibrittle Materials*; CRC Press: Boca Raton, FL, USA, 1998; p. 640.
31. Oesch, T. Investigation of Fiber and Cracking Behavior for Conventional and Ultra-High Performance Concretes Using X-ray Computed Tomography. Ph.D. Thesis, University of Illinois, Urbana, IL, USA, 2015.
32. Ehrig, K.; Goebbels, J.; Meinel, D.; Paetsch, O.; Prohaska, S.; Zobel, V. Comparison of Crack Detection Methods for Analyzing Damage Processes in Concrete with Computed Tomography. In Proceedings of the International Symposium on Digital Industrial Radiology and Computed Tomography, Berlin, Germany, 20–22 June 2011.
33. Lebbink, M.N.; Geerts, W.J.C.; van der Krift, T.P.; Bouwhuis, M.; Hertzberger, L.O.; Verkleij, A.J.; Koster, A.J. Template matching as a tool for annotation of tomograms of stained biological structures. *J. Struct. Biol.* **2007**, *158*, 327–335. [CrossRef]
34. Paetsch, O. Possibilities and Limitations of Automatic Feature Extraction shown by the Example of Crack Detection in 3D-CT Images of Concrete Specimen. In Proceedings of the 9th Conference on Industrial Computed Tomography, Padova, Italy, 13–15 February 2019.
35. Sato, Y.; Westin, C.F.; Bhalerao, A.; Nakajima, S.; Shiraga, N.; Tamura, S.; Kikinis, R. Tissue classification based on 3D local intensity structures for volume rendering. *IEEE Trans. Vis. Comput. Gr.* **2000**, *6*, 160–180. [CrossRef]
36. Watanabe, S.; Ohtake, Y.; Nagai, Y.; Suzuki, H. Detection of Narrow Gaps Using Hessian Eigenvalues for Shape Segmentation of a CT Volume of Assembled Parts. In Proceedings of the 8th Conference on Industrial Computed Tomography, Wels, Austria, 6–9 February 2018.
37. Picard, D.; Lauzon-Gauthier, J.; Duchesne, C.; Alamdari, H.; Fafard, M.; Ziegler, D.P. Crack Detection Method Applied to 3D Computed Tomography Images of Baked Carbon Anodes. *Metals* **2016**, *6*, 272. [CrossRef]
38. Cinar, A.F.; Hollis, D.; Tomlinson, R.A.; Marrow, T.J.; Mostafavi, M. Application of 3D phase congruency in crack identification within materials. In Proceedings of the 11th International Conference on Advances in Experimental Mechanics, Exeter, UK, 5–7 September 2016.
39. Cinar, A.F.; Barhli, S.M.; Hollis, D.; Flansbjer, M.; Tomlinson, R.A.; Marrow, T.J.; Mostafavi, M. An autonomous surface discontinuity detection and quantification method by digital image correlation and phase congruency. *Opt. Laser Eng.* **2017**, *96*, 94–106. [CrossRef]
40. Barhli, S.M.; Saucedo-Mora, L.; Jordan, M.S.L.; Cinar, A.F.; Reinhard, C.; Mostafavi, M.; Marrow, T.J. Synchrotron X-ray characterization of crack strain fields in polygranular graphite. *Carbon* **2017**, *124*, 357–371. [CrossRef]
41. Powierza, B.; Gollwitzer, C.; Wolgast, D.; Staude, A.; Bruno, G. Fully experiment-based evaluation of few digital volume correlation techniques. *Rev. Sci. Instrum.* **2019**, *90*, 115105. [CrossRef] [PubMed]
42. Kovesi, P. Phase congruency: A low-level image invariant. *Psychol. Res.* **2000**, *64*, 136–148. [CrossRef] [PubMed]
43. Bhandarkar, S.M.; Luo, X.Z.; Daniels, R.; Tollner, E.W. Detection of cracks in computer tomography images of logs. *Pattern Recogn. Lett.* **2005**, *26*, 2282–2294. [CrossRef]

44. DAfStb. *Vorbeugende Maßnahmen Gegen Schädigende Alkalireaktion im Beton (Alkali-Richtlinie)*; Deutscher Ausschuss für Stahlbeton; Beuth Verlag GmbH: Berlin, Germany, 2013.
45. DIN. *Tests for Geometrical Properties of Aggregates—Part 1: Determination of Particle Size Distribution—Sieving Method*; German Version EN 933-1:2012; Deutsches Institut für Normung; Beuth Verlag GmbH: Berlin, Germany, 2012.
46. Latief, F.D.; Fauzi, U.; Irayani, Z.; Dougherty, G. The effect of X-ray micro computed tomography image resolution on flow properties of porous rocks. *J. Microsc.* **2017**, *266*, 69–88. [CrossRef]
47. Krumm, A.; Means, B.; Bienkowski, M. *Learning Analytics Goes to School: A Collaborative Approach to Improving Education*; Routledge, Taylor & Francis Group: New York, NY, USA, 2018.
48. Heideklang, R.; Shokouhi, P. Multi-sensor image fusion at signal level for improved near-surface crack detection. *Ndt E Int.* **2015**, *71*, 16–22. [CrossRef]
49. The Mathworks. *MATLAB, R2017b*; The Mathworks: Natick, MA, USA, 2018.
50. Young, I.T.; Gerbrands, J.J.; van Vliet, L.J. *Fundamentals of Image Processing*, 2.2 ed.; Delft University of Technology: Delft, The Netherlands, 1998.
51. Zack, G.W.; Rogers, W.E.; Latt, S.A. Automatic measurement of sister chromatid exchange frequency. *J. Histochem. Cytochem.* **1977**, *25*, 741–753. [CrossRef]
52. Pringle, K.K. Visual Perception by a computer. In *Automatic Interpretation and Classification of Images*; Grasselli, A., Ed.; Academic Press: New York, NY, USA, 1969; pp. 277–284.
53. Brunauer, S.; Emmett, P.H.; Teller, E. Adsorption of gases in multimolecular layers. *J. Am. Chem. Soc.* **1938**, *60*, 309–319. [CrossRef]
54. Langmuir, I. The Adsorption of Gases on Plane Surfaces of Glass, Mica and Platinum. *J. Am. Chem. Soc.* **1918**, *40*, 1361–1403. [CrossRef]
55. Cooper, R.C.; Bruno, G.; Onel, Y.; Lange, A.; Watkins, T.R.; Shyam, A. Young's modulus and Poisson's ratio changes due to machining in porous microcracked cordierite. *J. Mater. Sci.* **2016**, *51*, 9749–9760. [CrossRef]
56. Mishurova, T.; Artzt, K.; Haubrich, J.; Requena, G.; Bruno, G. New aspects about the search for the most relevant parameters optimizing SLM materials. *Addit. Manuf.* **2019**, *25*, 325–334. [CrossRef]

materials

MDPI

Article

On the Response of a Micro Non-Destructive Testing X-ray Detector

Dionysios Linardatos [1], Vaia Koukou [1], Niki Martini [1], Anastasios Konstantinidis [2], Athanasios Bakas [3], George Fountos [1], Ioannis Valais [1,*] and Christos Michail [1]

1 Radiation Physics, Materials Technology and Biomedical Imaging Laboratory, Department of Biomedical Engineering, University of West Attica, Ag. Spyridonos, 12210 Athens, Greece; dlinardatos@uniwa.gr (D.L.); koukou@uniwa.gr (V.K.); mmartini@uniwa.gr (N.M.); gfoun@uniwa.gr (G.F.); cmichail@uniwa.gr (C.M.)

2 Radiological Sciences Group, Department of Medical Physics, Queen Alexandra Hospital, Portsmouth Hospitals University NHS Trust, Portsmouth PO6 3LY, UK; Anastasios.Konstantinidis@porthosp.nhs.uk

3 Department of Biomedical Sciences, University of West Attica, Ag. Spyridonos, 12210 Athens, Greece; abakas@uniwa.gr

* Correspondence: valais@uniwa.gr; Tel.: +30-210-5385-371

Abstract: Certain imaging performance metrics are examined for a state-of-the-art 20 μm pixel pitch CMOS sensor (RadEye HR), coupled to a Gd_2O_2S:Tb scintillator screen. The signal transfer property (STP), the modulation transfer function (MTF), the normalized noise power spectrum (NNPS) and the detective quantum efficiency (DQE) were estimated according to the IEC 62220-1-1:2015 standard. The detector exhibits excellent linearity (coefficient of determination of the STP linear regression fit, R^2 was 0.9978), while its DQE peaks at 33% and reaches 10% at a spatial frequency of 3 cycles/mm, for the measured with a Piranha RTI dosimeter (coefficient of variation CV = 0.03%) exposure value of 28.1 μGy DAK (detector Air Kerma). The resolution capabilities of the X-ray detector under investigation were compared to other commercial CMOS sensors, and were found in every case higher, except from the previous RadEye HR model (CMOS—Gd_2O_2S:Tb screen pair with 22.5 μm pixel pitch) version which had slightly better MTF. The present digital imager is designed for industrial inspection applications, nonetheless its applicability to medical imaging, as well as dual-energy is considered and certain approaches are discussed in this respect.

Keywords: CMOS; imaging; Gd_2O_2S:Tb; ZnSe:Te; non-destructive testing; DQE; scintillators; IEC 62220-1-1:2015

Citation: Linardatos, D.; Koukou, V.; Martini, N.; Konstantinidis, A.; Bakas, A.; Fountos, G.; Valais, I.; Michail, C. On the Response of a Micro Non-Destructive Testing X-ray Detector. *Materials* **2021**, *14*, 888. https://doi.org/10.3390/ma14040888

Academic Editor: Giovanni Bruno

Received: 7 January 2021
Accepted: 10 February 2021
Published: 13 February 2021

1. Introduction

The last decades have witnessed an unprecedented growth in the development of X-ray digital imaging technologies, to the extent that it is yet difficult to imagine the modern radiology in their absence. Digital imagers have revolutionized the otherwise classic procedures such as radiography, but more importantly have provided advanced techniques like digital breast tomosynthesis (DBT) [1–3], digital subtraction angiography (DSA) [4], thin tissue autoradiography [5,6], X-ray phase contrast imaging (XPCi) [7], etc. In the industrial field, this technology finds application in non-destructive testing (NDT) as well. In this quality control method, real-time radiography (RTR) is used for projection imaging and inspection of components from a production line. Material characterization and evaluation of substance properties are further applications under the term non-invasive inspection techniques. X-ray refraction, for instance, utilizes digital imagers for the characterization of well-defined geometry samples (fibers, capillaries) or bulks of micro-particles [8–10].

Digital imagers in their indirect-detection version, i.e., coupled to a scintillator material, have turned up in various forms; amorphous silicon on thin-film-transistor panels (a-Si-TFT) and their high-purity counterparts, charge-coupled devices (CCD) and complementary metal-oxide semiconductors (CMOS), either active or passive pixel sensor (APS or PPS) [11,12].

CCD detectors exhibit high resolution, good response linearity, low dark signal, low read noise and high sensitivity, making them superior at low exposures [12,13]. Initially their active surface used to be limited to less than 7 cm^2 [14], but today's designs have reached 85 cm^2 [15], albeit at a high cost. Those small-sized CCDs necessitate demagnification solutions for applications that require larger fields of view [16]. This inserts an additional source of noise and degrades image quality, up to the point of determining the whole system's performance (quantum sink) [11,17,18]. Something which, along with the specialized manufacturing process that these devices demand, adds to the complexity and gives rise to the overall cost. Moreover, CCDs are prone to radiation damage [19,20].

CMOS imagers take advantage of their sharing a highly developed fabrication process with the semiconductor industry, a field that progresses in a frantic pace and continuously curtails costs. Hence, they are presently a cost-effective solution that offers very high frame rates, X-Y addressability and low power consumption [21]. Their detecting area can reach more than 160 cm^2, while they can be designed 2–3 sides buttable [2,14], allowing placement in a tiled fashion and the building of a large area imager. The need for the (detrimental for the image quality) demagnification is eliminated and a direct coupling of the scintillator material is thus permitted. Regarding the attainable resolution, sub – 2 μm pixel sizes have been reported [22,23].

Initially, CMOS's main disadvantage over CCD was the higher electronic noise [24]. Today's APS designs, by integrating advanced functions like buffering and amplification on-chip and in-pixel, outperform in this respect. At the same time, they can support much higher read-out speeds because of their massively parallel architecture [7,25], that can be obtained, along with lower bandwidth–lower noise ADCs (analog to digital converters). Frame rates above 600 fps (frames per second) have been achieved for a general-purpose imager [26], while speeds of the order of MHz (Megahertz) have been realized using pixel-based storage for an ultra-high speed (UHS) one [27].

Apparently, certain of the novel medical imaging applications of the digital imagers can benefit from the distinguishable merits of the CMOS detectors. For example, a digital mammography detector has to be large enough for full field coverage, as well as fast enough in acquiring images, without lag and baseline drifts [21,28]. Similarly, in the industrial field, requirements include high resolution, physical ruggedness, high frame rates, cost effectiveness and portability for in-field applications. All the specifications above can be met by CMOS technology [10].

The aim of the present study was to examine the imaging performance of a state-of-the-art indirect detection CMOS X-ray detector for imaging applications, such as dual-energy (DE), in which the small pixel size could be beneficial in the detection of micro calcifications and masses, indicative of breast cancer, as well as industrial inspection applications requiring increased resolution to detect sub-millimeter structures [29]. Imaging performance metrics, such as the signal transfer property (STP), the modulation transfer function (MTF) and the normalized noise power spectrum (NNPS) were measured, to finally estimate the detective quantum efficiency (DQE) of the imaging system, following the International Electrotechnical Commission (IEC) procedures.

2. Materials and Methods

The detector under investigation is the Remote RadEye HR CMOS (Teledyne DALSA, Waterloo, ON, Canada) [30] photodiode pixel array (N-well diffusion on p-type epitaxial silicon), that features an active area of 33.0 × 24.9 mm^2, with 1650 × 1246 pixels at 20 μm pitch. The detector can be used in industrial applications, since it is small-sized and can reach tight spaces with its 12-bit USB interface. The Carestream Min-R 2190 scintillator screen (gadolinium oxysulfide activated with terbium: Gd_2O_2S:Tb of thickness 85 μm and screen coating weight of 33.91 mg/cm^2) is in direct contact with the CMOS active area, while a carbon-fiber window shields from the ambient light and provides mechanical protection [30,31].

The nominal conversion gain of the detector is 155 electrons per pixel value (e^-/PV). The nominal electronic noise (root mean square; r.m.s.) is around 310 e^- and the dynamic range is 66 dB. Hence, the calculated saturation charge is around 620×10^3 e^-. The maximum frame rate is 0.7 fps and the nominal average dark current is 930 e^-/s at 23 °C. A two-meter shielded cable is included to connect the sensor head to the electronics module, where the analog video signal is processed, digitized using 12-bit ADCs and transferred to a PC [30]. The previous model of this CMOS X-ray detector was studied by Konstantinidis [32], with 60 e^-/PV conversion gain (with a high gain option 2× available), 120 e^- electronic noise (r.m.s.), 72 dB dynamic range, 0.5 fps maximum frame rate, 3000 e^-/s average dark current (at 23 °C), one meter shielded cable and 12 bit-depth ADC [32].

Due to the variety of digital imaging detector configurations (sensors and scintillators combinations) it is necessary to establish standard methods to facilitate intercomparisons between various systems, in order to juxtapose their overall performance, in terms of widely used imaging metrics, such as the MTF, NNPS and DQE [33]. To this aim, the International Electrotechnical Commission (IEC) has established a standard method (IEC 62220-1:2003 [34]; referred to as IEC 2003 for brevity) which was amended in 2015 (IEC 62220-1-1:2015 [35]; also referred to as IEC 2015). The modifications between the two protocols were examined previously in [33]. In the present work, these metrics were estimated following the 2015 version, except for the MTF that was calculated additionally according to the 2003 version, in order to compare with the MTF values provided in the manufacturer's datasheet. The COQ plugin (Verison 2.6) for the ImageJ suite (Version 1.52a) [X]was used for the MTF, NNPS, DQE calculations [36,37].

The Del Medical Eureka radiographic system (Harisson, NY, USA) was used for the experiments. It features a rotating tungsten (W) anode with a focal spot size of 0.6 mm ("small" focal spot size selected) and an inherent filtration equivalent to 3 mm aluminum (Al). The RQA-5 radiation quality was used throughout the experiments, i.e., 70 kVp tube voltage and 6.8 mm half value layer (HVL). It was found that, in addition to the tube's inherent filtration, another 21 mm Al (type 1100, purity 99%) must be added in order to achieve this HVL. The Al filter was placed as close to the tube as possible and the source to detector distance (SDD) was 156 cm.

The STP or X-ray sensitivity of the detector is the relationship between the mean pixel value (MPV) and Air Kerma at Detector's surface (DAK). To obtain the DAK, the CMOS was removed altogether, and a calibrated RTI Piranha X-ray dosimeter (Mölndal, Sweden) was placed at the same position. In line with the IEC standard's recommendations regarding the reduction of backscatter to $< 0.5\%$, a 4 mm thick lead foil was placed at 45 cm behind the dosimeter [35]. A sequence of flat-field images was acquired at seven different exposure levels, each one consisting of five repetitions, for averaging reasons. The MPV was sampled from a 1×1 cm^2 region of interest (ROI) and the system's response curve was fitted using a linear equation of the form:

$$\mathrm{MPV} = \mathrm{a}\, \mathrm{K_a} + \mathrm{b} \quad (1)$$

where a is the detector's gain factor (G) [38] and b is the pixel offset at zero DAK [39].

The MTF of the detector is the variation of the output contrast as a function of the spatial frequency, and is normalized to the input contrast [40,41]. Following the IEC standard's procedures [34,35], the MTF was measured using the slanted edge technique with the PTW Freiburg tungsten edge device, which consists of a 1 mm thick tungsten edge plate (100×75 mm^2) fixed on a 3 mm thick lead plate [33,42]. The edge device was placed in contact with the detector's entrance window at a shallow angle (1.5°–3°) with respect to the pixel rows or columns. Images were obtained at 28.1 μGy and 48.3 μGy exposures.

For the edge spread function (ESF), a 2×2 cm^2 ROI was drawn, with the edge roughly at the center. The ROI's size, smaller than the suggested by the IEC protocol, is mandated by the detector's size. According to the IEC 62220-1:2003, the final MTF is determined by averaging the multiple MTFs obtained from the individual groups of N consecutive lines along the edge [34]. On the other hand, in the IEC 62220-1-1:2015 the final MTF is obtained by averaging the oversampled ESFs [35,43], by fitting of a modified Fermi–Dirac (F-D) distribution function of the form:

$$\mathrm{Fermi}(x) = \left(\frac{c}{e^{\frac{x-a}{b}} + 1}\right) + d \quad (2)$$

The oversampled ESFs resulted from the pixel values of the linearized data (using the inverse of the STP curve to get DAK values) of N consecutive lines across the edge. With differentiation of the fitted ESF, the line spread function (LSF) is obtained which, after the application of a Hanning filter (window width 2048 pixels) is Fourier-transformed to provide the MTF [35].

The NPS of the detector expresses the statistical variance of the output signal as a function of the spatial frequency. It was determined using flat-field images at the two exposure levels (28.1 μGy and 48.3 μGy coefficient of variation CV = 0.03%) measured with the RTI Piranha X-ray dosimeter. The normalized noise power spectrum (NNPS) tends to provide an estimation of the output noise, independent of gross exposure variations over the detector area (e.g., heel effect) [41,44]. This is achieved by fitting and then subtracting a two dimensional (2-D) second order polynomial fit S(x,y) to the original image data I(x,y) after converting to K_a units (see linearization process), using the inverse of the STP linear equation [35]. The average 2-D NPS is given by:

$$\mathrm{NPS}(u,v) = \frac{\Delta x \Delta y}{M N_x N_y} \sum_{m=1}^{M} \left|\mathrm{FFT}\{I(x_i, y_i) - S(x_i, y_i)\}\right|^2 \quad (3)$$

where u and v denote the x and y–axis spatial frequency, respectively; Δx and Δy are the x and y—axis pixel pitches; N_x and N_y express the ROI size in the x and y axes (256 pixels according to the IEC); M is the number of ROIs used in the ensemble average; FFT denotes the fast Fourier transform operation [2].

From this, the horizontal and vertical NPS were extracted, by averaging seven rows and seven columns on each side of each axis. Axes themselves were excluded, as the IEC standard suggests, since they may contain remnant column-to-column and/or row-to-row fixed-pattern noise (FPN). These 1-D NPS were divided by the square of the averaged K_a [33,45], also known as large area signal, in order to obtain the NNPS in both orientations, which were then combined to obtain the radial average NNPS.

The detector's DQE expresses its efficiency in transferring the signal to noise ratio square (SNR^2) impinging on it, towards the output [41,46]. It is given by:

$$\mathrm{DQE}(u) = \frac{\mathrm{MTF}^2(u)}{K_a \cdot q \cdot \mathrm{NNPS}(u)} \quad (4)$$

where q is the fluence per Air Kerma ratio, i.e., the number of X-ray photons per unit Air Kerma (in μGy) per mm^2. According to the IEC 62220-1-1:2015 protocol, the value of 29653 was used for the utilized X-ray beam quality (RQA-5) [35].

3. Results and Discussion

The MPV as a function of DAK is drawn in Figure 1 for the seven exposure levels and a linear interpolation is calculated.

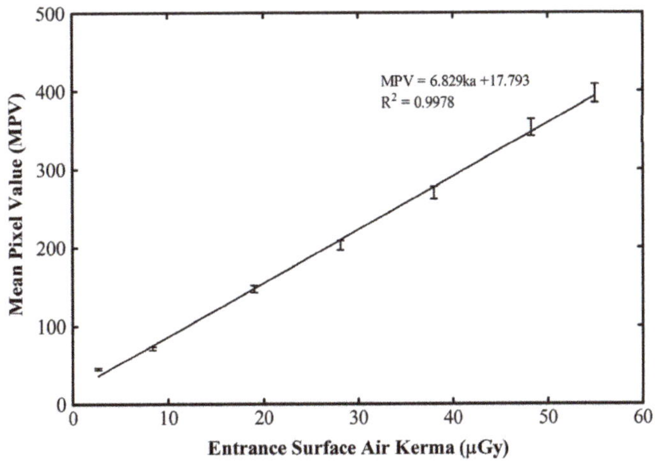

Figure 1. STP curve of the CMOS X-ray detector.

The detector's excellent linearity in the examined exposure range is obvious, since the coefficient of determination (R^2) is 0.9978. The gain factor (G) is found 6.829 digital units (DU) per μGy. It is a higher value compared to the 5.487 measured for the previous (discontinued) version of the Remote RadEye HR detector, studied by Konstantinidis [32], at almost the same beam quality (74 kVp). It is higher as well, in comparison to the 3.829 gain at 52 kVp, of the same study [32]. In that situation, the detector was also coupled to a Gd_2O_2S:Tb screen of mass thickness 33.91 mg/cm^2. The dynamic range of the present (new version) detector is practically double (up to 55 μGy, as opposed to 28 μGy of the old one), even though no loss of linearity is observed up to the exposure maxima in any case. There is also an offset of ~18 DU at zero input.

Despite the fact that two different detectors are being considered, each one with its own characteristics, increased G with increasing mean X-ray energy is an expected behavior. As described by Marshall [47] and Konstantinidis [2], with increasing mean energy of the spectrum, the number of X-ray photons per unit DAK increases as well. Every X-ray photon of higher energy causes the emission of more optical photons by the scintillator. Furthermore, higher energy X-ray photons have a greater penetration within the scintillator material, shifting the depth of interaction closer to the CMOS optical sensor and improving the detection probability of the secondary generated optical photons.

The oversampled ESF as a function of position across the edge device (in terms of number of pixels) is depicted in Figure 2a at 28.1 μGy and in Figure 3a at 48.3 μGy. Figures 2b and 3b show the corresponding averaged and Fermi-fitted ESFs. Signal fluctuations are mostly prominent in the area where the beam is not attenuated by the edge device, due to the Poisson distribution of input X-rays in spatial and temporal domains. Contrarily, in the "dark" area the fluctuations are mostly due to the electronic noise. The resulting LSFs after the fast Fourier transform (FFT) are demonstrated in Figures 2c and 3c, respectively, while the MTF curves according to the IEC 62220-1-1:2015 standard are shown in Figures 2d and 3d, respectively.

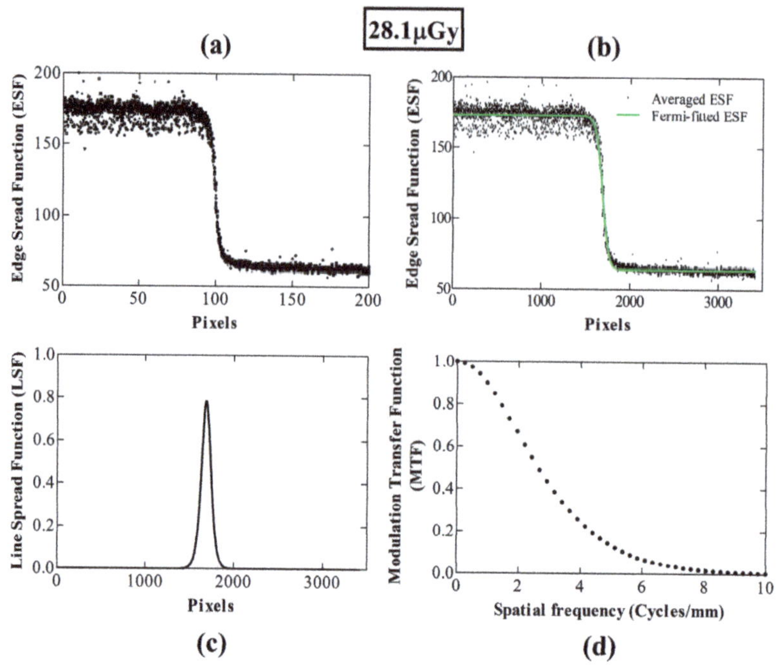

Figure 2. MTF calculation steps at an exposure level of 28.1 μGy, based on the IEC 2015; (**a**) all ESFs; (**b**) averaged and Fermi-fitted ESF; (**c**) LSF; (**d**) MTF.

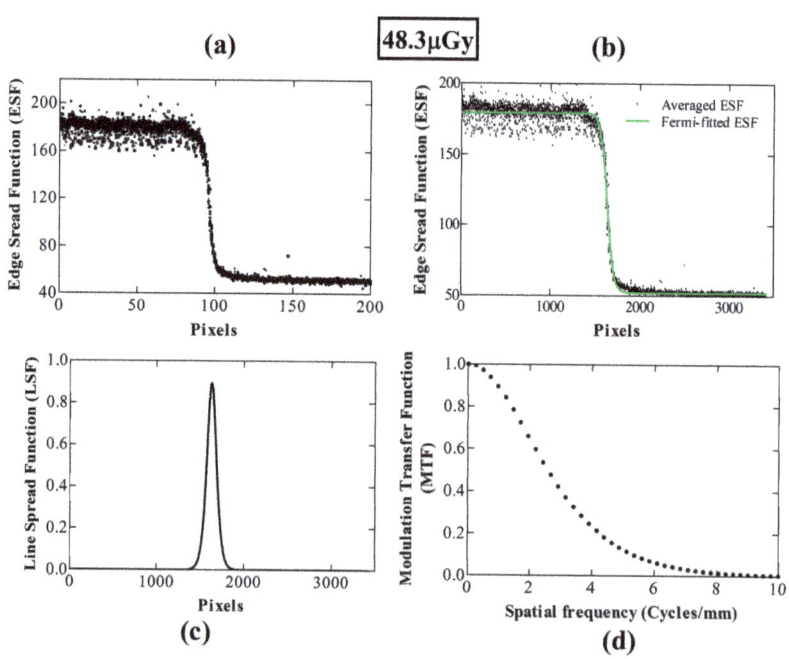

Figure 3. MTF calculation steps at an exposure level of 48.3 μGy, based on the IEC 2015; (**a**) all ESFs; (**b**) averaged and Fermi-fitted ESF; (**c**) LSF; (**d**) MTF.

For ease of comparison, these two MTF curves, along with the two MTFs according to the IEC 62220-1:2003 standard and the one extracted from the detector's datasheet, also calculated with the 2003 protocol [30], are illustrated in Figure 4. Regarding the two IEC 62220-1-1:2015 curves, there is a complete agreement between the different exposure levels. As for the two IEC 62220-1:2003 curves, the differences are within 4%; apart from the 12% at 10 cycles/mm. Hence, regardless of the exposure value, the MTFs of the 2003 and the 2015 versions of the IEC protocol show an agreement within 10%, up to the 3 cycles/mm. After that spatial frequency they start to diverge, with the curves originating from the 2003 protocol lying higher (up to ~17% difference roughly in the range 4.5–7 cycles/mm). This difference should be attributed to the averaging method of the 2003 protocol, which influences the F-D fitting and leads to an overestimation of the MTF [33]. On the other hand, the MTF curve extracted from the detector's datasheet follows within 6% agreement the two IEC 62220-1:2003 curves, except for the 13% difference to the 28.1 μGy curve at 10 cycles/mm. Presumably, an algorithm close to the IEC 2003 standard was used for the datasheet MTF curve calculation.

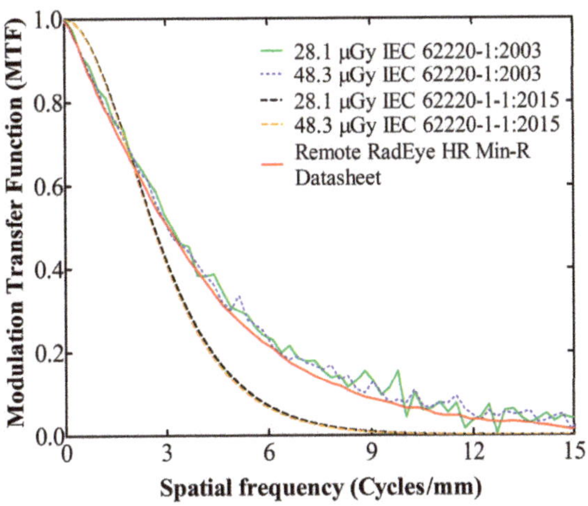

Figure 4. MTF comparison between measured values according to IEC 2003 and IEC 2015, along with manufacturer data [30].

Taking the IEC 62220-1-1:2015 MTF curve as a reference, it is noted that it falls to 50% (MTF50) at 2.6 cycles/mm and to 10% (MTF10) at 5.4 cycles/mm, which marks the limiting resolution of the detector [48].

For comparison purposes, the MTF values at 50% and 10% of the examined detector [49,50] are shown in the Table 1 (1st column), along with the corresponding values of other commercial detectors, all calculated following the IEC 2003 standard. In the 2nd column, the previous (discontinued) version of the Remote RadEye HR CMOS photodiode array is shown, with an active area of 2.7 × 3.6 cm^2, manufactured by Rad-icon Imaging Corp. (USA) with 1200 × 1600 pixels and a pixel pitch of 22.5 μm [32]. In the 3rd and 4th columns, the Dexela flat panel CMOS X-ray detector with a pixel pitch of 74.8 μm and 1944 × 1536 pixels, resulting in a 14.5 × 11.5 cm^2 active area. It is configured with two different Cesium Iodide (CsI) scintillator screens, with thicknesses 150 and 600 μm, respectively. In the 5th column the Hamamatsu C9732DK, which is a CMOS X-ray detector with 2400 × 2400 pixels and 50 μm pixel pitch, corresponding to an active photodiode area of 12 × 12 cm^2 [51]. Finally, in the 6th column, the Large Area Sensor (LAS) that contains 1350 × 1350 pixels at 40 μm pitch (photodiode area 5.4 × 5.4 cm^2) [52]. As it can be seen from Table 1, the RadEye HR, in both versions, retained the highest MTF values due to the smallest pixel pitch and the gold standard 85 μm Gd_2O_2S:Tb screen. However, the new

RadEye HR version appears to have slightly worse MTF than the previous version (4.8% and 5.5% differences at 10% and 50% of the MTF, respectively).

Table 1. MTF values at 10% and 50% for various CMOS X-ray detectors [32].

Detector	RadEye HR 20.0 μm Pitch	RadEye HR 22.5 μm Pitch	Dexela 150 μm CsI	Dexela 600 μm CsI	Hamamatsu C9732DK	LAS
Pixel Pitch (μm)	20.0	22.5	74.8	74.8	50	40
Dimensions (pixels)	1650 × 1246	1200 × 1600	1944 × 1536	1944 × 1536	2400 × 2400	1350 × 1350
Dimensions (cm^2)	2.49 × 3.3	2.7 × 3.6	14.5 × 11.5	14.5 × 11.5	12 × 12	5.4 × 5.4
MTF@10% (cycles/mm)	10.4	10.9	7.7	4.4	9	4.1
MTF@50% (cycles/mm)	3.6	3.8	2.7	1.3	2.7	1.3

Horizontal and vertical 1D NNPS values, along with the radial average NNPS are depicted in Figure 5 for 28.1 μGy exposure, and in Figure 6 for 48.3 μGy. The points' isotropy (i.e., similarity of the magnitudes between axes) for both exposures varies through the whole range of spatial frequencies; from negligible differences, up to an order of magnitude at 10.1–10.6 cycles/mm, where the vertical component exhibits a peak, probably due to remnant row-to-row FPN.

As expected, there is a tendency of NNPS decrease as the exposure increases. This is due to the fact that, with higher exposures the signal increase exceeds the noise increase (due to Poisson distribution in the detection of input X-rays). For example, at the exposure level of 28.1 the absolute NNPS value is 3.45×10^{-6} at 0.78 cycles/mm, whereas the corresponding value for the exposure level of 48.3 is 1.84×10^{-6}.

In Figure 7 is drawn the DQE in both exposures of our experiments. Given that the MTF between 28.1 μGy and 48.3 μGy did not show any great differences, whereas the NNPS showed a tendency to decrease with increasing exposure, an analogous behavior is expected for the DQE curve, i.e., increased values with higher exposures. This is confirmed by the graph. Besides, the curves have similar shape, since the frequency composition is generally not affected by the exposure [53]; any shape discrepancies should be attributed to the inherent non-linearity of CMOS APS detectors, remnant FPN and electronic noise.

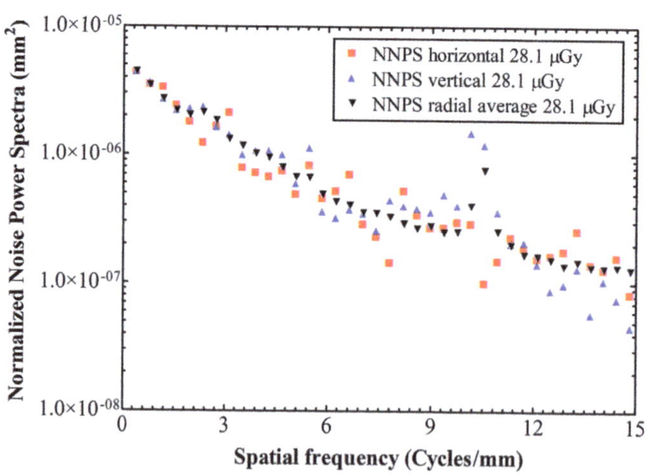

Figure 5. NNPS values at an exposure level of 28.1 μGy.

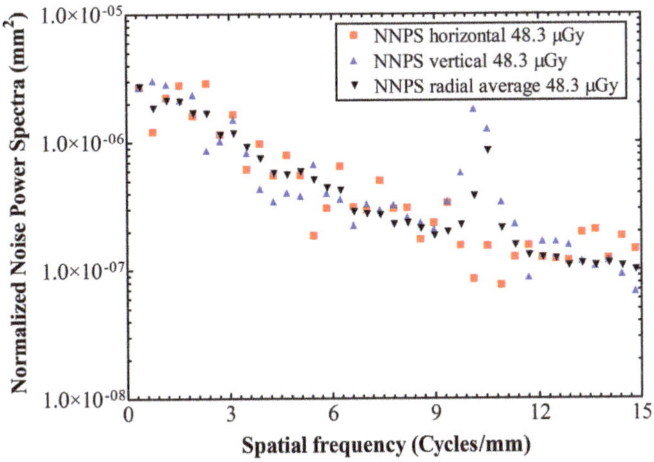

Figure 6. NNPS values at an exposure level of 48.3 μGy.

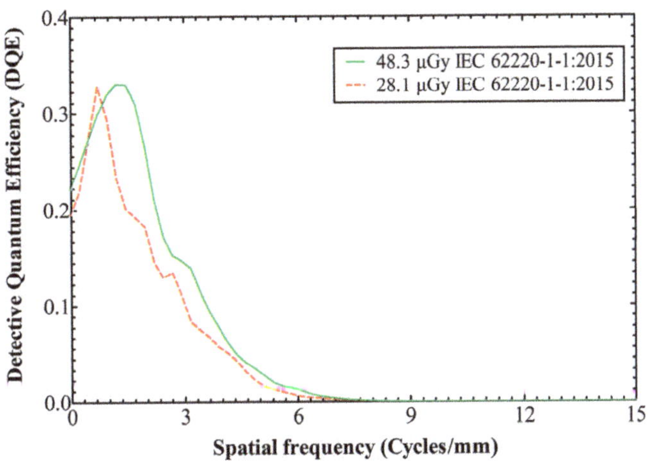

Figure 7. DQE curves for both the examined exposures.

Other than that, the DQE peaks at ~1 cycle/mm and then drops with spatial frequency, indicating that the SNR impinging on the detector is transferred less efficiently towards its output, as the spatial frequency increases. Under ~1 cycle/mm the NNPS levels are high enough to restrain the DQE values. Subsequently, the NNPS levels fall rapidly (Figures 5 and 6; logarithmic y-axes), thus letting the DQE values to build up. For the 48.3 μGy exposure, at the lowest spatial frequency limit the DQE starts from DQE(0) = 0.22, then peaks at DQE(1.3) = 0.33 and falls to 0.1 at 3.6 cycles/mm. As for the 28.1 μGy exposure, the DQE starts from DQE(0) = 0.19, then peaks at DQE(0.7) = 0.33 and falls to 0.1 at 3.0 cycles/mm.

As an indication of the present detector's imaging characteristics in the NDT context, a sample radiography is displayed in Figure 8. The filament coil (0.2 mm thickness) is clearly discernible. The sample lamp (halogen type; Geyer G4 12 V 28 W, Chalkida, Greece) was placed on the detector's surface and an exposure was taken at 70 kVp 20 mAs, using the same SDD as of the measurements, i.e., practically without geometric magnification.

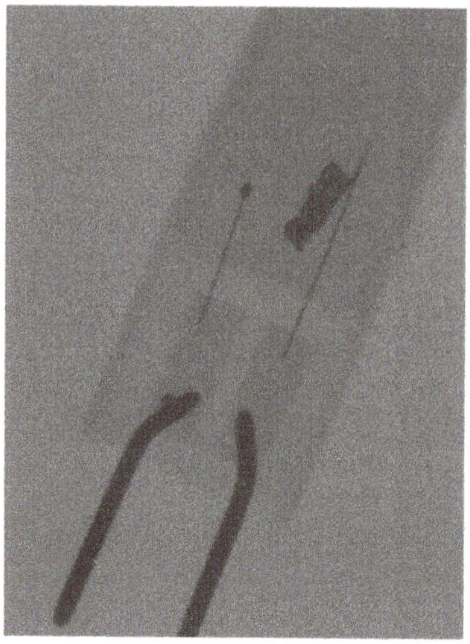

Figure 8. Halogen lamp radiography on the Remote RadEye HR (70 kVp 20 mAs).

Potential Application for Dual-Energy Imaging

The Min-R screen of the present work is made of Gd_2O_2S:Tb, one of the most commonly used scintillating materials coupled to digital imagers. Of its advantageous characteristics are the large effective atomic number (Z_{eff} = 60) and its density (7.3 g/cm^3), that both favor X-ray absorption. It also exhibits a high light conversion efficiency (19%) and a high light yield (60,000 photons/MeV) [54].

In dual-energy detectors, X-rays of different energies are absorbed selectively by two scintillators of the appropriate properties, thus improving the capability to image different composition materials. DE mammography for instance, has been proven to suppress the contrast between overlapping tissues, enhancing in this way the detectability of micro-calcifications and masses [29,55,56]. In the industrial context on the other hand, dual-energy imaging can provide quantitative information of the interior structure of composite materials. Examples are flaw detection of welds or complex geometry parts, luggage inspection, food industry samples [8–10].

The pair ZnSe:Te and $CdWO_4$ is commonly used in such arrays, with ZnSe:Te (low Z_{eff} and density) being the first layer to absorb the lower-energy photons and $CdWO_4$ (high Z_{eff} and density) detecting the X-rays that traversed the first layer. Comparing the properties of Gd_2O_2S:Tb and $CdWO_4$, one is becoming aware of the certain properties they have in common. Namely, $CdWO_4$ has effective atomic number 64.2, density 7.9 g/cm^3 and light yield 19,700 photons/MeV [57]. Its emission spectrum peaks at 495 nm, therefore it is well differentiated from that of ZnSe:Te (640 nm peak). Both spectra though, stand within the present imager's spectral sensitivity range (0.97 matching factor with ZnSe:Te) [58,59]. These results indicate high potential for the two phosphors, Gd_2O_2S:Tb and ZnSe:Te, to be considered as the high- and low-energy elements, respectively, of a dual-energy flat imager.

Such a material combination has been used by Altman et al. [60] for a dual-energy computed tomography (DECT) application. Nevertheless, in this approach the sensing photodetectors are placed next to every row of phosphor pair, i.e., the light detection takes place through the scintillators' side surfaces. This means an insensitive area of 0.125 mm thickness next to each phosphor row, as well as the necessity to radiation-shield these photodetectors [60].

The previous version of the examined detector, also having Gd_2O_2S:Tb phosphor, has been used in dual-energy breast imaging resulting in the detectability of 150 μm thick calcifications. The small pixel pitch allowed post processing in the final images and thus a 93 μm thick calcification was visible [55]. Considering that systems used in digital mammography can detect calcifications as small as 130 μm [61], such a detector has the potential to be used in dual-energy imaging. The same detector was also used in a study for characterization between malignant and benign calcifications. Based on the results of the study, such characterization could be accomplished for calcification thicknesses of 300 μm or higher [29]. Although compared to the previous version of RadEye HR the new version somehow has lower resolution (4.8% and 5.5% differences at 10% and 50% of the MTF, respectively; please see Table 1), such a slight reduction cannot substantially affect its diagnostic performance.

In conclusion, for those applications requesting low to medium diagnostic X-ray energies, the scintillator pair Gd_2O_2S:Tb and ZnSe:Te with the present CMOS imager could be considered as a possible DE array. Diverse approaches have been proposed, yet none of which utilizing these materials in a 2-D panel configuration, to the best of our knowledge. One could be similar to the dual-energy dual-color approach of Maier et al. [62]; the two phosphors are layered together, and the signal is differentiated by means of their different emission wavelengths and an optical layout consisting of a dichroic mirror and lenses. Another could be analogous to the Han et al. [63] solution of a sandwich detector; two Gd_2O_2S:Tb phosphors of different thicknesses are separated by an optically-opaque, radio-translucent foil and a CMOS is coupled on each side.

4. Conclusions

The present NDT CMOS-based detector exhibits excellent linearity across the examined exposure range and a higher gain factor compared to the older 22.5 μm CMOS. The MTF calculated according to the IEC 2003 standard is in close agreement with the CMOS manufacturer's MTF curve, whereas the MTF calculated with IEC 2015 is lower in the higher frequency range, due to the averaging method of IEC 2003 that leads to an overestimation of the MTF. MTF values are higher compared to other commercial CMOS detectors. However, the previous, discontinued, version of the detector under investigation performs slightly better in terms of the MTF, despite its coarser pixel pitch (22.5 μm instead of 20.0 μm). Regarding the detector's phosphor material, Gd_2O_2S:Tb, and based on its properties' pertinence to those of $CdWO_4$, dual-energy application is discussed. Specifically, the present CMOS with its Gd_2O_2S:Tb scintillator, in conjunction with ZnSe: Te, appears as a promising candidate for a dual-energy flat imager aimed at the NDT and medical fields.

Author Contributions: Conceptualization, C.M. and G.F.; methodology, C.M., D.L., I.V., V.K., A.B., and N.M.; software, V.K., and N.M.; validation, C.M. and A.K.; formal analysis, C.M., D.L., and G.F.; investigation, C.M., G.F., D.L., A.K., and I.V.; resources, A.B.; data curation, C.M., D.L., N.M., and V.K.; writing—original draft preparation, D.L. and C.M.; writing—review and editing, D.L., C.M., A.K., and I.V.; visualization, I.V.; supervision, C.M.; project administration, C.M., and G.F.; funding acquisition, I.V. All authors have read and agreed to the published version of the manuscript.

Funding: The APC was funded by the Special Account for Research Grants, of the University of West Attica, Greece.

Institutional Review Board Statement: Not applicable.

Informed Consent Statement: Not applicable.

Data Availability Statement: Data is contained within the article..

Conflicts of Interest: The authors declare no conflict of interest.

References

1. Zhao, C.; Kanicki, J.; Konstantinidis, A.C.A.C.; Patel, T. Large area CMOS active pixel sensor X-ray imager for digital breast tomosynthesis: Analysis, modeling, and characterization. *Med. Phys.* **2015**, *42*, 6294–6308. [CrossRef]
2. Konstantinidis, A.C.; Szafraniec, M.B.; Speller, R.D.; Olivo, A. The Dexela 2923 CMOS X-ray detector: A flat panel detector based on CMOS active pixel sensors for medical imaging applications. *Nucl. Instrum. Methods Phys. Res. A* **2012**, *689*, 12–21. [CrossRef]

3. Zhao, C.; Konstantinidis, A.C.; Zheng, Y.; Anaxagoras, T.; Speller, R.D.; Kanicki, J. 50 μm pixel pitch wafer-scale CMOS active pixel sensor X-ray detector for digital breast tomosynthesis. *Phys. Med. Biol.* **2015**, *60*, 8977–9001. [CrossRef]
4. Zhao, C.; Kanicki, J. Amorphous In-Ga-Zn-O thin-film transistor active pixel sensor X-ray imager for digital breast tomosynthesis. *Med. Phys.* **2014**, *41*, 091902. [CrossRef]
5. Cabello, J.; Bailey, A.; Kitchen, I.; Clark, A.; Crooks, J.; Halsall, R.; Key-Charriere, M.; Martin, S.; Prydderch, M.; Turchetta, R.; et al. Digital autoradiography using CCD and CMOS imaging technology. In Proceedings of the 2006 IEEE Nuclear Science Symposium Conference Record (NSS/MIC 2006), San Diego, CA, USA, 29 October–4 November 2006; Volume 4, pp. 2607–2612.
6. Pham, T.N.; Finck, C.; Marchand, P.; Brasse, D.; Boisson, F.; Laquerriere, P. 18F autoradiography using pixelated CMOS technology. In Proceedings of the 2017 15th IEEE International New Circuits and Systems Conference (NEWCAS 2017), Strasbourg, France, 25–28 June 2017; pp. 41–44.
7. O'Connell, D.W.; Morgan, K.S.; Ruben, G.; Schaff, F.; Croton, L.C.P.; Buckley, G.A.; Paganin, D.M.; Uesugi, K.; Kitchen, M.J. Photon-counting, energy-resolving and super-resolution phase contrast X-ray imaging using an integrating detector. *Opt. Express* **2020**, *28*, 7080–7094. [CrossRef]
8. Kupsch, A.; Hentschel, M.; Lange, A.; Bruno, G.; Müller, B. Direct X-ray Refraction of Micro Structures. In Proceedings of the 11th European Conference on Non-Destructive Testing (ECNDT 2014), Prague, Czech Republic, 6–10 October 2014.
9. Mishurova, T.; Artzt, K.; Haubrich, J.; Requena, G.; Bruno, G. New aspects about the search for the most relevant parameters optimizing SLM materials. *Addit. Manuf.* **2019**, *25*, 325–334. [CrossRef]
10. Prosch, A.; Larson, B. Real Time Radiography NDT Course Booklet. Available online: https://www.nde-ed.org/TeachingResources/Downloadable/Real-Time%20Radiography/Real-time%20Radiography%20Booklet.htm (accessed on 12 February 2021).
11. Nikl, M. Scintillation detectors for X-rays. *Meas. Sci. Technol.* **2006**, *17*, R37–R54. [CrossRef]
12. A Elbakri, I.; McIntosh, B.J.; Rickey, D.W. Physical characterization and performance comparison of active- and passive-pixel CMOS detectors for mammography. *Phys. Med. Biol.* **2009**, *54*, 1743–1755. [CrossRef] [PubMed]
13. Michail, C.M.; Spyropoulou, V.A.; Fountos, G.P.; Kalyvas, N.I.; Valais, I.G.; Kandarakis, I.S.; Panayiotakis, G.S. Experimental and Theoretical Evaluation of a High Resolution CMOS Based Detector Under X-ray Imaging Conditions. *IEEE Trans. Nucl. Sci.* **2010**, *58*, 314–322. [CrossRef]
14. Konstantinidis, A.; Anaxagoras, T.; Esposito, M.; Allinson, N.; Speller, R. DynAMITe: A prototype large area CMOS APS for breast cancer diagnosis using X-ray diffraction measurements. *SPIE Med Imaging* **2012**, *8313*, 83135. [CrossRef]
15. Teledyne CCD290-99 Product Datasheet. Available online: https://www.teledyne-e2v.com/markets/space/astronomy-imaging/ccd290-99/ (accessed on 12 February 2021).
16. E Bohndiek, S.; Cook, E.J.; Arvanitis, C.D.; Olivo, A.; Royle, G.J.; Clark, A.T.; Prydderch, M.L.; Turchetta, R.; Speller, R.D. A CMOS active pixel sensor system for laboratory- based X-ray diffraction studies of biological tissue. *Phys. Med. Biol.* **2008**, *53*, 655–672. [CrossRef]
17. Rowlands, J.A.; Yorkston, J. Flat Panel Detectors for Digital Radiography (Part I, Chapter 4). In *Handbook of Medical Imaging. Volume 1: Physics and Psychophysics*; Beutel, J., Kundel, H.L., Van Metter, R.L., Eds.; SPIE Press: Bellingham, WA, USA, 2000; Volume 1, pp. 223–327. ISBN 978-0-8194-7772-9.
18. Konstantinidis, A.C.; Szafraniec, M.B.; Rigon, L.; Tromba, G.; Dreossi, D.; Sodini, N.; Liaparinos, P.F.; Naday, S.; Gunn, S.; McArthur, A.; et al. X-ray Performance Evaluation of the Dexela CMOS APS X-ray Detector Using Monochromatic Synchrotron Radiation in the Mammographic Energy Range. *IEEE Trans. Nucl. Sci.* **2013**, *60*, 3969–3980. [CrossRef]
19. Gopinath, D.; Soman, M.; Holland, A.; Keelan, J.; Hall, D.; Holland, K.; Colebrook, D. Soft X-ray radiation damage in EM-CCDs used for Resonant Inelastic X-ray Scattering. *J. Instrum.* **2018**, *13*, C02027. [CrossRef]
20. Adams, R.; Zboray, R. Gamma radiography and tomography with a CCD camera and Co-60 source. *Appl. Radiat. Isot.* **2017**, *127*, 82–86. [CrossRef]
21. Arvanitis, C.D.; Bohndiek, S.E.; Royle, G.; Blue, A.; Liang, H.X.; Clark, A.; Prydderch, M.; Turchetta, R.; Speller, R. Empirical electro-optical and X-ray performance evaluation of CMOS active pixels sensor for low dose, high resolution X-ray medical imaging. *Med. Phys.* **2007**, *34*, 4612–4625. [CrossRef]
22. Ahn, J.; Moon, C.-R.; Kim, B.; Lee, K.; Kim, Y.; Lim, M.; Lee, W.; Park, H.; Moon, K.; Yoo, J.; et al. Advanced Image Sensor Technology for Pixel Scaling down toward 1.0 μm (Invited). In Proceedings of the 2008 IEEE International Electron Devices Meeting, San Francisco, CA, USA, 15–17 December 2008; pp. 1–4.
23. Asakura, K.; Hayashida, K.; Hanasaka, T.; Kawabata, T.; Yoneyama, T.; Okazaki, K.; Ide, S.; Noda, H.; Matsumoto, H. X-ray imaging polarimetry with a 2.5-μm pixel CMOS sensor for visible light at room temperature. *J. Astron. Telesc. Instrum. Syst.* **2019**, *5*, 035002. [CrossRef]
24. Zhang, L.; Jin, Y.; Lin, L.; Li, J.; Du, Y. The Comparison of CCD and CMOS Image Sensors. In Proceedings of the SPIE, Proceedings of 2008 International Conference on Optical Instruments and Technology: Advanced Sensor Technologies and Applications, Beijing, China, 16–19 November 2008; Wang, A., Liao, Y., Song, A., Ishii, Y., Fan, X., Eds.; Curran Associates, Inc.: Red Hook, NY, USA, 2009.
25. Olbinado, M.P.; Just, X.; Gelet, J.-L.; Lhuissier, P.; Scheel, M.; Vagovic, P.; Sato, T.; Graceffa, R.; Schulz, J.; Mancuso, A.; et al. MHz frame rate hard X-ray phase-contrast imaging using synchrotron radiation. *Opt. Express* **2017**, *25*, 13857–13871. [CrossRef]
26. Teledyne Lince11M CMOS Sensor. Available online: https://imaging.teledyne-e2v.com/products/2d-cmos-image-sensors/lince/lince-11m/ (accessed on 23 November 2020).

27. Kuroda, R.; Sugawa, S. Cameras with On-chip Memory CMOS Image Sensors. In *The Micro-World Observed by Ultra High-Speed Cameras*; Springer International Publishing: New York, NY, USA, 2018; pp. 103–124.
28. Koukou, V.; Martini, N.; Fountos, G.; Michail, C.; Bakas, A.; Oikonomou, G.; Kandarakis, I.; Nikiforidis, G. Application of a dual energy X-ray imaging method on breast specimen. *Results Phys.* **2017**, *7*, 1634–1636. [CrossRef]
29. Martini, N.; Koukou, V.; Fountos, G.; Michail, C.; Bakas, A.; Kandarakis, I.; Speller, R.; Nikiforidis, G. Characterization of breast calcification types using dual energy X-ray method. *Phys. Med. Biol.* **2017**, *62*, 7741–7764. [CrossRef] [PubMed]
30. Teledyne Remote RadEye HR. Available online: https://www.teledynedalsa.com/en/products/imaging/industrial-X-ray/remote-radeye-hr/ (accessed on 31 January 2021).
31. Cho, M.K.; Kim, H.K.; Graeve, T.; Yun, S.M.; Lim, C.H.; Cho, H.; Kim, J.-M. Measurements of X-ray Imaging Performance of Granular Phosphors With Direct-Coupled CMOS Sensors. *IEEE Trans. Nucl. Sci.* **2008**, *55*, 1338–1343. [CrossRef]
32. Konstantinidis, A. Evaluation of Digital X-ray Detectors for Medical Imaging Applications. PhD Thesis, University College London, London, UK, 2011.
33. Michail, C.; Valais, I.; Martini, N.; Koukou, V.; Kalyvas, N.; Bakas, A.; Kandarakis, I.; Fountos, G. Determination of the detective quantum efficiency (DQE) of CMOS/CsI imaging detectors following the novel IEC 62220-1-1:2015 International Standard. *Radiat. Meas.* **2016**, *94*, 8–17. [CrossRef]
34. *International Electrotechnical Commission IEC 62220-1:2003 Medical Electrical Equipment-Characteristics of Digital X-ray Imaging Devices-Part 1: Determination of the Detective Quantum Efficiency*; IEC: Geneva, Switzerland, 2003.
35. *International Electrotechnical Commission IEC 62220-1-1:2015 Medical Electrical Equipment: Characteristics of Digital X-ray Imaging Devices. Part 1-1*; IEC: Geneva, Switzerland, 2015; ISBN 978-2-8322-2389-5.
36. Donini, B.; Rivetti, S.; Lanconelli, N.; Bertolini, M. Free software for performing physical analysis of systems for digital radiography and mammography. *Med. Phys.* **2014**, *41*, 051903. [CrossRef]
37. ImageJ, An Open Platform for Scientific Image Analysis. Available online: https://imagej.net/Welcome (accessed on 12 February 2021).
38. Neitzel, U.; Günther-Kohfahl, S.; Borasi, G.; Samei, E. Determination of the detective quantum efficiency of a digital X-ray detector: Comparison of three evaluations using a common image data set. *Med. Phys.* **2004**, *31*, 2205–2211. [CrossRef]
39. Samei, E.; Flynn, M.J.; Reimann, D.A. A method for measuring the presampled MTF of digital radiographic systems using an edge test device. *Med. Phys.* **1998**, *25*, 102–113. [CrossRef]
40. Kandarakis, I.; Cavouras, D.; Sianoudis, I.; Nikolopoulos, D.; Episkopakis, A.; Linardatos, D.; Margetis, D.; Nirgianaki, E.; Roussou, M.; Melissaropoulos, P.; et al. On the response of $Y_3Al_5O_{12}$:Ce (YAG:Ce) powder scintillating screens to medical imaging X-rays. *Nucl. Instrum. Methods Phys. Res. A* **2005**, *538*, 615–630. [CrossRef]
41. Dobbins III, J.T. Image Quality Metrics for Digital Systems (Part I, Chapter 3). In *Handbook of Medical Imaging. Volume 1: Physics and Psychophysics*; Beutel, J., Kundel, H.L., Van Metter, R.L., Eds.; SPIE Press: Bellingham, WA, USA, 2000; Volume 1, pp. 161–222. ISBN 978-0-8194-7772-9.
42. Michail, C.; Valais, I.; Seferis, I.; Kalyvas, N.; Fountos, G.; Kandarakis, I. Experimental measurement of a high resolution CMOS detector coupled to CsI scintillators under X-ray radiation. *Radiat. Meas.* **2015**, *74*, 39–46. [CrossRef]
43. Fountos, G.P.; Michail, C.M.; Zanglis, A.; Samartzis, A.; Martini, N.; Koukou, V.; Kalatzis, I.; Kandarakis, I.S. A novel easy-to-use phantom for the determination of MTF in SPECT scanners. *Med. Phys.* **2012**, *39*, 1561. [CrossRef] [PubMed]
44. Michail, C. Image Quality Assessment of a CMOS/Gd_2O_2S:Pr,Ce,F X-ray Sensor. *J. Sens.* **2015**, *2015*, 1–6. [CrossRef]
45. Seferis, I.; Michail, C.; Valais, I.; Fountos, G.; Kalyvas, N.; Stromatia, F.; Oikonomou, G.; Kandarakis, I.; Panayiotakis, G. On the response of a europium doped phosphor-coated CMOS digital imaging detector. *Nucl. Instrum. Methods Phys. Res. A* **2013**, *729*, 307–315. [CrossRef]
46. Chan, H.-P.; Doi, K. Radiation dose in diagnostic radiology: Monte Carlo simulation studies. *Med Phys.* **1984**, *11*, 480–490. [CrossRef]
47. Marshall, N.W. Detective quantum efficiency measured as a function of energy for two full-field digital mammography systems. *Phys. Med. Biol.* **2009**, *54*, 2845–2861. [CrossRef]
48. Howansky, A.; Mishchenko, A.; Lubinsky, A.R.; Zhao, W. Comparison of CsI:Tl and Gd_2O_2S:Tb indirect flat panel detector X-ray imaging performance in front- and back-irradiation geometries. *Med. Phys.* **2019**, *46*, 4857–4868. [CrossRef]
49. Choi, S.; Seo, C.-W.; Cha, B.K. Effect of Filtered Back-Projection Filters to Low-Contrast Object Imaging in Ultra-High-Resolution (UHR) Cone-Beam Computed Tomography (CBCT). *Sensors* **2020**, *20*, 6416. [CrossRef]
50. Schulze, R.K.W.; I Doering, C. Simple computation of the approximated modulation transfer function (MTF) using spreadsheet-software: Method and evaluation in five maxillofacial CBCT-devices. *Dentomaxillofacial Radiol.* **2019**, *48*, 20180350. [CrossRef]
51. Hamamatsu Flat Panel Sensor C9732DK-11. Available online: https://www.hamamatsu.com/eu/en/product/type/C9732DK-11/index.html (accessed on 31 January 2021).
52. Bohndiek, S.E.; Blue, A.; Cabello, J.; Clark, A.T.; Guerrini, N.; Evans, P.M.; Harris, E.J.; Konstantinidis, A.; Maneuski, D.; Osmond, J.; et al. Characterization and Testing of LAS: A Prototype 'Large Area Sensor' With Performance Characteristics Suitable for Medical Imaging Applications. *IEEE Trans. Nucl. Sci.* **2009**, *56*, 2938–2946. [CrossRef]
53. Monnin, P.; Gutierrez, D.; Bulling, S.; Guntern, D.; Verdun, F.R. A comparison of the performance of digital mammography systems. *Med. Phys.* **2007**, *34*, 906–914. [CrossRef]

54. Litichevskyi, V.; Galkin, S.; Lalayants, O.; Voronkin, E.; Breslavskiy, I.; Tretiak, S.; Kosinov, N. Scintillation Panels Based on Zinc Selenide and Oxide Scintillators. *Funct. Mater.* **2011**, *18*, 391–397.
55. Koukou, V.; Martini, N.; Fountos, G.; Michail, C.; Sotiropoulou, P.; Bakas, A.; Kalyvas, N.; Kandarakis, I.; Speller, R.; Nikiforidis, G. Dual energy subtraction method for breast calcification imaging. *Nucl. Instrum. Methods Phys. Res. A* **2017**, *848*, 31–38. [CrossRef]
56. Martini, N.; Koukou, V.; Michail, C.; Fountos, G. Dual Energy X-ray Methods for the Characterization, Quantification and Imaging of Calcification Minerals and Masses in Breast. *Cryst.* **2020**, *10*, 198. [CrossRef]
57. Lecoq, P.; Gektin, A.; Korzhik, M. *Inorganic Scintillators for Detector Systems: Physical Principles and Crystal Engineering;* Springer: Berlin, Germany, 2006; ISBN 978-3-540-27766-8.
58. Linardatos, D.; Konstantinidis, A.; Valais, I.; Ninos, K.; Kalyvas, N.; Bakas, A.; Kandarakis, I.; Fountos, G.; Michail, C. On the Optical Response of Tellurium Activated Zinc Selenide ZnSe:Te Single Crystal. *Crystals* **2020**, *10*, 961. [CrossRef]
59. Seferis, I.E.; Michail, C.; Zeler, J.; Kalyvas, N.; Valais, I.; Fountos, G.; Bakas, A.; Kandarakis, I.; Zych, E.; Panayiotakis, G.S. Detective quantum efficiency (DQE) of high X-ray absorption Lu2O3:Eu thin screens: The role of shape and size of nano- and micro-grains. *Appl. Phys. A* **2018**, *124*, 604. [CrossRef]
60. Altman, A.; Shapiro, O.; Levene, S.; Wainer, N. Double Decker Detector for Spectral CT. US Patent US7968853B2, 28 June 2011.
61. Bick, U.; Diekmann, F. *Digital Mammography;* Springer: Heidelberg, Germany, 2010; ISBN 978-3-540-78449-4.
62. Maier, D.S.; Schock, J.; Pfeiffer, F. Dual-energy micro-CT with a dual-layer, dual-color, single-crystal scintillator. *Opt. Express* **2017**, *25*, 6924–6935. [CrossRef] [PubMed]
63. Han, J.C.; Kim, H.K.; Kim, N.W.; Yun, S.; Youn, H.; Kam, S.; Tanguay, J.; Cunningham, I.A. Single-shot dual-energy X-ray imaging with a flat-panel sandwich detector for preclinical imaging. *Curr. Appl. Phys.* **2014**, *14*, 1734–1742. [CrossRef]

materials

Article

Non-Destructive X-ray Characterization of a Novel Joining Method Based on Laser-Melting Deposition for AISI 304 Stainless Steel

Muhammad Arif Mahmood [1], Diana Chioibasu [1], Sabin Mihai [1,2], Mihai Iovea [3], Ion N. Mihailescu [1] and Andrei C. Popescu [1,*]

1 National Institute for Laser, Plasma and Radiation Physics (INFLPR), Magurele, 077125 Ilfov, Romania; arif.mahmood@inflpr.ro (M.A.M.); diana.chioibasu@inflpr.ro (D.C.); sabin.mihai@inflpr.ro (S.M.); ion.mihailescu@inflpr.ro (I.N.M.)
2 Faculty of Industrial Engineering and Robotics, University Politehnica of Bucharest, 060042 Bucharest, Romania
3 Accent Pro 2000 SRL, Nerva Traian Nr. 01 K6, Ap. 26, Sector 3, 031041 Bucharest, Romania; mihai.iovea@accent.ro
* Correspondence: andrei.popescu@inflpr.ro; Tel.: +40-214574550 (ext. 2423/2033)

Abstract: In this study, an application of the laser-melting deposition additive manufacturing technique as a welding method has been studied for the laser welding (LW) of AISI 304 stainless steel, specifically 0.4 mm and 0.5 mm thick sheets. The welding was carried out without and with filler material. Inconel 718 powder particles were used as filler material in the second case. A series of experiments were designed by changing the process parameters to identify the effect of operating conditions on the weld width, depth, and height. The welds were examined through metallographic experiments performed at various cross-sections to identify the defects and pores. All the deposited welds were passed through a customized mini-focus X-ray system to analyze the weld uniformities. The optimal operating conditions were determined for 0.4 mm and 0.5 mm sheets for the LW with and without filler material. It was found that laser power, laser scanning speed, powder flow rate, and helium to argon gases mixture-control the weld bead dimensions and quality. X-ray analyses showed that the optimal operating conditions gave the least peak value of non-uniformity in the laser welds. This study opens a new window for laser welding via additive manufacturing with X-ray monitoring.

Keywords: laser welding; laser-melting deposition; AISI 304 stainless steel sheets; process optimization; X-ray imaging characterization

Citation: Mahmood, M.A.; Chioibasu, D.; Mihai, S.; Iovea, M.; Mihailescu, I.N.; Popescu, A.C. Non-Destructive X-ray Characterization of a Novel Joining Method Based on Laser-Melting Deposition for AISI 304 Stainless Steel. *Materials* **2021**, *14*, 7796. https://doi.org/10.3390/ma14247796

Academic Editor: Giovanni Bruno

Received: 1 November 2021
Accepted: 14 December 2021
Published: 16 December 2021

1. Introduction

Laser welding (LW) combines metallic or thermoplastic pieces via a laser beam [1]. The laser beam delivers the focused heat, thus allowing for narrow and deep welds. The LW success majorly relies on careful process parameter consideration [2]. A set of optimum laser processing parameters can result in laser welds without pore and crack formation [3]. The LW of any material requires a complete understanding of its thermal, mechanical, and physical properties [4].

The operating parameters, especially the speed of the temperature increment, affect the weld bead dimensions, gas flow rate, welding velocity, specimen hardness, and microstructure formation. To this purpose, various studies have been carried out. The laser welding of AISI 304 stainless steel 1.6 mm thick sheets were carried out using the continuous wave (CW) solid-state Nd: YAG laser [5]. The effect of the laser power and welding speed was studied on weld depth and width. A direct correlation was found between laser power, weld depth, and width. However, welding speed showed an inverse relation with weld depth and width. These two operating conditions were identified as the parameters controlling the weld quality and efficiency. The LW process's heat-affected zone (HAZ) size was identified based on the laser scanning speed and laser power [6]. The HAZ maximum width and depth decreased by 32% and 62%, respectively, when the laser scanning speed increased from 10 to 100 mm/min by keeping the laser power constant (=10 W). There is no residual stress formation in the HAZ if the critical temperature is less than 840–890 °C. A study reported texture formation in aluminum alloys (AA5182-O and AA6111-T4) welded using the Nd: YAG laser source [7]. An electron backscattering diffraction (EBSD) technique was used to assess the texture. The measurements were carried out as a function of sample thickness. The results presented that the laser welds can develop a substantial texture, particularly the columnar grains entering the base plate into the laser weld. Additionally, they had a 001 texture along with the growth direction. The LW of aluminum-lithium sheets was carried out with a 3 kW CO_2 laser beam [8]. The influence of various gases was evaluated on the weld appearance, fusion regime dimensions, solute vaporization, hardness, tensile characteristics, and pore formation. It was found that the energy density input plays a direct role in defining the characteristics mentioned above. The effect of laser pulse frequency, laser input energy, and scanning speed was examined on low carbon steels' mechanical and metallurgical properties [9]. A series of experiments were designed using the Taguchi technique. It was determined that the effective pulse energy is a governing factor in determining the laser-welded joints' strength. AISI 304 stainless steel sheets laser welding was carried out using a CW CO_2 laser [10]. They determined that the absorbance of CO_2 in conduction and keyhole welding was 15% and 65%, respectively. Furthermore, the steel surface pre-oxidation resulted in 30–50% better CO_2 absorbance.

One of the best ways to analyze the weld quality, using non-destructive techniques, is to combine the welding process with X-ray analyses. In this regard, a high-speed X-ray imaging system was used to observe the keyhole variabilities concerning the defects formation during the LW of copper sheets [11]. Breakthrough results were presented, explaining the bubble generation at the capillary tip. LW penetration as a function of welding speed was studied [12]. X-ray imaging was compiled with the welding process to study the melt pool surface and ejected plume behavior. A wide range of laser scanning speeds was utilized (0–50 m/min) for observation. An inverse correlation was found between laser scanning speed and penetration depth for higher laser scanning speeds. In contrast, when the welding speed is decreased, the weld pool dynamics and plume interaction disturb the keyhole generation stability. At this point, the weld penetration depth showed a random behavior, thus breaking the prior-defined trend. The effect of laser beam pulse shape was analyzed on the keyhole formation via an in situ micro-focused X-ray system [13]. The AISI 304 stainless steel sheet experimentation was irradiated using the Nd: YAG laser beam (pulse duration = 1.1 ms and 4.6 kW peak power = 4.6 kW). It was determined that the keyhole generation started at 0.6–0.7 ms, became deep at 1.5 ms, and finally collapsed at 1 ms. It was also analyzed that the pores were formed at the beam profile, where the peak laser power declined rapidly. The laser welding and weldability phenomena were interpreted concerning the association of physical and metallurgical points of view [14]. Additionally, a micro-focused X-ray in situ monitoring system was used to observe keyhole, melt-flow, bubbles, and pore formation. It was determined that the input parameters, such as laser power, scanning speed, and gas flow rate, play a key role in defining the stability of keyhole, melt-flow, and bubbles and pores generation. The dynamics of keyhole generation and melt pool formation in the LW process was studied by combining the X-ray imaging system with LW [15]. Additionally, the real cause of pores formation was also studied. It was found that during the LW process, a laser hole (keyhole) is generated in the melt pool due to an intense evaporation recoil pressure. This keyhole formation leads to a penetration depth with a high aspect ratio. This characteristic is the most beneficial feature of high-energy-density laser beams (HEDLB). The keyhole in the liquid is unstable, causing cavities or pores in the solidified weld, which is a severe problem of HEDLB.

Errico et al. [16] carried out the LW of AISI 304 stainless steel with AISI 316 L filler powder. To achieve defect-free weld seams, the influence of process parameters, including laser power, laser scanning speed, powder feeding rate, gas flow rate, and laser beam spot size, was investigated on the geometry, microstructure, and porosity inside the weld seams. It was identified that the change in the laser beam spot size is critical in producing various welding regimes. The profile of the laser keyhole is responsible for generating the porosity in the welded region. Prabakaran and Kannan [17] used the Taguchi-based grey relational analysis to optimize the laser welding process parameters for austenitic stainless steel (AISI316) and low carbon steel (AISI1018) materials. Butt joint testing was performed after utilizing a 3.5 kW diffusion-cooled slab CO2 laser and the laser power, welding speed, and focal distance were varied. The authors determined that the ideal operating conditions are laser power = 2600 W, laser scanning speed = 1.5 m/min, and focal distance = 20 mm with the tensile strength = 475.112 MPa. Landowski [18] presented the results for the microstructure of laser beam-welded stainless steel under various welding conditions. Welded joints were produced using an Ytterbium fiber laser without using filler material. The test material was 2205 ferritic-austenitic duplex stainless-steel plates with an 8 mm thickness. It was discovered that the optimum focusing position is one of the most important parameters for shaping high-quality welds. A focus position above the specimen surface (+3, +6 mm) is preferred for high-power fiber laser welding of thick plates. Ding et al. [19] performed the laser welding of heterogeneous metals. Different thermal gradients were attained by varying the operating conditions, including laser welding speed, frequency, and pulse width. The results showed that the melt pool was asymmetric due to the melting of the brass alloy. The measured temperature and melt volume were higher due to the brass alloy's lower melting temperature and high heat transfer rate. Intermetallic complexes were also present in the melt pool microstructure. The melt pool dimensions can be increased by increasing the peak power and decreasing the welding speed.

This study presents a novel technique for AISI 304 stainless steel, with 0.4 mm and 0.5 mm thick sheets and laser welding. The laser melting deposition (LMD) technique, an additive manufacturing method, has been implemented for sheets' LW. The welding was carried out without and with filler material. For the second category, Inconel 718 powder particles were added coaxially. In the case of LW without filler, the laser scanning speed, argon gas flow rate, and robot axis angle were varied, while laser power was kept fixed. In contrast, regarding LW with filler, the laser power, laser scanning speed, powder flow rate, and helium and argon gases mixture were changed. A series of experiments were designed by changing the operating conditions to identify their effect on the weld width, depth, and height (with filler material only). The metallographic experiments were conducted at various sections for the deposited welds to analyze the defects and pores in the cross-section. All the deposited welds were subjected to the customized mini-focus X-ray system to analyze the weld uniformities. Last but not least, a set of optimal parameters was produced in the case of LW without and with filler material based on the information generated via experimental analyses.

2. Materials and Methods

This study focused on determining the optimal laser welding parameters for AISI 304 stainless steel (SS) sheets' butt-welding. These sheets were provided by NIKO AUTO COM S.R.L., a metal parts producer in Romania, with the following dimensions: length = 100 mm, width = 50 mm, and thickness = 0.4 mm and 0.5 mm. The chemical composition of AISI 304 SS sheets is presented in Table 1.

Table 1. Chemical composition of AISI 304 stainless steel sheets [20].

Elements	(%)
Carbon	0.08
Manganese	2.0
Silicon	0.75
Phosphorus	0.045
Sulfur	0.03
Chromium	18.0–20.0
Nickel	8.0–10.5
Iron	Balance

Laser welding was executed using a Yb: YAG laser source (TruDisk 3001, Trumpf, Ditzingen, Germany) emitting in continuous mode with wavelength $\lambda = 1030$ nm, integrated with a KUKA robot and a 3-beam powder nozzle. The laser beam's focused spot was 800 μm with a top-hat shape. The deposition optics in the LMD system were mounted on the robotic arm integrated and expressing 6 degrees of freedom through an electromagnetic plate. The laser beam was guided and focused on the workpiece through a system of lenses and mirrors, and the powder stream was blown into the laser focal spot via a 3-beam nozzle. Helium and argon carrier gases transported the metallic powder. The welding experiments were carried out with and without the powder addition by varying the laser power, scanning speed, laser focal point, and carrier gases. Inconel 718 powder particles in the range of 45–90 μm were used as a filler material. The chemical composition of Inconel 718 is provided in Table 2.

Table 2. Chemical composition of Inconel 718 powder particles [21].

Elements	(%)
Carbon	0.08
Manganese	0.35
Phosphorus	0.015
Sulfur	0.015
Silicon	0.35
Chromium	17–21
Nickel	50–55
Molybdenum	2.80–3.30
Columbium	4.75–5.50
Titanium	0.65–1.15
Aluminum	0.20–0.80
Cobalt	1.0
Boron	0.006
Copper	0.30
Tantalum	0.05
Iron	Balance

Before carrying out the welding experiments, all the sheets were cleaned with ethyl alcohol to eliminate oil and grease particles. Then, these samples were fixed in position using support fixtures. The experimental assembly used for the welding is shown in Figure 1.

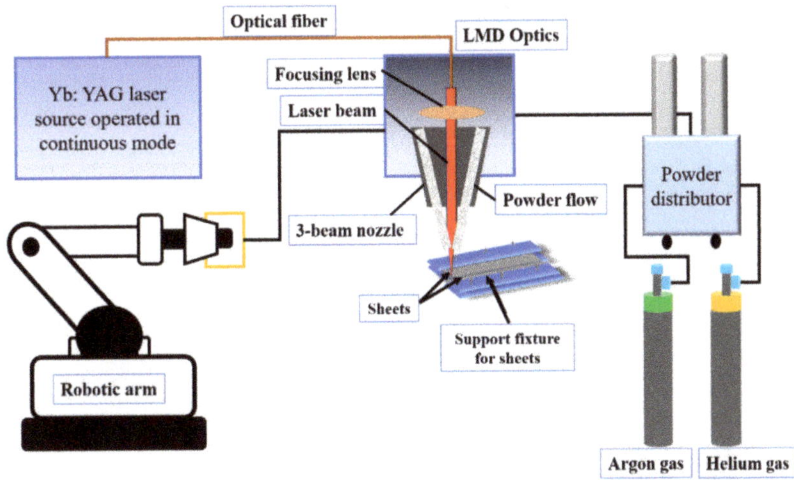

Figure 1. Laser welding via LMD set-up.

Table 3 presents the list of parameters used for the laser welding optimization without the powder addition. The main objective is to obtain fully penetrated welds without defects. During the laser welding, priority was given to (a) obtaining continuous and flawless weld seams and (b) streamlining the welding process in terms of processing time, energy consumption, and total costs. For this purpose, in the case of 0.5 mm thick sheets, the laser power was kept constant (=1000 W) while the laser scanning speed (P1-P4), robot axis angle (P5–P8), and gas flow rate (P9–P12) were varied to determine the optimized set of parameters. The optimized set of parameters was used as the starting point for 0.4 mm thick sheets and only the laser scanning speed was changed to speed up the welding process (W1 and W2).

Table 4 shows the process parameters in stainless-steel samples' laser welding with a filler material (Inconel 718). The filler metal powder was transported from the powder distributor via a three-jet powder nozzle using a mixture of helium and argon gases. In this case, the laser power, scanning speed, powder flow, and carrier gas flow rates were varied. Initially, the process optimization was conducted for a 0.4 mm thick sheet. The optimized parameters, attained from the welding of 0.4 mm sheets, were used for 0.5 mm sheet welding, except the laser scanning speed was diminished progressively to achieve a better beam penetration into the material.

Table 3. Laser welding operating parameters without powder addition.

Sr. No	Sheet Thickness (mm)	Laser Power (P; W)	Laser Scanning Speed (V; m/min)	Argon Gas Flow (Ag; slpm)	Robot Axis Angle (θ; °)
P1			6.0		
P2			4.8		0
P3			3.6		
P4					
P5				6.0	5
P6	0.5	1000			10
P7			2.4		15
P8					20
P9				5.0	
P10				7.0	0
P11				9.0	
P12				11.0	

Sr. No	Sheet Thickness (mm)	Laser Power (W)	Laser Scanning Speed (m/min)	Argon Gas Flow (slpm)	Robot Axis Angle (θ; °)
W1	0.4	1000	2.4	5.0	0
W2	0.4	1000	3.0	5.0	0

∵ P(Sr. No) = sheets with 0.5 mm thickness; W(Sr. No) = sheets with 0.4 mm thickness.
∵ number of scanning lines = 1.0.

Table 4. Laser welding operating parameters with filler material (Inconel 718).

<table>
<tr><th>Sr. No.</th><th>Sheet Thickness (mm)</th><th>Laser Power (P; W)</th><th>Laser Scanning Speed (V; m/min)</th><th>Powder Flow Rate (Pf; g/min)</th><th>Helium Gas Flow Rate (Hg; slpm)</th><th>Argon Gas Flow Rate (Ag; slpm)</th></tr>
<tr><td>E1</td><td rowspan="20">0.4</td><td>700</td><td rowspan="3">0.6</td><td rowspan="8">4.0</td><td rowspan="13">4.0</td><td rowspan="16">5.0</td></tr>
<tr><td>E2</td><td>800</td></tr>
<tr><td>E3</td><td>900</td></tr>
<tr><td>E4</td><td rowspan="19">1000</td><td>0.6</td></tr>
<tr><td>E5</td><td>1.2</td></tr>
<tr><td>E6</td><td>1.8</td></tr>
<tr><td>E7</td><td>2.4</td></tr>
<tr><td>E8</td><td>3.0</td></tr>
<tr><td>E9</td><td rowspan="12">2.4</td><td>3.0</td></tr>
<tr><td>E10</td><td>6.0</td></tr>
<tr><td>E11</td><td>9.0</td></tr>
<tr><td>E12</td><td>12.0</td></tr>
<tr><td>E13</td><td rowspan="10">6.0</td></tr>
<tr><td>E14</td><td>5.0</td></tr>
<tr><td>E15</td><td>6.0</td></tr>
<tr><td>E16</td><td>7.0</td></tr>
<tr><td>E17</td><td rowspan="6">4.0</td><td>10.0</td></tr>
<tr><td>E18</td><td>12.0</td></tr>
<tr><td>E19</td><td>14.0</td></tr>
<tr><td>E20</td><td>16.0</td></tr>
<tr><td>F1</td><td rowspan="2">0.5</td><td>1.2</td><td rowspan="2">10.0</td></tr>
<tr><td>F2</td><td>2.4</td></tr>
</table>

∵ E(Sr. No) = sheets with 0.4 mm thickness; F(Sr. No) – sheets with 0.5 mm thickness
∵ number of scanning lines = 1.0.

Figure 2a,b show the schematic and nomenclature of a typical laser weld in the case of with and without filler material, and can be divided into three sections: (a) weld height, (b) weld width, and (c) weld depth. The effect of primary operating conditions on these three features was studied.

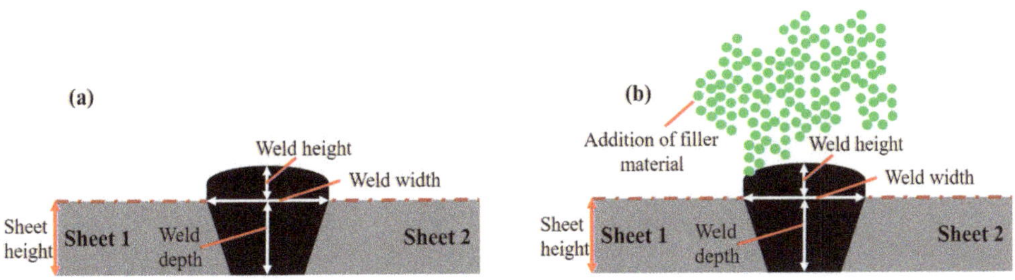

Figure 2. Schematic and nomenclature of weld cross-section: (**a**) without filler and (**b**) with filler material.

After performing the laser-welding experiments, the welds were subjected to radiographic analyses via customized mini-focus X-ray equipment developed by ACCENT PRO, Romania, as shown in Figure 3.

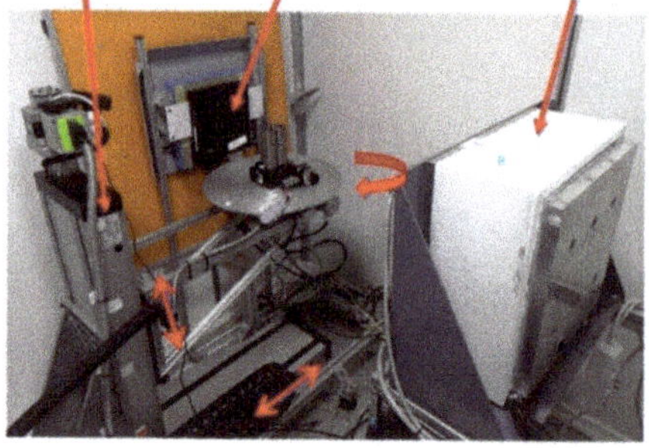

Figure 3. Mini-focus X-ray equipment for radiographic analyses.

Table 5 presents the X-ray setup description applied to conduct the X-ray experimentation.

Table 5. X-ray setup description used to perform X-ray analyses.

Item	Value
X-ray source	
Type	Gilardoni Monoblock AION
Peak voltage	160 kVp
Current	3 mA
Power	500 W
Focal spot size	1.2 mm
Cone beam angle	40°
X-ray detector	
Detection Technology	Flat panel type X-Panel 1512
Size	150 mm × 120 mm
Pixel size	100 μ
ADC type	14 bits
Peak voltage	225 kVp
Geometry	
Source-detector distance (SDO)	800 mm
Object-detector distance (ODD)	40 mm
Acquisition parameters	
Peak voltage	150 kVp
Current	2 mA
Integration time per frame	50 ms
Average frames acquired	25 frames

The samples images are given in attenuation values (*Att*) computed for each detector's pixel by using the following formula:

$$Att = ln\left[\frac{I_o - I_{off}}{I_s - I_{off}}\right], \quad (1)$$

where I_o is the image taken with X-ray source ON and without any object/sample in the beam, I_s is the image taken with X-ray source ON and with the sample placed in the beam, and I_{off} is the image taken with X-ray source OFF.

X-ray analysis was performed on the whole sample to analyze the weld quality. However, three different sections, namely the (1) start, (2) middle, and (3) end of the weld seam, were chosen to calculate the weld uniformity, as presented in Figure 4. These analyses were carried out orthogonal to the weld seam, resulting in weld seam discontinuities. X-rays are electromagnetic beams and work on electromagnetic radiation's principles [22]. Electromagnetic radiation, similar to radio waves, visible light, or microwaves, uses waves and photons to carry energy through space, also known as radiant energy. When the interaction between X-rays and a given material occurs, they lose a certain amount of energy based on the material's absorption behavior. Besides various other parameters, this absorption coefficient depends on the surface uniformity/regularities in such a way that a uniform surface yields a higher absorption coefficient value [23]. X-ray absorption, in general, quantifies the loss of energy at the incident region. Hence, the attained image contrast can assist in quantifying the weld quality.

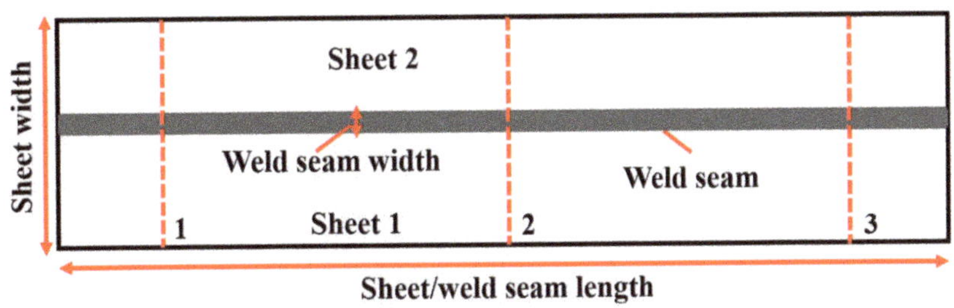

Figure 4. X-ray analyses' schematic on the weld seam at various locations: (1) at the beginning, (2) in the center, and (3) at the end of the sheet.

3. Results and Discussion

Figure 5a–c exhibit the influence of the laser scanning speed (V), robot axis angle (θ), and argon gas flow rate (Ag) on the laser welds' depth (WD) and width (WW). It should be noted that laser power (P) was kept constant. For a given P, Ag, and θ, an increment in V causes a declination in the WD and WW values. The V plays a vital role in defining the laser–material interaction time. A higher V defines less laser–material interaction time, resulting in a melt pool formation with smaller dimensions. In Figure 5a, a curve fitting was performed to analyze the comprehensive and deterministic outcomes, resulting in a polynomial expression that can predict the different outputs based on given inputs. Figure 5b shows a correlation between θ, WD, and WW. "θ" is defined as the inclination angle provided by the robot axis. In the current study, θ did not show any significant impact over WD and WW dimensions. Additionally, a random behavior was observed in the datasets. A curve fitting was performed to identify the trend in the generated data points. An increase in θ causes a decrease in WD and an increase in WW, keeping the P, Ag, and V constant. When the robot arm is perpendicular to the metallic sheets welded, the laser spot focuses on a particular circular area. However, with an increment in the robot arm axis angle, the laser spot no longer remains circular; instead, an elliptical shape is attained, resulting in higher WW and lesser WD than the last case. Furthermore, Ag presented an inverse relation with WD and direct relation with WW for a given P, V, and θ, as shown in Figure 5c. It can be attributed to the increase in argon content in the weld pool and the formation of a thick protective layer on the weld surface, reducing the WD. These findings are in good correlation with the one provided in Reference [24]. The data trend can be analyzed via a curve fitting, as shown in Figure 5. Furthermore, error bars have been plotted in Figure 5 that show the standard deviation in the data attained by varying the operating conditions. In other words, the short error bars demonstrate that the concentration of the values is high compared to the long error bars.

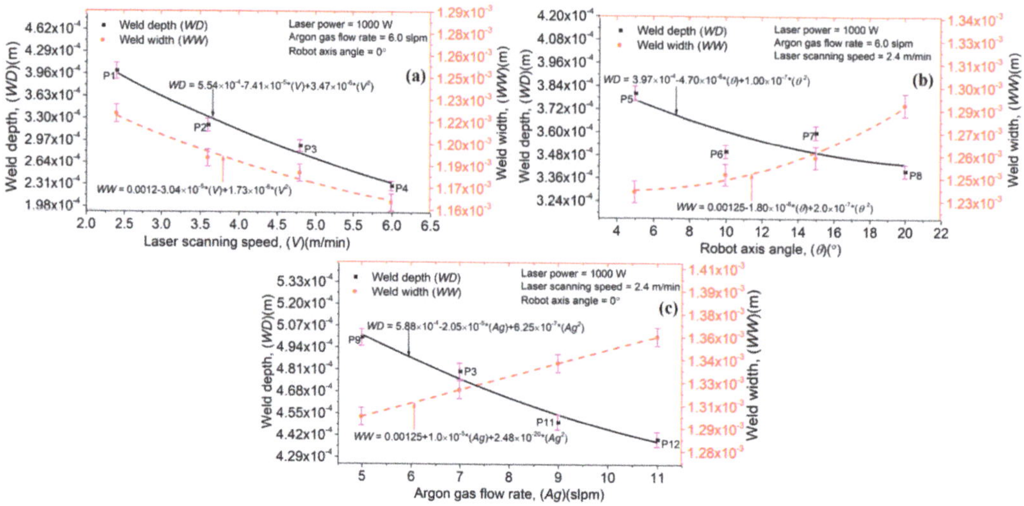

Figure 5. Laser welds' depth and width by changing: (**a**) laser scanning speed, (**b**) robot axis angle, and (**c**) argon gas flow rate.

Using the operating conditions defined in Table 3, several laser welds combined with metallography experiments were carried out. The optimal parameters for laser welding of AISI 304 stainless steel sheets without filler material, with a thickness of 0.5 mm, are: (a) laser power = 1000 W, (b) laser scanning speed = 2.4 m/min, (c) argon gas flow rate = 5 slpm, and (d) robot tilt angle = 0°. Figure 6a,b show the optical images of the top view and cross-section of the welded sheets by applying optimal laser processing parameters. It can be observed that the weld is uniform, without pores and cracks. When the V increases beyond 3 m/min, the laser–material interaction time decreases, ultimately reducing the laser energy density transfer into the material, thus leading to narrower weld seams which can lead to incomplete penetration across the sheet thickness. When the laser energy density cannot travel across the sheet thickness, it results in a lack of fusion between the two provided sheets, thus giving incomplete laser weld penetration that can be analyzed near the bottom end of the sheets with reference to the laser–material interaction surface.

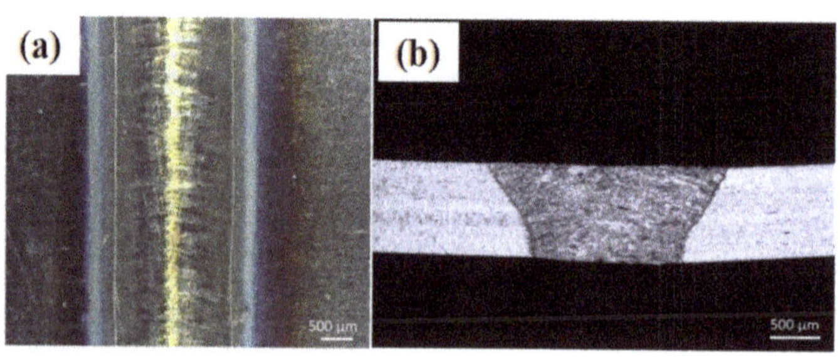

Figure 6. Optical images of laser-welded 0.5 mm thick sheets' cross-sections without powder addition and with laser power = 1000 W, laser scanning speed = 2.4 m/min, gas flow rate = 5 slpm, and robot tilt angle = 0°: (**a**) top view and (**b**) cross-section.

To determine the laser welding operating conditions for AISI 304 stainless-steel sheets with 0.4 mm thickness, the optimal conditions identified for the 0.5 mm thickness sheets were initially used, as displayed in Figure 7a. As the sheet thickness (=0.4 mm) is smaller than 0.5 mm, the laser scanning speed was increased from 2.4 to 3.0 m/min, as presented in Figure 7b. It can be observed that the weld width decreases with the increase in scanning speed, indicating a complete and flawless welding penetration. However, beyond 3.0 m/min, an incomplete laser depth was achieved. Hence, the optimal parameters for the laser welding of AISI 304 stainless steel sheets without filler material, with a thickness of 0.4 mm, are: (a) laser power = 1000 W, (b) laser scanning speed = 3.0 m/min, (c) argon gas flow rate = 5 slpm, and (d) robot tilt angle = 0°.

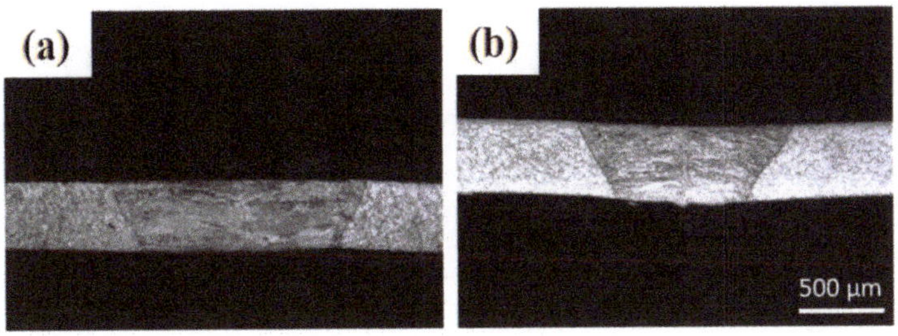

Figure 7. Optical images of 0.4 mm thick sheets' laser weld cross-sections without powder addition and with laser power = 1000 W, gas flow rate = 5 slpm, robot tilt angle = 0° and (**a**) laser scanning speed = 2.4 m/min and (**b**) laser scanning speed = 3.0 m/min.

For weld joints, good positioning of the sheets is essential to obtain completely penetrated welds. A larger spot size results in easier sheet positioning for laser welding. Therefore, the laser beam was defocused to increase the spot diameter. In LMD equipment, laser defocusing decreases the energy density at the sheet surface. Figure 8a,b show the results of a defocused laser beam with (focal plane + 5 mm) and (focal plane + 10 mm) distances, respectively. The spot diameter was increased from 0.8 to 1 mm (focal plane + 5 mm) and 1.15 mm (focal plane + 10 mm). Acceptable weld seams were achieved via a defocused laser beam, thus relaxing the exact positioning of the laser spot at the two plates' junction. However, this strategy can only be implemented using a "top-hat" laser spot in which the laser intensity is uniformly distributed. Extending this technique to other distributions (Gaussian or various non-uniformities) is difficult and often impossible.

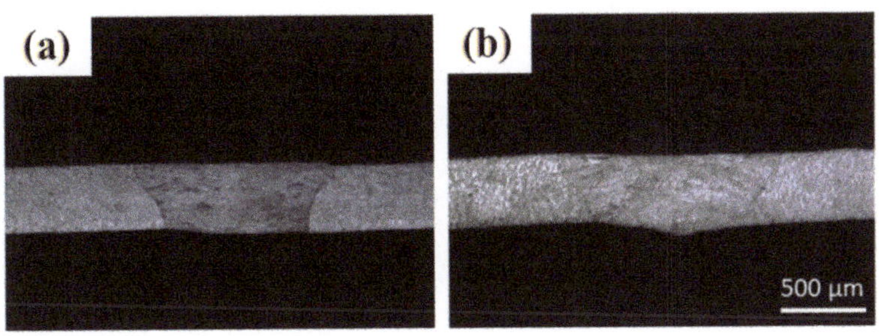

Figure 8. Optical images of 0.4 mm thick sheets' laser weld cross-sections without powder addition and with laser power = 1000 W, gas flow rate = 5 slpm, robot tilt angle = 0°, and laser scanning speed = 3.0 m/min: (**a**) focal plane + 5 mm and (**b**) focal plane + 10 mm.

After determining the optimal laser welding parameters without filler materials, experiments were performed to determine the operating conditions with filler material. To this purpose, a mixture of helium and argon gases was used to transport powder particles from the powder feeder to the AISI 304 stainless steel sheets to be welded. Initially, the optimization was carried out for 0.4 mm thick sheets. Figure 9 explains the effect of the laser power (*P*) on the weld width (*WW*), depth (*WD*), and height (*WH*). A direct relation can be observed between *P*, *WW*, *WD*, and *WH* by keeping the laser scanning speed (*V*), powder flow rate (*Pf*), and a mixture of helium (*Hg*) and argon (*Ag*) gases constant. With a rise in *P*, the energy density for a given area also increases, forming a melt pool with higher width and depth. The LMD technique includes the coaxial inclusion of powder particles which pass across the laser beam, resulting in laser beam attenuation. The laser beam is accountable for particles' melting, thus causing weld formation. However, not all the powder particles passing across the laser beam contribute to weld generation and give powder utilization efficiency, defined as the ratio between the total powder provided from the powder feeder to the powder used in layer formation. With an increase in P, the laser energy density rises and causes an improved powder particle melting. It, in return, increases the weld height. Error bars have been provided in Figure 9 to identify the effect of *P* on *WW*, *WD*, and *WH*. The short error bars show that the change in *P* is not significantly affected and vice versa for long error bars.

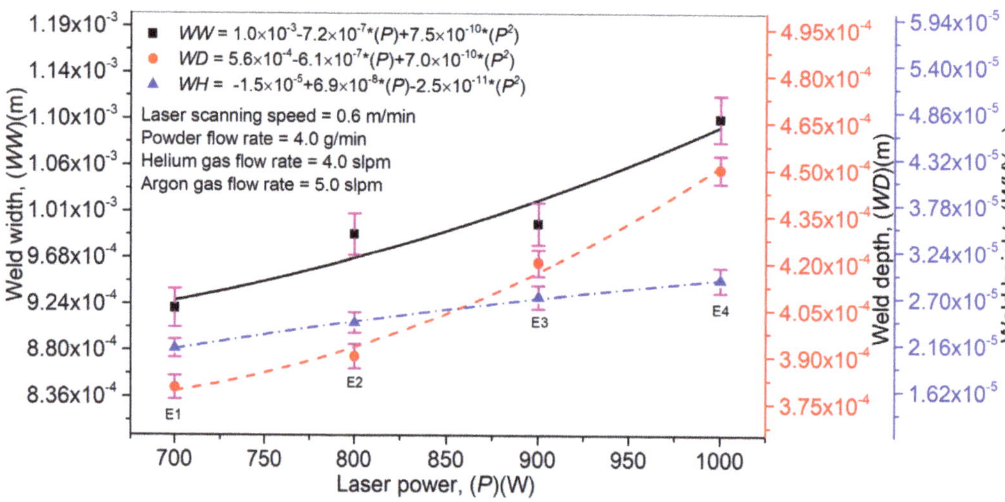

Figure 9. Effect of laser power on laser weld width, depth, and height in laser welding with filler material (Inconel 718).

The influence of *V* on *WW*, *WD*, and *WH* is highlighted in Figure 10. Additionally, *P*, *Pf*, and a mixture of *Hg* and *Ag* were kept constant. It can be observed that an increment in *V* decreases the laser–material interaction time, which in turn reduces the melt pool dimensions and powder particles' melting, thus diminishing the *WW*, *WD*, and *WH* values. An optimum *V* value is necessary to determine as a prolonged laser–material time usually causes residual stress formation in the weld and disturbs the welded part's mechanical properties. Figure 10 shows the error bars plotted to analyze the effect of *V* on *WW*, *WD*, and *WH*. It can be seen that an increase in *V* affects *WW* significantly compared to the *WD* and *WH*.

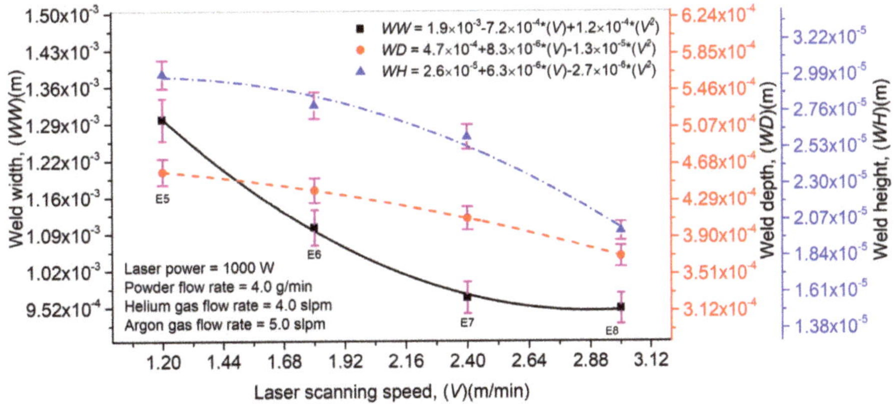

Figure 10. Effect of laser scanning speed on laser weld width, depth, and height in laser welding with filler material (Inconel 718).

An influence of *Pf* on *WW*, *WD*, and *WH* is plotted in Figure 11. A direct correlation can be analyzed between *Pf*, *WW*, and *WH*. In contrast, *Pf* presented an inverse relationship with *WD*. As explained above, laser beam attenuation is an inherited phenomenon in the LMD technique, which prevails in the deposition process with the increase in *Pf*. Excess laser beam energy reaches the substrate's surface and does not find enough time to arrive at the depth. It causes an elevation in *WW* and *WH* while *WD* declines steadily. Error bars have been drawn in Figure 11 to illustrate the influence of *Pf* on *WW*, *WD*, and *WH*. A short error bar means that the concentration of values is high and the average value is more certain, while it is the opposite for long error bars.

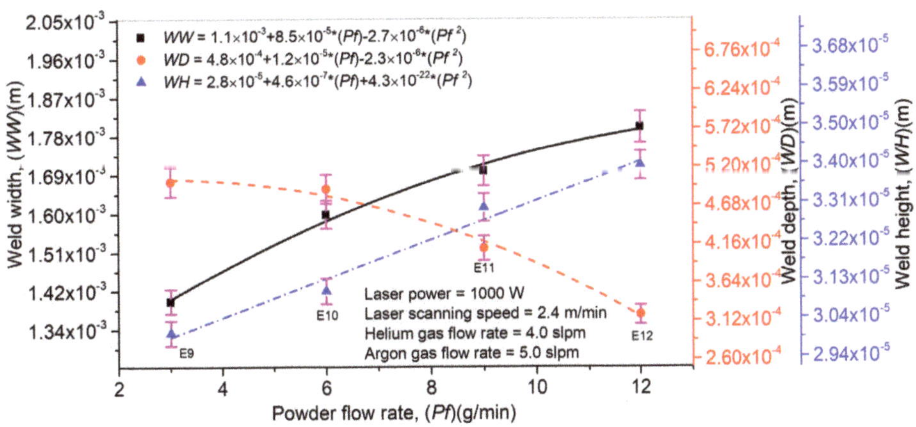

Figure 11. Effect of powder flow rate on laser weld width, depth, and height in laser welding with filler material (Inconel 718).

Figure 12 displays a link between the *Hg*/*Ag* ratio on *WW*, *WD*, and *WH*. For a given set of *P*, *V*, and *Pf*, the *Hg*/*Ag* ratio showed a direct link with *WW* and *WH*, while an inverse link was presented between the *Hg*/*Ag* ratio and *WD*. With an increase in the gas mixture ratio, a thick protective layer is formed at the melt pool surface, not allowing the laser beam energy to travel in the substrate's depth. It causes the entire powder deposition at the substrate's surface. It is the possible explanation behind the behaviors presented by *WW*, *WD*, and *WH*. In Figure 12, error bars have been presented to display the effect of the

Hg/Ag ratio on *WW*, *WD*, and *WH* using standard deviation. It can be analyzed that the Hg/Ag ratio affects *WD* significantly compared to *WH* and *WW* quantities.

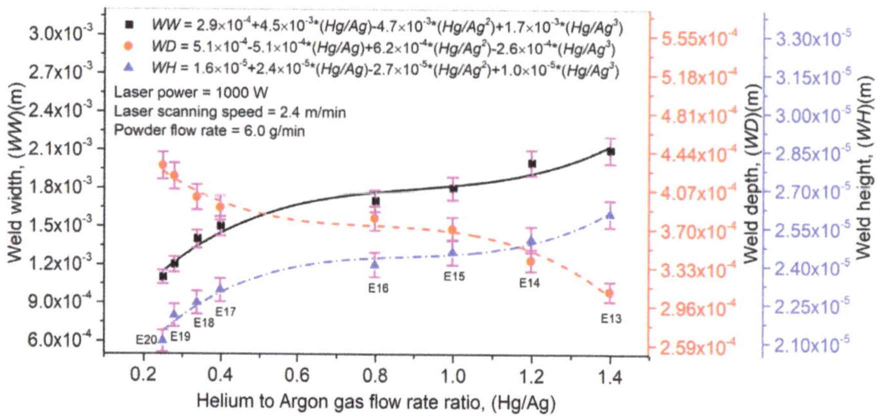

Figure 12. Effect of helium to argon gas flow rate ratio on laser weld width, depth, and height in laser welding with filler material (Inconel 718).

After carrying out various experiments with metallography analysis, the following optimal laser processing parameters for the 0.4 mm thick sheet were determined: (a) laser power = 1000 W, (b) laser scanning speed = 2.4 m/min, (c) powder flow rate = 6 g/min, (d) helium gas flow rate = 4 slpm, (e) argon gas flow rate = 10 slpm, and (f) robot tilt angle = 0°. The optical imaging and metallography analysis are shown in Figure 13a,b. To identify the difference between the weld seam without filler material, the results in Figure 13 can be compared with those provided in Figure 7. In Figure 13a, both sheets are positioned precisely. However, in Figure 13b, both sheets are placed in perfect position and adjusted at a difference of 100 μm with respect to height. Hence, one can conclude that using the filler material can provide a reasonable solution to the difficulties that arise from improper positioning in the case of laser welding without filler material. The same operating conditions as for 0.4 mm thick sheets were implemented for 0.5 mm thick sheets, except the laser scanning speed was varied by (i) 1.2 m/min and (ii) 2.4 m/min. However, a thorough and defect-free weld was achieved with a 2.4 m/min scanning speed. Hence, the optical parameters were kept the same as for the 0.4 mm thick sheet.

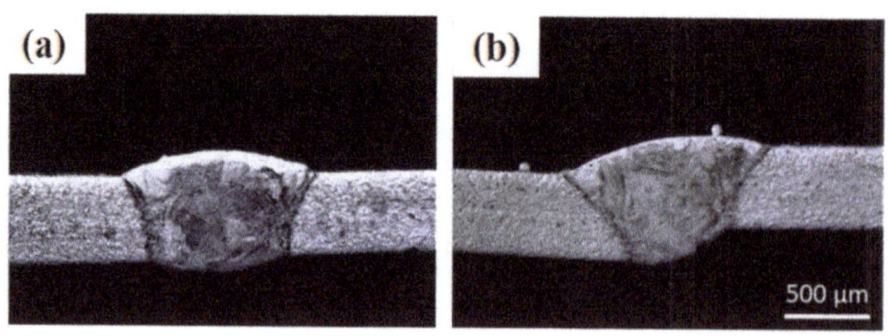

Figure 13. Optical images of 0.4 mm welded plates using Inconel 718 filler material with laser power = 1000 W, laser scanning speed = 2.4 m/min, powder flow rate = 6 g/min, helium gas flow rate = 4 slpm, argon gas flow rate = 10 slpm, and robot tilt angle = 0°: (**a**) perfect positioning and (**b**) positioning with a height difference of ~100 μm.

To analyze the effect of weld quality, the seams were subjected to X-ray analyses. However, only the P2 sample is shown here due to space limitations. Figure 14a,b show the image of a laser weld using the LMD equipment and the corresponding radiographic image, respectively. The weld uniformities were plotted between the scan length and thickness (weld) at three different locations (1, 2, and 3), as shown in Figure 14c. These surface uniformities explain the irregularities within the weld seam based on the X-ray absorption coefficient at the incident regime so that a uniform surface yields a higher absorption coefficient value [23]. Hence, the X-rays can be utilized to quantify the weld seams. It is worth mentioning that X-rays present several advantages over a traditional profilometer setup. One of the advantages is that X-rays can be applied for in situ measurement and characterization of the weld seam. In addition, X-rays can pass across the weld seams, just providing analysis across the samples. As highlighted with green circles, these weld uniformities were used to identify the weld quality. Furthermore, from the graphs, the peak ($T1$) and valley values were calculated to compute the peak value of non-uniformity in the weld seams (percentage), as given in Equation (2):

$$Peak\ value\ of\ non-uniformity\ in\ the\ weld\ (\%) = \left(\frac{T1 - T2}{T1}\right) \times 100 \quad (2)$$

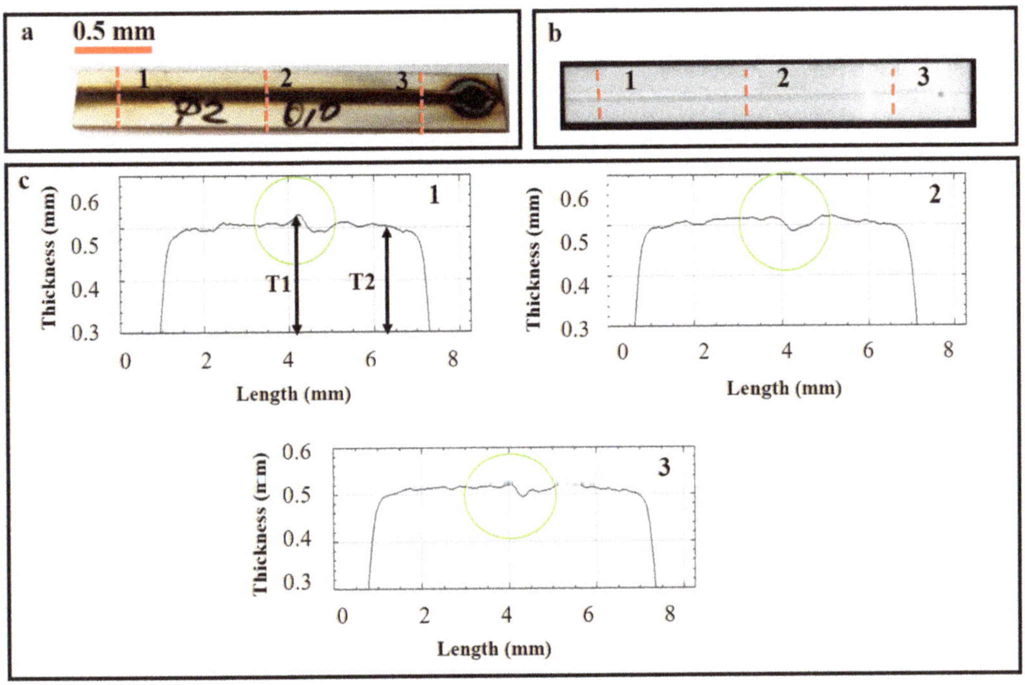

Figure 14. P2 sample (**a**) laser weld by the LMD, (**b**) X-ray analysis of weld using mini-focus X-ray equipment, and (**c**) graphical results of various scans at different positions (1, 2, and 3).

After utilizing Equation (1), the peak value of non-uniformity in the weld seams (percentage) without filler material was calculated for different locations (1, 2, and 3). An average was calculated, as presented in Figure 15. The optimal weld parameters, given by sample P9 for 0.5 mm thick sheets, gave the least non-uniformity value (=3.8%). Similarly, in the case of 0.4 mm sheets, the optimal parameters, shown by the W2 sample, presented the least non-uniformity value (=3.5%).

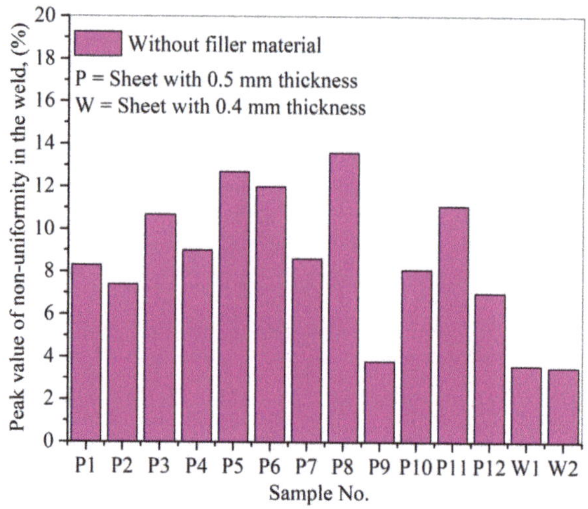

Figure 15. The peak value of non-uniformity was calculated via X-ray analyses for laser welds without filler material.

Figure 16 collects the peak value of non-uniformity in the welds (percentage) in the case of laser welding with filler material (Inconel 718). In the case of E17 (0.4 mm thick sheets) and F2 (0.5 mm thick sheets), the optimal weld parameters presented the least non-uniformity values equal to 7.3% and 8.5%, respectively. From the above results, one can conclude that X-ray analysis is an efficient non-destructive technique for characterizing weld seams.

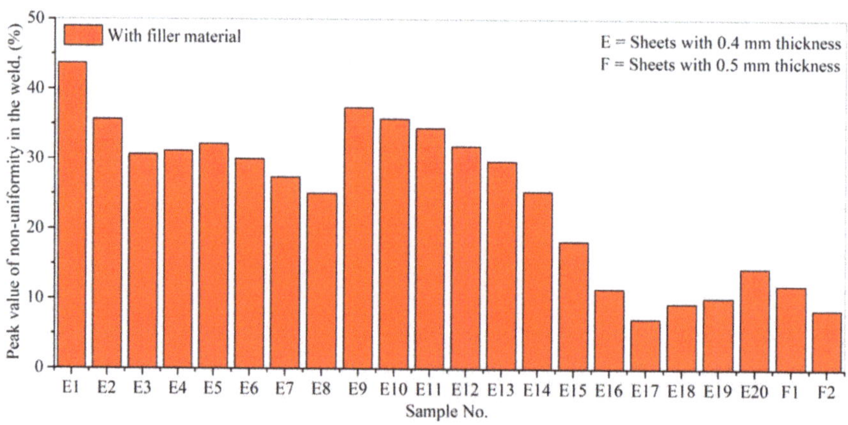

Figure 16. The peak value of non-uniformity was calculated via X-ray analyses for laser welds with filler material (Inconel 718).

Practically, the weld non-uniformity can be defined in terms of the weld quality, surface morphology characteristics, and melting situation of a given material. Each weld presented a non-uniformity value (percentage); however, it primarily depends on the set of operating conditions. Furthermore, the welds without filler showed fewer non-uniformities than those with filler material. The powder particles (filler) are micro-grains fed simultaneously along with the laser beam responsible for particles' melting and melting pool formation. These melted particles define a new layer in correspondence with the base material. Furthermore, the quality of the melted layer is entirely dependent on the melting situation defined by the operating conditions. The two phenomena, namely the (a) inclusion of a new layer and (b) melting situation of the particles, are responsible for giving high non-uniformity (percentage) compared to the welding without material addition. To achieve weld with minimum non-uniformities, it is desirable to determine optimum process parameters. It is worth mentioning that each set of process parameters, even when repeating the same operating condition, will present a value of non-uniformity (percentage) as the deposited material will have a certain morphology and surface characteristics. These values can be lessened but not eliminated.

4. Conclusions

This study carried out laser welding of AISI 304 stainless steel, specifically 0.4 mm and 0.5 mm thick sheets, via the laser-melting deposition (LMD) technique. The welding was carried out without and with filler material. Inconel 718 powder particles were used as filling material in the joints welded. Metallographic experiments were conducted to analyze the defects and pores in the cross-sections. These welds were subjected to X-ray analyses to identify the weld quality in weld uniformities. Based on the study, the following conclusions are deduced:

- In laser welding without filler material, an increment in the laser scanning speed causes a decrease in the weld width and depth. An increment in the robot axis angle decreases the weld depth and increases the weld width. Additionally, the argon gas flow rate presented an inverse relation with weld depth and a direct relation with weld width.
- The optimal parameters for the laser welding of AISI 304 stainless steel sheets without filler material, with a thickness of 0.5 mm, are: (a) laser power = 1000 W, (b) laser scanning speed = 2.4 m/min, (c) argon gas flow rate = 5 slpm, and (d) robot tilt angle = 0°. All the parameters are the same for 0.4 mm thick sheets, except the laser scanning speed increases from 2.4 to 3.0 m/min due to inferior thickness compared to the 0.5 mm thick sheet.
- Good positioning of the sheets is necessary to achieve entirely penetrated welds. A larger spot size results in easier sheet positioning for laser welding. The laser beam was defocused with focal plane + 5 mm and focal plane + 10 mm distances. Acceptable welds were achieved through the defocused laser beam, thus comforting the exact positioning of the laser spot at the sheet area to be welded. This strategy can be implemented with a "top-hat" laser beam profile and loses application using a "Gaussian" intensity distribution.
- For laser welding with Inconel 718 filler material, a direct relation was found between laser power and weld width, depth, and height. Increasing the laser scanning speed decreases the laser–material interaction time, thus diminishing the weld width, depth, and height.
- A direct correlation was analyzed between the powder flow rate and weld width and height, while an inverse relationship was presented between the powder flow rate and weld depth. Laser beam attenuation is an inherited phenomenon in the LMD process caused by laser beam energy absorption by the powder particles. These powder particles melt and an excess of material is deposited on the surface, while the remaining laser energy (unabsorbed by the particles) reaching the surface is not sufficient for completely penetrating and melting the material to be welded. The deposition of molten particles on the surface of the joint causes an elevation in the weld width and height while weld depth declines steadily.
- By utilizing filler material, the optimized set of parameters for 0.4 mm and 0.5 mm sheets are: (a) laser power = 1000 W, (b) laser scanning speed = 2.4 m/min, (c) powder flow rate = 6 g/min, (d) helium gas flow rate = 4 slpm, (e) argon gas flow rate = 10 slpm, and (f) robot tilt angle = 0°.

- X-ray analyses presented a quantitative value for the peak value of non-uniformity in laser welds. In welding without filler material, the optimal weld parameters, given by sample P9 for 0.5 mm thick sheets, gave the least non-uniformity value (=3.8%). Similarly, for 0.4 mm sheets, the optimal parameters, shown by the W2 sample, presented the least non-uniformity value (=3.5%). Furthermore, the optimal weld parameters for welding with filler material in the case of E17 (0.4 mm thick sheets) and F2 (0.5 mm thick sheets) presented the least non-uniformity values equal to 7.3% and 8.5%, respectively.

Author Contributions: Conception, M.A.M., D.C., S.M., M.I., I.N.M. and A.C.P.; literature study, M.A.M., D.C. and S.M.; experiments, D.C., S.M., M.I. and A.C.P.; writing—review and editing, M.A.M., M.I., I.N.M. and A.C.P.; funding acquisition and project administration, I.N.M. The authors agree to publish the current version of the manuscript. All authors have read and agreed to the published version of the manuscript.

Funding: M.A.M. received financial support from the European Union's Horizon 2020 (H2020) research and innovation program under the Marie Skłodowska–Curie grant agreement (no. 764935). This research study has been conducted in the framework of POC-G, contract no. 135/2016, and UEFISCDI 45/2021. This work was also supported by grants from the Romanian Ministry of Education and Research, CNCS-UEFISCDI, project numbers PN-III-P4-ID-PCE-2020-1634 and PN-III-P2-2-1-PED-2019-3953, within PNCDI III, and the Romanian Ministry of Education and Research under the Romanian National Nucleu Program LAPLAS VI, contract no. 16N/2019.

Institutional Review Board Statement: Not applicable.

Informed Consent Statement: Not applicable.

Data Availability Statement: Not applicable.

Acknowledgments: The authors want to thank the NIKO Auto S.R.L company for providing the metal sheets for the welding experiments.

Conflicts of Interest: The authors declare no conflict of interest.

References

1. Liao, Y.C.; Yu, M.H. Effects of laser beam energy and incident angle on the pulse laser welding of stainless steel thin sheet. *J. Mater. Process. Technol.* **2007**, *190*, 102–108. [CrossRef]
2. Zhang, Y.M.; Kovacevic, R.; Li, L. Characterization and real-time measurement of geometrical appearance of the weld pool. *Int. J. Mach. Tools Manuf.* **1996**, *36*, 799–816. [CrossRef]
3. Khan, M.M.A.; Romoli, L.; Fiaschi, M.; Dini, G.; Sarri, F. Experimental design approach to the process parameter optimization for laser welding of martensitic stainless steels in a constrained overlap configuration. *Opt. Laser Technol.* **2011**, *43*, 158–172. [CrossRef]
4. Yoshiaki, A.; Isamu, M. Theoretical Analysis of Weld Penetration Due to High Energy Density Beam. *Trans. JWRI* **1972**, *1*, 11–16.
5. Balasubramanian, K.R.; Buvanashekaran, G.; Sankaranarayanasamy, K. Modeling of laser beam welding of stainless steel sheet butt joint using neural networks. *CIRP J. Manuf. Sci. Technol.* **2010**, *3*, 80–84. [CrossRef]
6. Singh, R.; Alberts, M.J.; Melkote, S.N. Characterization and prediction of the heat-affected zone in a laser-assisted mechanical micromachining process. *Int. J. Mach. Tools Manuf.* **2008**, *48*, 994–1004. [CrossRef]
7. Hector, L.; Chen, Y.; Agarwal, S.; Briant, C. *Texture Characterization of Autogenous Nd: YAG Laser Welds in AA5182-O and AA6111-T4 Aluminum Alloys*; Elsevier: Amsterdam, The Netherlands, 2003; Volume 356.
8. Lee, M.F.; Huang, J.C.; Ho, N.J. Microstructural and Mechanical Characterization of Laser-Beam Welding of a 8090 Al-Li Thin Sheet. *J. Mate. Sci.* **1996**, *35*, 1455–1468. [CrossRef]
9. Masoumi, M.; Marashi, S.P.H.; Pouranvari, M. Metallurgical and mechanical characterization of laser spot welded low carbon steel sheets. *Steel Res. Int.* **2010**, *81*, 1144–1150. [CrossRef]
10. Nath, A.K.; Sridhar, R.; Ganesh, P.; Kaul, R. Laser Power Coupling Efficiency in Conduction and Keyhole Welding of Austenitic Stainless Steel. In *Sadhan*; Indian Academy of Sciences: Bengaluru, India, 2002; Volume 27, pp. 383–392. [CrossRef]
11. Heider, A.; Sollinger, J.; Abt, F.; Boley, M.; Weber, R.; Graf, T. High-speed X-ray analysis of spatter formation in laser welding of copper. In Proceedings of the Physics Procedia; Elsevier: Amsterdam, The Netherlands, 2013; Volume 41, pp. 112–118.
12. Fabbro, R. Melt pool and keyhole behaviour analysis for deep penetration laser welding. *J. Phys. D. Appl. Phys.* **2010**, *43*, 445501. [CrossRef]
13. Fujinaga, S.; Takenaka, H.; Narikiyo, T.; Katayama, S.; Matsunawa, A. Direct observation of keyhole behaviour during pulse modulated high-power Nd:YAG laser irradiation. *J. Phys. D. Appl. Phys.* **2000**, *33*, 492–497. [CrossRef]

14. Katayama, S.; Kawahito, Y.; Mizutani, M. Collaboration of physical and metallurgical viewpoints for understanding and process development of laser welding. In Proceedings of the 26th International Congress on Applications of Lasers and Electro-Optics, ICALEO 2007—Congress Proceedings, Laser Institute of America, Orlando, FL, USA, 29 October—1 November 2007; Volume 2007, p. 701.
15. Matsunawa, A.; Kim, J.-D.; Seto, N.; Mizutani, M.; Katayama, S. Dynamics of keyhole and molten pool in laser welding. *J. Laser Appl.* **1998**, *10*, 247–254. [CrossRef]
16. Errico, V.; Campanelli, S.L.; Angelastro, A.; Mazzarisi, M.; Casalino, G. On the feasibility of AISI 304 stainless steel laser welding with metal powder. *J. Manuf. Process.* **2020**, *56*, 96–105. [CrossRef]
17. Prabakaran, M.P.; Kannan, G.R. Optimization of laser welding process parameters in dissimilar joint of stainless steel AISI316/AISI1018 low carbon steel to attain the maximum level of mechanical properties through PWHT. *Opt. Laser Technol.* **2019**, *112*, 314–322. [CrossRef]
18. Landowski, M. Influence of parameters of laser beam welding on structure of 2205 duplex stainless steel. *Adv. Mater. Sci.* **2019**, *19*, 21–31. [CrossRef]
19. Ding, H.; Ma, J.; Zhao, C.; Zhao AFFILIATIONS, D. Effect of welding speed, pulse frequency, and pulse width on the weld shape and temperature distribution in dissimilar laser welding of stainless steel 308 and brass alloy. *J. Laser Appl.* **2021**, *33*, 22009. [CrossRef]
20. Stainless Steel-Grade 304 (UNS S30400). Available online: https://www.azom.com/article.aspx?ArticleID=965 (accessed on 20 May 2021).
21. Nickel Alloy Inconel 718—Properties and Applications by United Performance Metals. Available online: https://www.azom.com/article.aspx?ArticleID=4459 (accessed on 20 May 2021).
22. Berger, M.; Yang, Q.; Maier, A. *X-ray Imaging*; Lecture Notes in Computer Science Book Series; Springer: Berlin/Heidelberg, Germany, 2018; Volume 11111, pp. 119–145. [CrossRef]
23. Niu, C.; Zhu, T.; Lv, Y. Influence of Surface Morphology on Absorptivity of Light-Absorbing Materials. *Int. J. Photoenergy* **2019**, *2019*, 1476217. [CrossRef]
24. Nakhaei, R.; Khodabandeh, A.; Najafi, H. Effect of Active Gas on Weld Shape and Microstructure of Advanced A-TIG-Welded Stainless Steel. *Acta Metall. Sin.* **2016**, *29*, 295–300. [CrossRef]

materials

MDPI

Article

X-ray Computed Tomography Procedures to Quantitatively Characterize the Morphological Features of Triply Periodic Minimal Surface Structures

Sergei Evsevleev [1,*], Tatiana Mishurova [1], Dmitriy Khrapov [2], Aleksandra Paveleva [2], Dietmar Meinel [1], Roman Surmenev [2], Maria Surmeneva [2], Andrey Koptyug [2,3] and Giovanni Bruno [1,4]

1 Bundesanstalt für Materialforschung und -prüfung (BAM), Unter den Eichen 87, 12205 Berlin, Germany; tatiana.mishurova@bam.de (T.M.); dietmar.meinel@bam.de (D.M.); giovanni.bruno@bam.de (G.B.)
2 Physical Materials Science and Composite Materials Centre, Research School of Chemistry & Applied Biomedical Sciences, National Research Tomsk Polytechnic University, 30 Lenina Avenue, 634050 Tomsk, Russia; dah8@tpu.ru (D.K.); aleksandra-paveleva@mail.ru (A.P.); rsurmenev@mail.ru (R.S.); surmenevamaria@mail.ru (M.S.); andrey.koptyug@miun.se (A.K.)
3 Department of Engineering and Sustainable Development, Mid Sweden University, Akademigatan 1, SE-831 25 Östersund, Sweden
4 Institute of Physics and Astronomy, University of Potsdam, Karl-Liebknecht-Str. 24-24, 14476 Potsdam, Germany
* Correspondence: sergei.evsevleev@bam.de

Citation: Evsevleev, S.; Mishurova, T.; Khrapov, D.; Paveleva, A.; Meinel, D.; Surmenev, R.; Surmeneva, M.; Koptyug, A.; Bruno, G. X-ray Computed Tomography Procedures to Quantitatively Characterize the Morphological Features of Triply Periodic Minimal Surface Structures. *Materials* **2021**, *14*, 3002. https://doi.org/10.3390/ma14113002

Academic Editor: George Kenanakis

Received: 20 April 2021
Accepted: 27 May 2021
Published: 1 June 2021

Abstract: Additively manufactured (AM) metallic sheet-based Triply Periodic Minimal Surface Structures (TPMSS) meet several requirements in both bio-medical and engineering fields: Tunable mechanical properties, low sensitivity to manufacturing defects, mechanical stability, and high energy absorption. However, they also present some challenges related to quality control, which can prevent their successful application. In fact, the optimization of the AM process is impossible without considering structural characteristics as manufacturing accuracy, internal defects, as well as surface topography and roughness. In this study, the quantitative non-destructive analysis of TPMSS manufactured from Ti-6Al-4V alloy by electron beam melting was performed by means of X-ray computed tomography (XCT). Several advanced image analysis workflows are presented to evaluate the effect of build orientation on wall thicknesses distribution, wall degradation, and surface roughness reduction due to the chemical etching of TPMSS. It is shown that the manufacturing accuracy differs for the structural elements printed parallel and orthogonal to the manufactured layers. Different strategies for chemical etching show different powder removal capabilities and both lead to the loss of material and hence the gradient of the wall thickness. This affects the mechanical performance under compression by reduction of the yield stress. The positive effect of the chemical etching is the reduction of the surface roughness, which can potentially improve the fatigue properties of the components. Finally, XCT was used to correlate the amount of retained powder with the pore size of the functionally graded TPMSS, which can further improve the manufacturing process.

Keywords: metamaterials; functionally graded porous structure; triply periodic minimal surface structures; roughness analysis; powder removal; deep learning segmentation

1. Introduction

Additive manufacturing (AM) techniques, such as electron beam melting (EBM) or laser powder bed fusion (LPBF) allow producing geometrically complex structures, including near net-shape ones. Triply Periodic Minimal Surface Structures (TPMSS) are one of the topologies, which can be produced only by AM technologies. TPMSS have mean curvature equal to zero at every point and exhibit periodicity in three independent space directions [1].

One of the most promising applications of TPMSS is tissue engineering, since they possess mechanical characteristics (e.g., Young's modulus and strength) that can be tuned to those of the bone [2]. It has been also consistently observed that the rate of tissue generation is proportional to the curvature (and the amount) of the surface [3] on which it grows. Thus, TPMSS can effectively be used as a porous scaffold thanks to their high surface-to-volume ratio. Moreover, TPMSS mechanical and physical properties provide possibilities in a variety of engineering applications: From lightweight construction elements and battery electrodes to photonic crystals [4,5]. These diverse structural and functional applications give the motivation for an accurate characterization of such metamaterials.

One of the challenges in the usage of TPMSS is the choice of a suitable geometry for a specific application. Yang et al. [6] have reported that the variation of the TPMSS unit cell size and wall thickness (WT) provides an opportunity to tune their elastic modulus, strength, and energy absorption. Moreover, in contrast to the strut-based lattice structures, this can be achieved without loss of structural integrity.

Another challenge is the production of defect-free, accurate TMPSS with low deviation from the intended geometrical models. This is particularly challenging in the case of metal PBF (powder bed fusion) manufacturing. The internal defects (gas pores, lack-of-fusion defects) as well as dimensional inaccuracies and excessive surface roughness are characteristic for EBM and LPBF processes [7,8]. It should be mentioned that overall surface topography together with physicochemical properties determine the biocompatibility of the implants at early stages of implant integration. Though commonly presented in the literature, averaged roughness parameters do not adequately represent the variety of the surface topography features important at different stages of the implant integration [9–14]. The detailed studies of the implant surface topography are commonly carried out using, for example, tactile profilometry, optical methods, or atomic force microscopy. However, the majority of the topography assessment methods are not suitable for lattices and other lightweight structures having the elements obscured from direct observation.

Finally, the inner structure of the TPMSS consists of curved channels. Therefore, the cleaning process of such structures from residual powder is also a highly challenging task. For example, removing residual powder from porous EBM parts is performed by a special powder recovery system (PRS) using precursor powder suspended in the jet of the compressed air. However, the efficiency of this method, namely the depth of powder blast penetration, is largely affected by the channel aspect ratio defined by the structure geometry, as well as by the size and type of the unit cells [15]. Other ways to treat the whole volume of the porous specimen are chemical and electrochemical etching (CE, ECE). These methods have been widely investigated by some researchers in order to improve some morphological characteristics (in particular surface roughness) of AM-samples [16,17]. However, such etching is often anisotropic and can lead to wall degradation, especially at the periphery of the samples. This can significantly affect the mechanical performance of the part.

X-ray Computed tomography (XCT) has already proved to be one of the best tools for non-destructive analysis of lattice and cellular structures [8,18,19]. It has been reported that XCT can detect the structural defects originated from the manufacturing process, as well as analyze the deflection of the geometry of manufactured components against their computer aided design (CAD) models [4]. Such knowledge can be further used to optimize the manufacturing process, reducing the defects, and leading to better geometry fidelity. Most importantly, XCT allows not only imaging but also extracting quantitative information about geometrical accuracy [20], manufacturing defects [21], surface texture [22–24], interconnectivity of the pores (see [25]), etc. However, there are few challenges in the analysis of the CT data of TPMSS. As mentioned above, the morphology of the structure even after the cleaning procedure can lead to retained powder (RP). The segmentation of the powder-containing regions is impossible with standard tools due to several reasons: (1) The identical gray level of the powder particles and wall regions limits the application of the global thresholding [26]; (2) the morphological filters or Watershed segmentation tools could modify the surface profile as well as the thickness of the walls, bringing additional error to the analysis.

Another challenge is the quantitative assessment of material loss caused by CE. The presence of RP inside TPMSS can hinder the uniform penetration of etchant, resulting in the gradient of the wall thickness (WT). Inhomogeneous CE can potentially cause anisotropy of the mechanical properties and unpredictability of the behavior under load condition. In this case, the mean WT value or the overall WT distribution determined by standard procedures are not informative and more sophisticated analysis is required.

In this work, we focus on XCT characterization of sheet-based TPMSS manufactured using EBM from Ti-6Al-4V. We present novel strategies and workflows for quantitative XCT image analysis of metallic TPMSS and functionally graded porous structures (FGPS). The procedures presented in this work allow obtaining information about the production and post-processing of AM structures. This information can be integrated into the optimization of the manufacturing process, and into the numerical simulations of the properties for the resulting structures [27]. The detailed investigation of the correlation between morphological characteristics, materials properties, and structure performance is not in the scope of this work and will be discussed in the follow-up publications.

2. Materials and Methods

2.1. Samples Design and Production

Two types of gyroid TPMSS (with constant and with functionally graded porosity) [4] were investigated. The gyroid surface was generated according to the following equation:

$$\sin(kx)\cos(ky) + \sin(ky)\cos(kz) + \sin(kz)\cos(kx) = 0 \quad (1)$$

where parameter k controls the unit cell size. In this study, k = 2 was chosen for TPMSS specimens with constant porosity. To make the gradient structure, k was defined as a function of one Cartesian coordinate, in our case z:

$$k(z) = 1 + jz \quad (2)$$

The scaling factor j was selected to be 0.05. The models of the designed structures are shown in Figure 1.

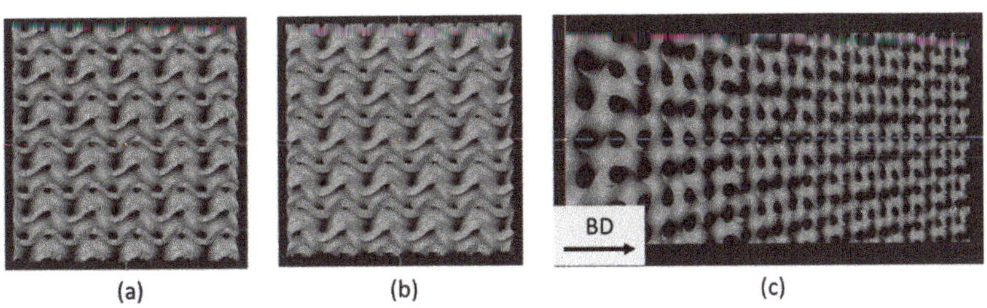

Figure 1. Models of the gyroid TPMSS with nominal WT of 0.4 mm (**a**), 0.6 mm (**b**), and gyroid-based functionally graded porous scaffold (FGPS) (**c**). BD indicates the build direction during manufacturing.

All specimens were manufactured from a Ti-6Al-4V alloy using an ARCAM A2 EBM machine by ARCAM, EBM (Mölnlycke, Sweden). The precursor powder was purchased from ARCAM EBM. The layer thickness was set to 50 μm and the process temperature was set to 720 °C. Standard process parameters for Ti-6Al-V from the ARCAM system library were used (the "Melt Theme" of the ARCAM EBM process library). The manufactured specimens (Figure 2a) were separated from the start plate and cleaned from the residual semi-sintered powder using a standard ARCAM Powder Recovery System.

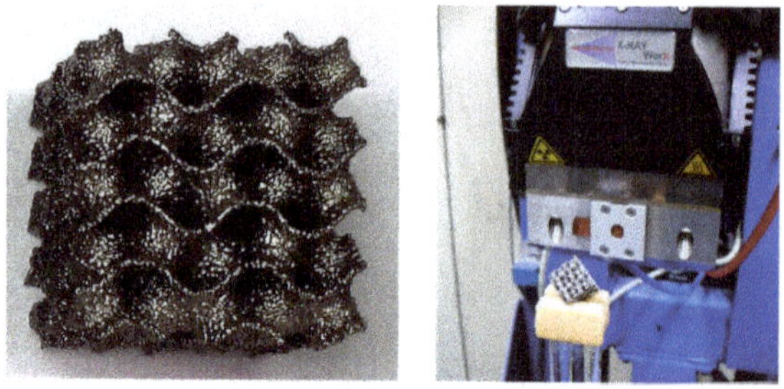

(a) (b)

Figure 2. Photographs of (**a**) specimen with nominal WT of 0.4 mm; (**b**) the XCT experimental setup.

Cubic specimens with constant porosity were designed with two different WT of 0.4 mm and 0.6 mm resulting in the specimens (15 mm × 15 mm × 15 mm) with relative material apparent density of 50% and 42%, respectively. In the following text, we name these specimens as 2 × 04 and 2 × 06, according to the parameter k and their nominal WT.

The specimen with gradient porosity was manufactured using a model with zero thickness and hence its WT was purely controlled by the electron beam application strategy (current, surface speed). The standard "Wafer Theme" from the ARCAM EBM process library was used, leading to the smallest WT achievable with EBM.

The removal of the precursor powder retained inside the structure after manufacturing is a critical step in the manufacturing chain. All specimens were subjected to a standard PRS cleaning for 10 min. Additionally, one specimen with constant porosity for each WT was chemically etched (CE) according to the following procedure: A 2 × 04 specimen was immersed in a 1% HF:10% HNO_3 water solution 14 times for 3 min each, without solution renewal. The 2 × 06 specimen was immersed into the solution of the same composition 12 times for 3 min. It is known that the intensity of the etching decreases with time due to the formation of stable and soluble titanium hexafluoro complexes [27]. Therefore, in the case of the 2 × 06 specimen, the chemical solution was renewed after 12 and 24 min of etching. After each immersion step, the specimens were rinsed first with demineralized water and then with ethanol, and finally, air-dried.

2.2. *X-ray Computed Tomography*

The X-ray computed tomography (XCT) measurements were performed using a microfocus X-ray tube XWT-225-SE (maximum voltage 225 kV, (X-RAY WorX, Garbsen, Germany) from X-Ray WorX GmbH and an XRD1620 detector (CsI scintillator, 2048 × 2048 pixel, from PerkinElmer Inc., Waltham, MA, USA) with in-house built case and cooling system (Figure 2b). The tube voltage and a tube current were set to 120 kV and 120 μA respectively. The effective voxel size was $(15.3\ \mu m)^3$. The reconstruction of 3D volumes from 2D projections was carried out by in-house software using a filtered back-projection algorithm. Specimens were scanned with the cube diagonal parallel to the rotation axis (see Figure 2b), in order to reduce the influence of cone-beam artifacts.

3. Results and Discussion

3.1. *Virtual Powder Removal*

Retained powder (RP) is one of the limitations for the application of cellular structures with small pore size and is especially critical in medical implants [16]. The efficiency of the loose powder removal techniques can be non-destructively investigated by XCT. An example of XCT reconstructed slice of 2 × 04 specimen (Figure 3a) shows RP particles inside the structure. Obviously, the amount of RP depends on the pore and structure sizes. As it can be expected, the efficiency of the narrow channel cleaning by PRS drops with the distance from the specimen's outer surface, leaving the powder residue mainly in the center of the specimen [15]. As stated above, it is challenging (and sometimes impossible) to remove all RP by conventional methods. The challenge is so formidable, that so far no perfect solution is available. However, vibration-assisted methods indicate that although damage can occur in beam-based lattice structures, such methods can be well suited for the TPMSS [7].

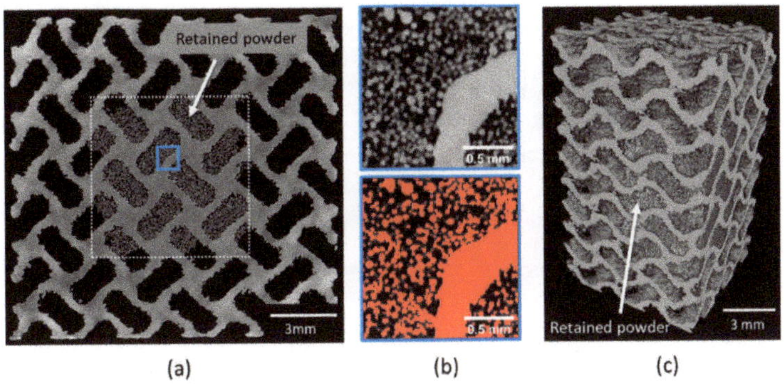

Figure 3. (**a**) Reconstructed XCT slice of the 2 × 04 specimen subjected to PRS showing the RP particles; (**b**) Magnified region indicated by blue rectangle in (**a**) and the result of the automatic thresholding using a standard Otsu's method; (**c**) 3D rendering of the reconstructed TPMSS with the RP inside.

The RP prevents precise analysis of the XCT data, as such data should be segmented prior to further evaluation. The apparent density of the powder is smaller than the density of the bulk material (due to inter-particle void volume and internal porosity). Furthermore, if scanned with the low resolution, it could have had a sufficient contrast to be segmented by automatic thresholding [28]. In our case, the segmentation of the powder-containing regions is a challenging task. The resolution achieved allows distinguishing the individual powder particles that have the same gray value distribution in the reconstructed slices as the scaffold (Figure 3b). This limits the application of standard tools such as global thresholding [26]. One of the solutions could be the downscaling of the data, which would simulate the CT scan with lower resolution. However, this would cause loss of information about the surface topography as well as lead to additional error in the WT calculation. Therefore, we used a deep learning (DL) algorithm to solve this segmentation problem.

For the segmentation of the scaffold, we employed a 2D fully convolutional neural network (CNN) [29] implemented as a plugin in Fiji ImageJ (Version 1.53c) [30]. To perform segmentation, the CNN must undergo an iterative training procedure. Eight slices (with and without RP) were used for the training session. On these slices, the areas with RP were manually segmented and the rest of the scaffold was segmented using an advanced surface determination algorithm implemented in VG Studio Max (Version 3.4.3, Volume Graphics GmbH, Heidelberg, Germany) [31]. The CNN was trained with 500 iterations. The number of iterations usually depends on the diversity of the datasets. By tracking the convergence of the validation and training errors, we found that 500 iterations were optimal for the investigated datasets.

The application of the 2D network only on XY slices resulted in under-segmentation of the scaffolds surface. In this case, the surface roughness of walls parallel to the XY plane (indicated by arrows on XZ slices, Figure 4a) appears as loose powder on the XY planes and is classified by the CNN as powder (Figure 4b). To overcome this issue, the segmentation procedure of every dataset was carried out in three planes (XY, XZ and YZ), followed by the logical operation AND on the three segmented volumes (Figure 4c). As a result, the scaffold was segmented with overall accuracy exceeding 95%. The accuracy was evaluated by pixelwise comparison of the ground truth (manually segmented) with the result of DL segmentation. The workflow is presented in Figure 5.

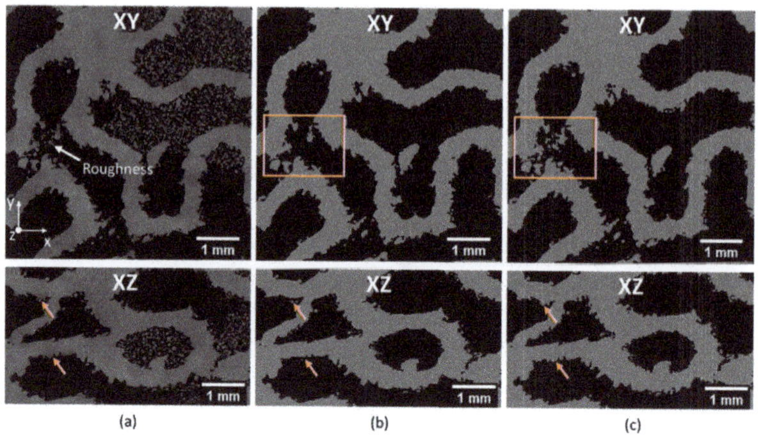

Figure 4. (**a**) A 2D reconstructed XY and XZ slices; (**b**) the result of the 2D DL segmentation of the scaffold only on XY planes; (**c**) the result of the combined segmentation on XY, XZ, and YZ planes. Orange boxes and arrows indicate the critical regions with surface roughness being wrongly segmented in (**b**).

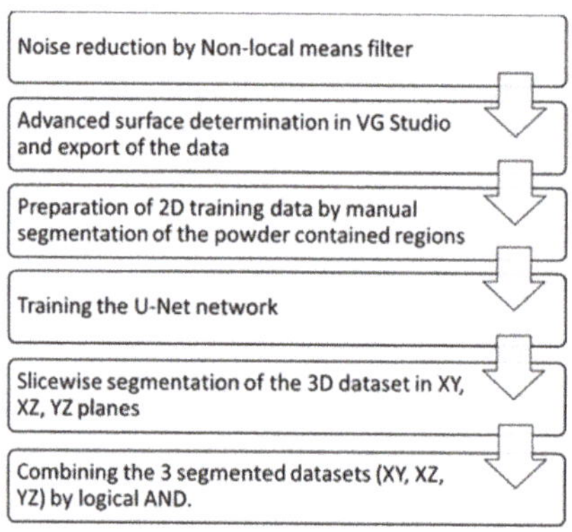

Figure 5. The description of the workflow used for segmentation of the RP.

It is highly important to note that the presented methodology of using a 2D CNN instead of 3D CNN allowed achieving a high segmentation accuracy at a low cost. Typically, the amount of required training data is significantly smaller for the 2D networks and their computational efficiency is much higher.

3.2. Effect of Wall Orientation

The thickness of a powder-based AM component can vary depending on its angle to the build direction. Therefore, a calculation of an average WT within the specimen would imply a loss of information. A different WT of horizontal (within the deposited layers) and vertical (along the build direction) elements in TPMSS can lead to unwanted (and anyway uncontrolled) anisotropic mechanical properties of the component.

XCT allows investigating the distribution of the WT for the elements printed at different angles to the build direction. In this case, the WT was quantitatively analyzed for all near-horizontal and near-vertical walls in both the 2 × 04 and 2 × 06 specimens (Figure 6) using the algorithm of volumetric local thickness [32] implemented in the BoneJ plugin [33] for Fiji ImageJ (Version 1.53c). The algorithm is based on fitting maximal spheres to every point of the structure. It is important to note that the used algorithm of local thickness evaluates the surface roughness of a wall as an intersection of objects of different size (the wall itself and the attached powder particles). The local thickness of such a joint structure is determined by the diameter of the largest sphere fitted to the wall region and smaller spheres fitted to the roughness regions. Consequently, the presence of partially molten powder particles attached to the surface can lead to an underestimation of the mean WT. Therefore, only local thickness values higher than 150 μm were included in the analysis. We experimentally found the value of 150 μm to be the optimal to exclude the contribution of those powder particles attached to the surface.

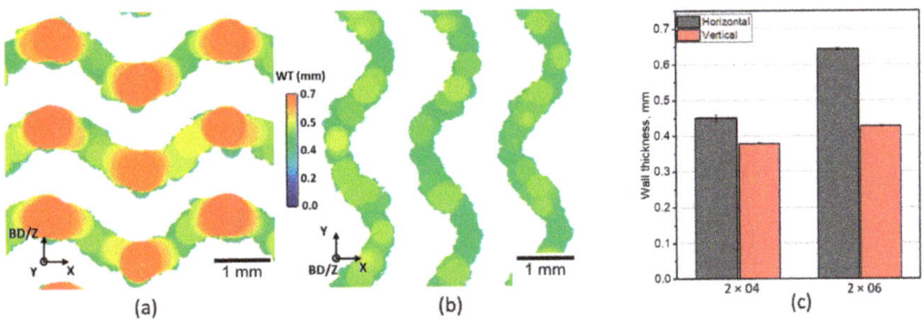

Figure 6. Comparison of the WT in the horizontal (printed normal to BD) and vertical (printed parallel to BD) sections: (**a**) Example of the CT-slice containing the horizontal wall of the 2 × 06 specimen; (**b**) example of the CT-slice containing the vertical wall of the 2 × 06 specimen; (**c**) quantitative comparison of WT in 2 × 04 and 2 × 06 specimens. BD indicates the build direction.

A significant difference in the thickness of horizontal and vertical walls was observed for both specimens (Figure 6c). In the 2 × 06 specimen, the vertical elements appear up to 180 μm (around 30%) thinner than the nominal WT while horizontal elements almost coincide in thickness with the nominal value.

The effect of the wall orientation (with respect to the build direction) on the WT and roughness in the PBF-AM has a known origin but is hard to control. The thickness of the printed walls not only depends on the CAD model, but also on the size of the melt pool and on the direction of heat dissipation. The "Melt Theme" in ARCAM EBM used in this work optimizes the parameters of the electron beam for bulky specimens i.e., considering the large cross-sections of the printed part within the build plane. The cross-section size of TPMSS vertical and horizontal walls (within the build plane) differs significantly, leading to different temperature gradients and hence different sizes of the melt pool. In the vertical walls, the melt pool is supported by the solid material in the layers below, leading to a fast heat dissipation and swells mainly in the horizontal plane, capturing some powder particles. This phenomenon leads to some waviness of the vertical wall sides (with characteristic "wavelengths" proportional to the used layer thickness) and irregular roughness caused by the partially fused powder particles. In the horizontal walls, the metal solidifies over underlying powder layers. While in EBM, such powder layers are pre-sintered, rather than loose as in LPBF, liquid metal from the melt pool is able to "leak" through the powder bed. This phenomenon results in excessive surface roughness and even stalactite-like structures.

It is important to consider the angular dependence of WT since it can cause anisotropic mechanical properties and unpredictable mechanical performance of TPMSS. One approach to reduce the influence of the temperature gradients and hence the difference in the WT is to build the cubic TPMSS tilted by 45° (i.e., to align the main cube diagonal with the build direction). Another approach is to directly compensate for the expected WT deviation in CAD models. This, however, requires detailed research to define the exact correlation between build angles of the walls and their thickness deviations.

3.3. *Wall Degradation after Etching*

Post-manufacture chemical etching (CE) is often used on PBF lattice structures to reduce the surface roughness [34], which usually results in improvement of the mechanical properties [35]. In the case of sheet-based TPMSS, in addition to surface modification, CE can also be applied as a cleaning procedure dissolving the RP. However, the penetration of the fresh etchant into the deep voids is restricted and the solution etches the specimen periphery faster than the interior. This is visible in Figure 7a,c. This preferential etching leads to a gradient of the WT towards the center of the specimen and can cause the unpredictability of the mechanical performance.

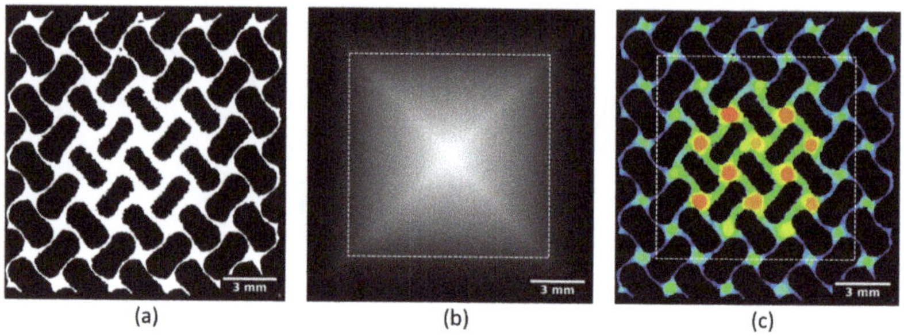

Figure 7. Schematic workflow generated for WT analysis of the gradient structures. (**a**) A binarized 2D slice of the 2 × 06 specimen after CE; (**b**) the Euclidean distance map; (**c**) the result of the WT analysis. The white square indicates one set of voxels having the same distance from the closest edge.

There is no ready solution for a XCT-based quantitative evaluation of the WT gradient, although it is highly important for the quality assessment as well as for the prediction of the mechanical behavior. Therefore, in this section, we present a method we developed to evaluate the wall degradation after CE. Figure 7a shows a binarized reconstructed XCT slice of the 2 × 06 specimen after CE (the RP is not shown). The WT clearly increases from the specimen periphery to its center. The overall mean WT determined by standard procedures is no longer relevant in such a gradient structure. Therefore, an adapted XCT image analysis workflow was developed to quantitatively evaluate the loss of material as a function of distance from the specimen periphery. This procedure will help to tune the CE parameters as a function of the porosity of the specimen. The procedure consists of the following steps (see Figures 7 and 8):

1. Prior to the WT analysis, the scaffold was segmented from the loose powder using the algorithm described in the Section 3.1.
2. The envelope structure defined by the outer region of the specimen was generated. For simple convex objects, this can be done by calculation of the convex hull, which is the smallest convex shape containing all points of an object. For more complex geometries, the envelope surface can be identified by sophisticated DL algorithms.
3. The 3D Euclidean distance map was calculated on the convex hull of the structure (Figure 7b). This allows assigning every point of the structure the value of the shortest distance between this point and the background (specimen's edge).
4. Finally, the average value of the local thickness was calculated for every set of voxels equidistantly located from the specimen's edges (with the same Euclidean distance value, as indicated by the white dashed line in Figure 7b,c). This combination of complementary information on the WT and the Euclidean distance allows a quantitative characterization of the WT gradient in a direction from the specimen edges to the center. Figure 8 summarizes the workflow.

In both 2 × 04 and 2 × 06 specimens, the CE could not completely remove the RP from the structure. A small amount of powder remained in the center of specimens (see Figure 9) in the region where WT curves of as-build and CE specimens coincide: At a distance of around 5 mm for 2 × 06 and of around 6 mm for 2 × 04. The region where some powder is retained shows a linear increase of the WT to the region where the pores are completely filled with powder. A strong gradient of WT is created by the CE in the case of the 2 × 06 specimen, showing a reduction in the mean WT from around 470 μm (in as-build condition) to 200 μm (after CE). The large material loss (also seen in Figure 7a) is caused by the aggressive etching strategy, i.e., the renewal of the chemical solution leads to a more intensive etching but still does not completely remove the RP.

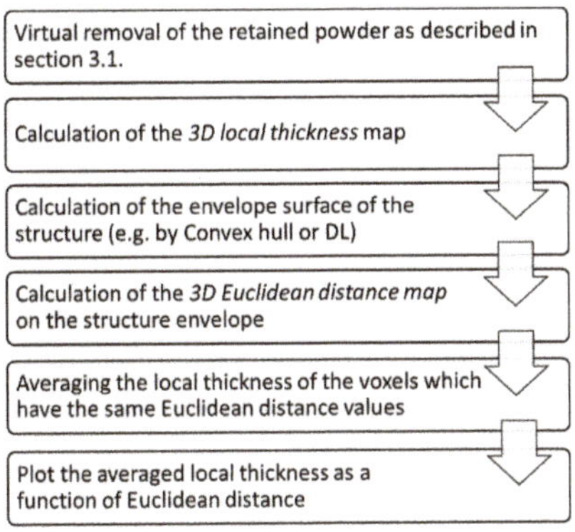

Figure 8. The description of the workflow used for the wall degradation analysis.

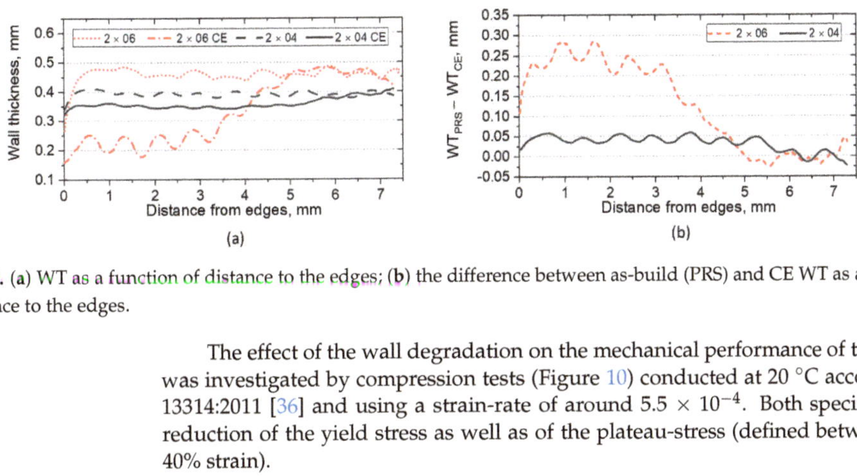

Figure 9. (**a**) WT as a function of distance to the edges; (**b**) the difference between as-build (PRS) and CE WT as a function of distance to the edges.

The effect of the wall degradation on the mechanical performance of the specimens was investigated by compression tests (Figure 10) conducted at 20 °C according to ISO 13314:2011 [36] and using a strain-rate of around 5.5 × 10^{-4}. Both specimens show a reduction of the yield stress as well as of the plateau-stress (defined between 20% and 40% strain).

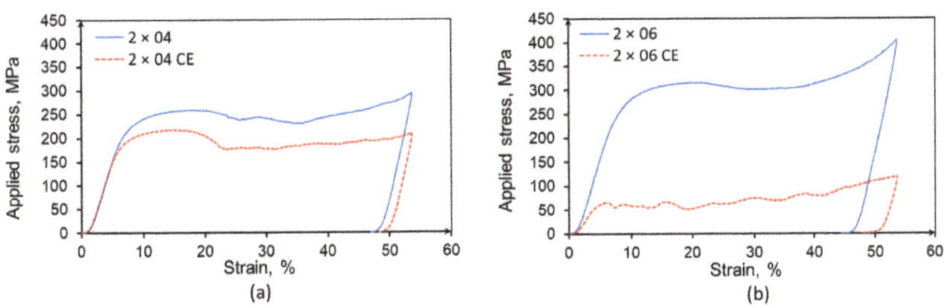

Figure 10. Stress–strain curves obtained during compression tests on 2 × 04 (**a**) and 2 × 06 (**b**) specimens in as-build condition and after CE.

3.4. Surface Roughness Analysis

Surface roughness is another critical parameter for the mechanical performance of TPMSS, especially under cyclic load [37,38]. XCT must be used to determine surface roughness in lattice structures, as no other tactile or laser technique can interrogate their interior. However, the XCT determination of the surface roughness on the curved surface is a challenging task. Therefore, we present a multistep procedure we developed to determine roughness of the walls based on the XCT data. In order to exclude the influence of building orientation and compare different specimens, only vertical walls (parallel to BD) in regions not containing the RP were analyzed. First, 2D cross-sections of the vertical walls were extracted for every specimen (Figure 11a). Then the advanced surface determination tool available in VG Studio Max [31] was employed to identify the surface of the wall. The segmented walls were then exported, and an in-house-developed Fiji ImageJ routine was used for further analysis. Wall profiles were created on both sides of the segmented walls (Figure 11a middle). To identify the base line, the profiles were processed with a low-pass filter (Figure 11a bottom represents the filtered profiles).

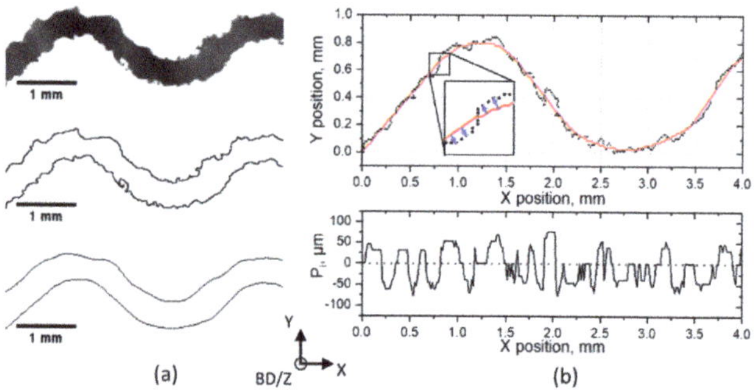

Figure 11. Evaluation of the surface roughness. (**a**) A 2D cross-section extracted from the vertical wall of 2 × 04 specimen (top), the profiles of both sides of the wall (middle), the profiles after low-pass filtering (bottom); (**b**) example showing the original (black points) and filtered profiles (continuous red line) and the resulting roughness profile (bottom). BD indicates the build direction (normal to the layer surfaces).

To obtain the roughness profile, a simple subtraction of the original and filtered profiles would not be accurate in the regions with non-zero slope. Instead, the distance between the baseline and the profile at each point was calculated along the normal to the filtered profile. This procedure is shown in the inset in Figure 11b top, where the red continuous curve represents the filtered profile, and the points are the (discrete) measured surface values. Re-entrant features were not considered in the analysis. The resulting roughness profile is shown at the bottom of Figure 11b.

The average roughness parameter P_a was calculated according to Equation (3), and the resulting values are presented for all specimens in Table 1.

$$P_a = \frac{1}{n}\sum_{i=1}^{n}|P_i(x)| \quad (3)$$

Table 1. Averaged roughness parameter P_a for as-build and CE specimens.

Specimen	2 × 04	2 × 04_CE	2 × 06	2 × 06_CE
P_a, µm	36.6 ± 3.1	30.8 ± 3.5	38.3 ± 4.6	16.4 ± 2.7

In order to get a statistically correct result and estimate a statistical error, the P_a value was calculated for ten 2D wall cross-sections, resulting in 20 roughness profiles for every specimen. The error is represented by the standard deviation of 20 measurements.

It is shown in literature that the arithmetic roughness (R_a) for EBM-manufactured vertical walls (or struts) ranges from 20 µm to 50 µm [39,40]. In our case, similar values (slightly below 40 µm) were obtained for the as-build 2 × 04 and 2 × 06 specimens. A slight reduction of the roughness was observed for the 2 × 04 CE specimen. In the case of the 2 × 06 CE specimen, the etching procedure could significantly reduce the roughness. However, as it was shown in the previous section, the periphery wall degradation, and at the same time the poor cleaning results, indicate the necessity of an improvement of the CE process. The combination of XCT data with advanced algorithms for surface topography determination provides a valuable method for the study (in 3D) and the improvement of the biocompatibility of lattice structures.

3.5. Analysis of the FGPS Lattice Structure

As mentioned above, one of the approaches to mimic the bone structure is the manufacture of functionally graded porous structures along a particular direction. In this case, the problem of RP is also present and needs to be considered when optimizing the size of the smallest pore (see Figure 12). A FGPS specimen with a porosity gradient in the build direction was investigated, combining the approaches described above. The pore size distribution was analyzed in the direction of the gradient.

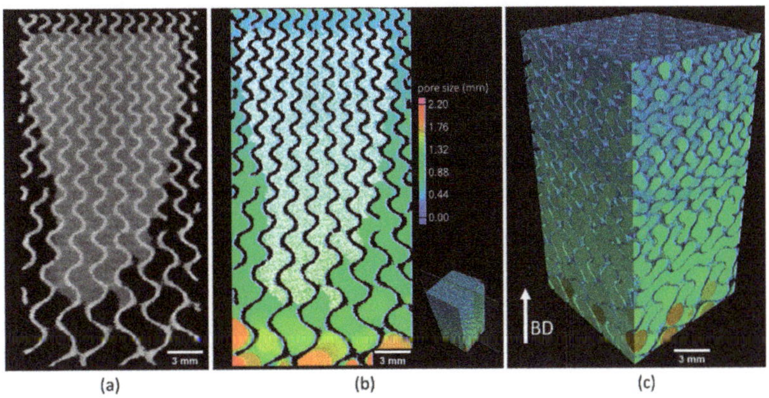

Figure 12. (**a**) 2D slice of a 3D reconstruction of a FGPS structure; (**b**) 2D slice showing the result of 3D pore size analysis (the white area represents the entrapped powder particles); (**c**) 3D rendering of the pore size distribution along the build direction (BD).

The volume fraction of RP was calculated as a ratio of the volume occupied by powder to the volume of the pores. The resulting mean and maximum pore size distribution as well as the volume fraction of RP are shown in Figure 13. It can be seen that the efficiency of the powder removal process is only satisfactory for pore sizes up to 1.6 mm. Lower porosity leads to the presence of RP, and, consequently, requires advanced cleaning procedures.

The theoretical pore size distribution was also analyzed using a model of the FGPS with a nominal WT of 0.25 mm. The theoretical distribution is in a good agreement (within a standard deviation) with the experimental one. Interestingly, no fluctuations of the pore size curve were observed, indicating no thickness difference between horizontal and vertical walls.

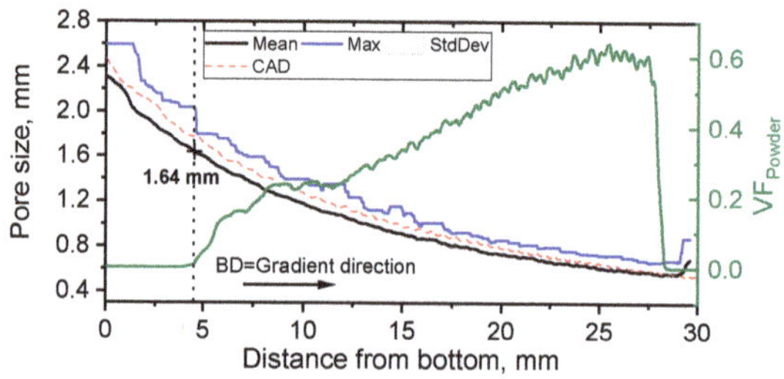

Figure 13. Pores size and volume fraction of entrapped powder (VF_{powder}) as a function of FGPS height.

To conclude, we can state that although more detailed studies are necessary to elucidate this aspect further, certain general criteria can be formulated for the design of FGPS manufactured by PBF AM: The smallest feasible cell size is determined not only by the manufacturing process and by the spatial resolution achievable with a particular machine, but also by the possibility to effectively remove the RP. In addition, powder cleaning efficiency depends on the overall dimensions of the porous sections and design of the implant. For example, the efficiency of the PRS system standard for EBM strongly falls as a function of depth in any structure: It is hard for the airflow to penetrate deep into the long channels filled with RP or solid sections of the component [15]. Multiple additional technologies supporting easier powder recovery should be applied with care, as some of them can negatively impact the mechanical properties of the structures (e.g., [7]). Until more reliable industry-accepted powder recovery methods are in place, the most reliable way to ensure thorough cleaning is to manufacture sets of structures with fixed TPMS cell properties and to use standard methods, such as EBM-PRS. Routines developed in this work for non-destructive characterization of the FGPS structures are an effective way to assess the quality of manufacturing and post-manufacturing processes, being a prerequisite for further technology improvement.

4. Conclusions

X-ray computed tomography (XCT) is the only tool for non-destructive and quantitative morphological characterization of additively manufactured triply periodic minimal surface structures (TPMSS). We showed that the challenge of segmenting the retained powder can be efficiently tackled by employing deep learning algorithms. The presented methodology of using a 2D convolutional neural network (CNN) instead of a 3D CNN allowed achieving high segmentation accuracy at a low cost (small amount of training data and high computational efficiency). This step allowed further metrological analysis of the TPMSS:

(1) The wall thickness showed a clear dependence on the wall orientation (with respect to the build direction): The horizontal walls (parallel to the build plane) appear up to 30% thicker than the vertical ones. Quantification of this known phenomenon is extremely important since it can cause a strong anisotropy of the mechanical properties.

(2) The chemical etching (CE) as a postprocessing technique was characterized assessing its efficiency in removing the retained powder, its effect on the wall thickness, and the surface roughness. The wall thickness gradient (from the edges to the specimen's center) was quantified using a developed methodology by combining the local thickness algorithm with the Euclidean distance transform. Strong gradient and large material loss were found in the specimen subjected to aggressive etching. In both specimen types, the powder could not be fully removed. This indicates that CE cannot be used as a standalone cleaning tool. Moreover, the material loss resulted in a reduction of the mechanical properties under compressive load. We found a positive effect of the CE on surface roughness. Our surface roughness analysis method allowed

investigating curved walls of the TPMSS and could disclose a decrease of the average roughness after the CE.

(3) Finally, the pore size distribution in functionally graded TPMSS was analyzed and correlated to the amount of retained powder. This procedure can be used to evaluate the cleaning routines and determine the minimum pore size compatible with a powder-free part, which is the premise for the optimization of the manufacturing process and for utilization of TPMSS in the biomedical sector.

Author Contributions: Conceptualization, S.E., T.M., D.K., R.S., M.S., A.K., and G.B.; methodology, S.E., T.M., D.K., A.P., and A.K.; CT measurements, D.M.; software, S.E., T.M., and D.K.; investigation, S.E., T.M., D.K., A.P. and D.M.; writing—original draft preparation, S.E.; writing—review and editing, S.E., T.M., D.K., D.M., R.S., M.S., A.K., A.P., and G.B. All authors have read and agreed to the published version of the manuscript.

Funding: This research was supported by the G-RISC program funded by the German Federal Foreign Office via the German Academic Exchange Service (DAAD) (Funding Decision No. M-2020a-6) and by Tomsk Polytechnic University Competitiveness Enhancement Program.

Institutional Review Board Statement: Not applicable.

Informed Consent Statement: Not applicable.

Data Availability Statement: The data presented in this study are available on request from the corresponding author. The data are not publicly available as they also form part of an ongoing study.

Conflicts of Interest: The authors declare no conflict of interest.

References

1. Shi, J.; Zhu, L.; Li, L.; Li, Z.; Yang, J.; Wang, X. A TPMS-based method for modeling porous scaffolds for bionic bone tissue engineering. *Sci. Rep.* **2018**, *8*, 7395. [CrossRef] [PubMed]
2. Fernandez-Yague, M.A.; Abbah, S.A.; McNamara, L.; Zeugolis, D.I.; Pandit, A.; Biggs, M.J. Biomimetic approaches in bone tissue engineering: Integrating biological and physicomechanical strategies. *Adv. Drug Deliv. Rev.* **2015**, *84*, 1–29. [CrossRef]
3. Zadpoor, A.A. Bone tissue regeneration: The role of scaffold geometry. *Biomater. Sci.* **2015**, *3*, 231–245. [CrossRef]
4. Al-Ketan, O.; Al-Rub, R.K.A. Multifunctional Mechanical Metamaterials Based on Triply Periodic Minimal Surface Lattices. *Adv. Eng. Mater.* **2019**, *21*, 1900524. [CrossRef]
5. Al-Ketan, O.; Rowshan, R.; Al-Rub, R.K.A. Topology-mechanical property relationship of 3D printed strut, skeletal, and sheet based periodic metallic cellular materials. *Addit. Manuf.* **2018**, *19*, 167–183. [CrossRef]
6. Yang, E.; Leary, M.; Lozanovski, B.; Downing, D.; Mazur, M.; Sarker, A.; Khorasani, A.; Jones, A.; Maconachie, T.; Bateman, S.; et al. Effect of geometry on the mechanical properties of Ti-6Al-4V Gyroid structures fabricated via SLM: A numerical study. *Mater. Des.* **2019**, *184*, 108165. [CrossRef]
7. Khrapov, D.; Koptyug, A.; Manabaev, K.; Léonard, F.; Mishurova, T.; Bruno, G.; Cheneler, D.; Loza, K.; Epple, M.; Surmenev, R.; et al. The impact of post manufacturing treatment of functionally graded Ti6Al4V scaffolds on their surface morphology and mechanical strength. *J. Mater. Res. Technol.* **2020**, *9*, 1866–1881. [CrossRef]
8. Echeta, I.; Feng, X.; Dutton, B.; Leach, R.; Piano, S. Review of defects in lattice structures manufactured by powder bed fusion. *Int. J. Adv. Manuf. Technol.* **2019**, *106*, 2649–2668. [CrossRef]
9. Koptyug, A.; Rännar, L.-E.; Bäckström, M.; Franzén, M.S.; Dérand, D. Additive manufacturing technology applications targeting practical surgery. *Int. J. Life Sci. Med Res.* **2013**, *3*, 15–24. [CrossRef]
10. Alla, R.K.; Ginjupalli, K.; Upadhya, N.; Mohammed, S.; Sekar, R.; Ravi, R. Surface Roughness of Implants: A Review. *Trends Biomater. Artif. Organs* **2011**, *25*, 112–118.
11. Albrektsson, T.; Wennerberg, A. On osseointegration in relation to implant surfaces. *Clin. Implant. Dent. Relat. Res.* **2019**, *21*, 4–7. [CrossRef] [PubMed]
12. Rosa, M.B.; Albrektsson, T.; Francischone, C.E.; Schwartz Filho, H.O.; Wennerberg, A. The influence of surface treatment on the implant roughness pattern. *J. Appl. Oral Sci.* **2012**, *20*, 550–555. [CrossRef] [PubMed]
13. Koptyug, A.; Rännar, L.-E.; Bäckström, M. Multiscale surface structuring of the biomedical implants manufactured in Electron Beam Melting technology: Demands, advances and challenges. In Proceedings of the iCAT-2014, International Conference on Additive Technologies, Vienna, Austria, 15–17 October 2014.
14. Novaes, A.B., Jr.; de Souza, S.L.; de Barros, R.R.; Pereira, K.K.; Iezzi, G.; Piattelli, A. Influence of implant surfaces on osseointegration. *Braz Dent. J.* **2010**, *21*, 471–481. [CrossRef]
15. Drescher, P.; Reimann, T.; Seitz, H. Investigation of powder removal of net–structured titanium parts made from electron beam melting. *Int. J. Rapid Manuf.* **2014**, *4*, 81–89. [CrossRef]

16. Wysocki, B.; Idaszek, J.; Buhagiar, J.; Szlązak, K.; Brynk, T.; Kurzydłowski, K.J.; Święszkowski, W. The influence of chemical polishing of titanium scaffolds on their mechanical strength and in-vitro cell response. *Mater. Sci. Eng. C* **2019**, *95*, 428–439. [CrossRef]
17. Hasib, H.; Harrysson, O.L.A.; West, H.A. Powder Removal from Ti-6Al-4V Cellular Structures Fabricated via Electron Beam Melting. *JOM* **2015**, *67*, 639–646. [CrossRef]
18. Blanquer, S.B.G.; Werner, M.; Hannula, M.; Sharifi, S.; Lajoinie, G.P.R.; Eglin, D.; Hyttinen, J.; Poot, A.A.; Grijpma, D.W. Surface curvature in triply-periodic minimal surface architectures as a distinct design parameter in preparing advanced tissue engineering scaffolds. *Biofabrication* **2017**, *9*, 025001. [CrossRef] [PubMed]
19. Plessis, A.D. X-ray tomography for the advancement of laser powder bed fusion additive manufacturing. *J. Microsc.* **2020**. [CrossRef]
20. Thompson, A.; Maskery, I.; Leach, R.K. X-ray computed tomography for additive manufacturing: A review. *Meas. Sci. Technol.* **2016**, *27*, 072001. [CrossRef]
21. Mishurova, T.; Artzt, K.; Haubrich, J.; Requena, G.; Bruno, G. New aspects about the search for the most relevant parameters optimizing SLM materials. *Addit. Manuf.* **2019**, *25*, 325–334. [CrossRef]
22. Suard, M.; Plancher, E.; Martin, G.; Dendievel, R.; Lhuissier, P. Surface Defects Sensitivity during the Unfolding of Corrugated Struts Made by Powder-Bed Additive Manufacturing. *Adv. Eng. Mater.* **2020**, *22*, 2000315. [CrossRef]
23. Kerckhofs, G.; Pyka, G.; Moesen, M.; Van Bael, S.; Schrooten, J.; Wevers, M. High-Resolution Microfocus X-ray Computed Tomography for 3D Surface Roughness Measurements of Additive Manufactured Porous Materials. *Adv. Eng. Mater.* **2013**, *15*, 153–158. [CrossRef]
24. Zanini, F.; Pagani, L.; Savio, E.; Carmignato, S. Characterisation of additively manufactured metal surfaces by means of X-ray computed tomography and generalised surface texture parameters. *CIRP Ann.* **2019**, *68*, 515–518. [CrossRef]
25. Bernard, D.; Léonard, F.; Plougonven, E.; Bruno, G. On the use of autocorrelation functions, permeability tensors, and computed tomography to characterise the anisotropy of diesel particulate filter materials. *Philos. Mag.* **2020**, *100*, 2802–2835. [CrossRef]
26. Hunter, L.W.; Brackett, D.; Brierley, N.; Yang, J.; Attallah, M.M. Assessment of trapped powder removal and inspection strategies for powder bed fusion techniques. *Int. J. Adv. Manuf. Technol.* **2020**, *106*, 4521–4532. [CrossRef]
27. Sefer, B.; Dobryden, I.; Almqvist, N.; Pederson, R.; Antti, M.-L. Chemical Milling of Cast Ti-6Al-4V and Ti-6Al-2Sn-4Zr-2Mo Alloys in Hydrofluoric-Nitric Acid Solutions. *Corrosion* **2016**, *73*, 394–407. [CrossRef]
28. Cooper, R.C.; Bruno, G.; Onel, Y.; Lange, A.; Watkins, T.R.; Shyam, A. Young's modulus and Poisson's ratio changes due to machining in porous microcracked cordierite. *J. Mater. Sci.* **2016**, *51*, 9749–9760. [CrossRef]
29. Falk, T.; Mai, D.; Bensch, R.; Cicek, O.; Abdulkadir, A.; Marrakchi, Y.; Bohm, A.; Deubner, J.; Jackel, Z.; Seiwald, K.; et al. U-Net: Deep learning for cell counting, detection, and morphometry. *Nat. Methods* **2019**, *16*, 67–70. [CrossRef]
30. Schindelin, J.; Arganda-Carreras, I.; Frise, E.; Kaynig, V.; Longair, M.; Pietzsch, T.; Preibisch, S.; Rueden, C.; Saalfeld, S.; Schmid, B.; et al. Fiji: An open-source platform for biological-image analysis. *Nat. Methods* **2012**, *9*, 676–682. [CrossRef]
31. Volume Graphics. Available online: https://www.volumegraphics.com/ (accessed on 19 April 2021).
32. Hildebrand, T.; Rüegsegger, P. A new method for the model-independent assessment of thickness in three-dimensional images. *J. Microsc.* **1997**, *185*, 67–75. [CrossRef]
33. Doube, M.; Kłosowski, M.M.; Arganda-Carreras, I.; Cordelières, F.P.; Dougherty, R.P.; Jackson, J.S.; Schmid, B.; Hutchinson, J.R.; Shefelbine, S.J. BoneJ: Free and extensible bone image analysis in ImageJ. *Bone* **2010**, *47*, 1076–1079. [CrossRef] [PubMed]
34. Pyka, G.; Kerckhofs, G.; Papantoniou, I.; Speirs, M.; Schrooten, J.; Wevers, M. Surface Roughness and Morphology Customization of Additive Manufactured Open Porous Ti6Al4V Structures. *Materials* **2013**, *6*, 4737. [CrossRef] [PubMed]
35. De Formanoir, C.; Suard, M.; Dendievel, R.; Martin, G.; Godet, S. Improving the mechanical efficiency of electron beam melted titanium lattice structures by chemical etching. *Addit. Manuf.* **2016**, *11*, 71–76. [CrossRef]
36. International Organization for Standartization. *Mechanical Testing of Metals—Ductility Testing—Compression Test for Porous and Cellular Metals*; ISO: Geneva, Switzerland, 2011.
37. Greitemeier, D.; Dalle Donne, C.; Syassen, F.; Eufinger, J.; Melz, T. Effect of surface roughness on fatigue performance of additive manufactured Ti–6Al–4V. *Mater. Sci. Technol.* **2016**, *32*, 629–634. [CrossRef]
38. Sanaei, N.; Fatemi, A. Analysis of the effect of surface roughness on fatigue performance of powder bed fusion additive manufactured metals. *Theor. Appl. Fract. Mech.* **2020**, *108*, 102638. [CrossRef]
39. Vayssette, B.; Saintier, N.; Brugger, C.; Elmay, M.; Pessard, E. Surface roughness of Ti-6Al-4V parts obtained by SLM and EBM: Effect on the High Cycle Fatigue life. *Procedia Eng.* **2018**, *213*, 89–97. [CrossRef]
40. Townsend, A.; Senin, N.; Blunt, L.; Leach, R.K.; Taylor, J.S. Surface texture metrology for metal additive manufacturing: A review. *Precis. Eng.* **2016**, *46*, 34–47. [CrossRef]

materials

MDPI

Article

Different Approaches for Manufacturing Ti-6Al-4V Alloy with Triply Periodic Minimal Surface Sheet-Based Structures by Electron Beam Melting

Dmitriy Khrapov [1], Maria Kozadayeva [1], Kayrat Manabaev [1], Alexey Panin [2], William Sjöström [3], Andrey Koptyug [3], Tatiana Mishurova [4], Sergei Evsevleev [4], Dietmar Meinel [4], Giovanni Bruno [4,5], David Cheneler [6], Roman Surmenev [1] and Maria Surmeneva [1,*]

1 Physical Materials Science and Composite Materials Centre, Research School of Chemistry & Applied Biomedical Sciences, National Research Tomsk Polytechnic University, 30 Lenina Avenue, 634050 Tomsk, Russia; dah8@tpu.ru (D.K.); mariakoz71@gmail.com (M.K.); mkk@tpu.ru (K.M.); rsurmenev@mail.ru (R.S.)
2 Institute of Strength Physics and Materials Science of Siberian Branch Russian Academy of Sciences (ISPMS SB RAS), pr. Akademicheskii 2/4, 634055 Tomsk, Russia; pav@ispms.tsc.ru
3 SportsTech Research Center, Mid Sweden University, Akademigatan 1, SE-831 25 Östersund, Sweden; william.sjostrom@miun.se (W.S.); andrey.koptyug@miun.se (A.K.)
4 Bundesanstalt für Materialforschung und -Prüfung (BAM), Unter den Eichen 87, 12205 Berlin, Germany; tatiana.mishurova@bam.de (T.M.); sergei.evsevleev@bam.de (S.E.); dietmar.meinel@bam.de (D.M.); giovanni.bruno@bam.de (G.B.)
5 Institute of Physics and Astronomy, University of Potsdam, Karl-Liebknecht-Str. 24-25, 14476 Potsdam, Germany
6 Engineering Department, Lancaster University, Lancaster LA1 4YW, UK; d.cheneler@lancaster.ac.uk
* Correspondence: surmenevamaria@mail.ru

Citation: Khrapov, D.; Kozadayeva, M.; Manabaev, K.; Panin, A.; Sjöström, W.; Koptyug, A.; Mishurova, T.; Evsevleev, S.; Meinel, D.; Bruno, G.; et al. Different Approaches for Manufacturing Ti-6Al-4V Alloy with Triply Periodic Minimal Surface Sheet-Based Structures by Electron Beam Melting. *Materials* **2021**, *14*, 4912. https://doi.org/10.3390/ma14174912

Academic Editor: Francesco Iacoviello

Received: 9 July 2021
Accepted: 20 August 2021
Published: 29 August 2021

Abstract: Targeting biomedical applications, Triply Periodic Minimal Surface (TPMS) gyroid sheet-based structures were successfully manufactured for the first time by Electron Beam Melting in two different production Themes, i.e., inputting a zero (Wafer Theme) and a 200 μm (Melt Theme) wall thickness. Initial assumption was that in both cases, EBM manufacturing should yield the structures with similar mechanical properties as in a Wafer-mode, as wall thickness is determined by the minimal beam spot size of ca 200 μm. Their surface morphology, geometry, and mechanical properties were investigated by means of electron microscopy (SEM), X-ray Computed Tomography (XCT), and uniaxial tests (both compression and tension). Application of different manufacturing Themes resulted in specimens with different wall thicknesses while quasi-elastic gradients for different Themes was found to be of 1.5 GPa, similar to the elastic modulus of human cortical bone tissue. The specific energy absorption at 50% strain was also similar for the two types of structures. Finite element simulations were also conducted to qualitatively analyze the deformation process and the stress distribution under mechanical load. Simulations demonstrated that in the elastic regime wall, regions oriented parallel to the load are primarily affected by deformation. We could conclude that gyroids manufactured in Wafer and Melt Themes are equally effective in mimicking mechanical properties of the bones.

Keywords: Electron Beam Melting; scaffold; lightweight structures; computed tomography; Finite Element Analysis

1. Introduction

The optimization of additively manufactured (AM) porous structures for biomedical applications aims at increasing fatigue life, enhancing mass transport properties for tissue regeneration, decreasing the occurrence of infections, minimizing powder release from the structures, and minimizing stress shielding. Stress shielding is caused by differences between Young's moduli of the bone and the implant, and can be prevented by adjusting

Young's modulus of the implant through manipulating its structure (porosity) and material [1]. The designed porosity in regular-geometry lattice systems primarily depends on the type of unit cell. Usually, beam-based or sheet-based cell elements are used. AM porous structures based on the beam-like elements are intensively studied [2] and are most commonly used for porous scaffolds design.

Triply Periodic Minimal Surfaces (TPMS) have recently gained interest as the new approach to the design of the sheet-based porous scaffolds for tissue engineering. TPMS attracts attention due to zero-mean curvature at every point that is admittedly a great advantage since it improves the structure load-bearing capacity simultaneously assisting bone cell ingrowth [3]. The well-known TPMS are Schwarz Gyroid (G), Schwarz Primitive (P), and Schwarz Diamond (D) [4].

There are different approaches of using TPMS geometries for designing porous structures. First approach utilizes beam-based TPMS, strut-based [4], network-based or skeletal structure designs [5,6]. They are used to overcome stress concentrators in the sharp turns of the metal that are typical for unit cells with straight beam-like struts and a polyhedral core. Porous structures using straight beams can experience severe stress concentrations under loading, especially in regions where beams are merging, or bending at acute angles. Severe overloading and increased fatigue-related failure in the stress-concentration zones can lead to a complete collapse of the corresponding structural elements [7]. Beam-based TPMS are designed to have smooth struts and smoother connections between horizontal and vertical elements as compared to conventional beam-based structures. Most of the beam-based TPMS were manufactured using struts with circular cross-section. Beam-based TPMS structures were manufactured by the stereolithography rapid prototyping [8,9], selective laser sintering (SLS) [10] and laser powder bed fusion (L-PBF) [11]. Beam-based TPMS structures were also manufactured by Electron Beam Melting (EBM) by Yánez et al. [12,13] and Ataee et al. [14].

A second approach to designing TMPS structures is by using sheet-based [4] or sheet-like [15] elements. Such structures are also referred to as list structures or matrix phase lattices [5]. They comprise a wall of solid material bounded by two unconnected void regions. The continuity of the sheet-based TPMS is supposed to provide higher strength and damage-tolerance through effective obstruction of crack propagation [16]. Crack propagation in continuous sheet-based porous structures requires more energy as compared to common strut-based ones. Sheet-based gyroid structures also have higher Young's modulus, peak stress, and toughness in comparison with beam-based gyroid structures. For example, Al-Ketan et al. [6] demonstrated that sheet-TPMS structures have superior mechanical properties in terms of Young's modulus in comparison with conventional strut-based and skeletal-TPMS porous structures. Among such TPMS structures, gyroid structures attract the attention of many scientists. Kapfer et al. [17] demonstrated that the sheet-based gyroid structures have higher stiffness than the beam-based ones with the same porosity and manufactured from the same material. Aremu et al. [18] noted that gyroid lattices, unlike several other lattice types, possess axisymmetric stiffness making them desirable candidates for applications where the exact nature and direction of the loads are not fully known or if such loads are subject to large variations. Sheet-based structures have been manufactured by selective laser sintering (SLS) [6,15], L-PBF [5,19] but so far not by EBM. It is well-known that EBM structures have lower resulting porosity [18] and lower residual stress as compared to similar L-PBF- and SLS-manufactured structures because of the preheating during manufacturing that acts as a stress relief heat treatment [20,21]. Internal defects of the EBM-manufactured objects affect their fatigue life [22]. To evaluate manufacturing quality and porosity, the size and the form of the defects process monitoring approaches for is investigated [23]. Moreover, manufacturing parameters' optimization leads to porosity reduction [24]. One way to optimize manufacturing parameters is EBM process simulation that helps to predict physical properties of AM [25,26]. To evaluate the behavior of the Ti-6Al-4V implant in human body, friction and wear performance of the wrought and EBM-manufactured Ti-6Al-4V in simulated body fluid solution were studied [27].

Studying EBM TPMS sheet structures is of great interest for biomedical applications as the manufacturing of implants with porous elements is one of the core applications of EBM technology. Sheet-based TPMS can be produced by EBM using different manufacturing parameter sets commonly known as Themes. "Melt" Themes originally designed for manufacturing solid structural elements require a 3D model with predetermined material thickness. The "Wafer" Theme originally designed for manufacturing different support structures uses zero-thickness 3D models [28]. Taking into consideration the complexity of TPMS, the Wafer Theme may become a new key to controlling specimen's porosity and preventing the stress shielding effect. The differences between the gyroid samples manufactured using Melt and Wafer Themes are the subject of this work.

Our initial assumption was that specimens based on 200 μm thick model manufactured using Melt Theme and specimens based on zero-thickness model manufactured using Wafer Theme would have identical sheet thickness and identical mechanical properties. This assumption was based on the fact that the beam spot diameter set in the ARCAM A2 machine is commonly 200 μm [28].

In this work, we investigate the mechanical properties of the TPMS porous specimens based on model with 200 μm thickness manufactured by EBM using the Melt Theme and specimens with an equivalent design but produced by EBM using the Wafer Theme, zero-thickness model. Samples were manufactured using the standard Themes provided by ARCAM EBM for Ti-6Al-4V alloy.

We address the relationship between structural performance and manufacturing modality, keeping porosity constant. The novelty of the research lies in combination of design methods of TPMS and EBM-manufacturing modalities. The aim of the current investigation was to evaluate the worthiness of Wafer Theme in comparison with the Melt Theme for TPMS structures' fabrication from the mechanical point of view.

2. Materials and Methods

2.1. TPMS File Preparation

Wolfram Mathematica (version 12) [29] was used to visualize the gyroids using the following equation:

$$\sin(kx)\cos(ky) + \sin(ky)\cos(kz) + \sin(kz)\cos(kx) = 0 \quad (1)$$

The limits of the surface were chosen from $-5/2\ \pi$ to $5/2\ \pi$ in all directions. So, the designed structure consists of three unit cells along each coordinate direction (X, Y, Z), resulting in a total size of 15 mm × 15 mm × 15 mm. The coefficient k controls the unit cell size. For this research $k = 1$ was chosen. Two different sets of the 3D models were implemented. The models of the first set with zero-thickness were exported from Mathematica as STL files with the default density of the polygon mesh, Figure 1a.

The second set of models was produced from the first one by assigning a thickness of 200 μm to all surfaces, Figure 1c and also was exported from Mathematica as STL files with the same conditions. Since the surface of gyroid consists of semicircular surfaces, the exported 3D models had a large number of vertices (more than 10^6), thus creating a high-poly model. The design files of large size can in some cases cause memory issues of ARCAM Build Assembler software. Thus, the high-poly meshes generated with random polygon size and shape distribution required additional mesh optimization. For this purpose, the MeshLab software, an open-source Mesh Processing Tool [30], was used. The number of the vertices was steeply reduced to 10,000 for zero-thickness model and to 44,000 for 200 μm thick model, preserving the boundaries and the topology of the mesh. Topological errors, such as non-manifold faces, self-intersections, duplicate faces, etc., were also removed using the MeshLab.

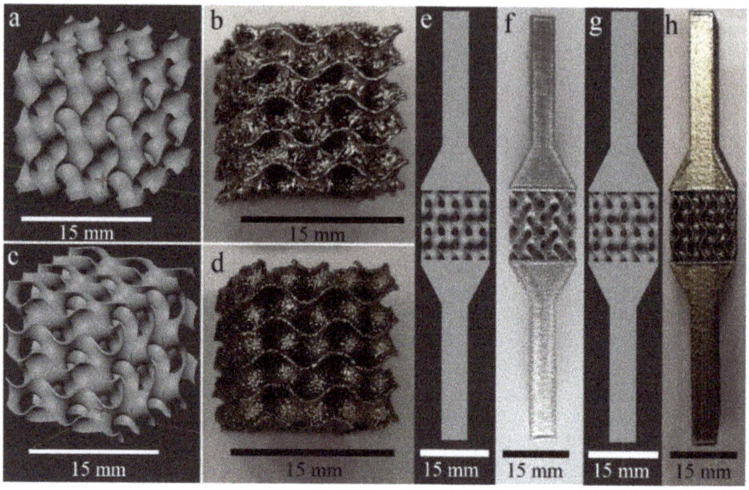

Figure 1. Samples for compression tests: (**a**) 3D model with zero thickness; (**b**) WT specimen; (**c**) 3D model with 200 μm thickness; (**d**) MT specimen. Samples for tension tests: (**e**) 3D model with zero thickness; (**f**) WT specimen; (**g**) 3D model with 200 μm thickness; (**h**) MT specimen.

For the tensile samples, two opposite 4 cm long tapering blocks were added to the gyroid structures of both types, Figure 1e,g. The total length of these specimens was 95 mm. The process of error correction was repeated in FreeCAD [31]. This software was used for STL to STEP file conversion (necessary for the FE analyses). The obtained STL files were used for designing all specimen models and for processing in the ARCAM Build Assembler—for EBM—manufacturing.

2.2. Finite Element Analysis

The aim of FE simulations was to qualitatively analyze the deformation process and the stress distribution. The models with thicknesses of 0.25 and 0.4 mm were imported to ANSYS Workbench (ANSYS, Canonsburg, PA, USA). The values of thicknesses were taken based on the experimental results (see Section 3.2). A tetrahedral mesh model was implemented. This method was convenient since the initial models obtained in Wolfram Mathematica consisted of triangle polygons. The total number of nodes and elements for the model with the thickness of 0.25 mm were 910,727 and 489,256, respectively. Using the physical properties of Ti-6Al-4V, the modeled specimen had an estimated mass of 1.53 g and a porosity of 85%. The total number of nodes and elements for the model with the thickness of 0.4 mm were 1,070,675 and 610,874, respectively. Using the physical properties of Ti-6Al-4V, the modeled specimen had an estimated mass of 3.18 g and a porosity of 75%. The von Mises failure criterion was chosen, and a yield strength of 970 MPa for Ti-6Al-4V was selected [32]. Boundary conditions were applied as follows: frictionless support was applied to the bottom face and 1 mm displacement was applied to the top. Only the elastic regime was simulated for both tension and compression tests.

2.3. Manufacturing

Scaffolds were manufactured from Ti-6Al-4V powder in ARCAM A2 EBM machine by ARCAM, EBM (Mölnlycke, Sweden) operating under firmware version 3.2 using standard Themes without modification of the default parameter settings. All samples were manufactured using 50 μm layer thickness and process temperature of 720 °C.

In beam-based structures, PBF AM specimens can be manufactured using different beam energy application strategies, influencing, in particular, melt pool dimensions, solidification rate and final material microstructure. Modern PBF AM machines are commonly operated using parameter sets provided by the manufacturer. An ARCAM "Theme" is a set of settings incorporating beam scanning strategy and process parameters. The standard Themes provided by the EBM machine manufacturer ARCAM EBM are called "Melt", "Net", and "Wafer". The "Melt" Theme is optimized for manufacturing of solid structural elements. The "Net" Theme is designed for the manufacturing of the beam-based porous structures. The "Wafer" Theme is designed for manufacturing of support structures (essential elements of the EBM process), where wafer-thin surfaces are required [28]. These supports are commonly used for stabilizing overhanging elements, and as spacers between the solid elements and the base plate. Such supports are removed after manufacturing. So, the "Wafer" Theme was designed to produce very thin structures to guarantee a small amount of waste material. Structures manufactured using Net and Wafer Themes are designed as zero-thickness geometry, and the cross-section of the elements is defined by the beam energy and its deposition rate.

The "Melt" Theme is more complicated than the "Wafer" one. The first path of the beam (called first contour) is shifted out from the CAD-defined element periphery by a certain value called the first contour offset (CO1). Next, the beam performs a second contour scan with a second contour offset (CO2) moving slightly inward from the CAD-defined cross-section periphery. After that, the beam melts the area enclosed by these two contours (hatching) using different types and strategies of the raster motion. Corresponding contours are performed by continuous beam spot motion over the whole periphery length, or in so-called 'Multibeam^{TM} mode'. In the latter case, the beam moves only through the short sector of the contour and 'jumps' away to melt another sector, repeating the operation many times to cover all needed contours. The main purpose for selecting single- or double- contours in continuous or Multibeam^{TM} mode is the optimization of the process to obtain the smallest possible roughness of side surfaces of the components. It is clear that when using any mode for the manufacturing of lightweight and porous structures, careful parameter selection is needed to guarantee that the resulting element cross-sections are as close to the CAD design as possible. When the elements with a small cross-sectional area are EBM-manufactured using Melt Theme, automated file preprocessing can reduce the number of contours (in an automatic fashion) until only the hatch is left, minimizing the increase in the dimensions of the manufactured elements.

The "Wafer" Theme, traditionally, is mainly used for the plane sheet supports with relatively small surface areas; our experience shows that it is also effective when used for quite complicated structures. Slicer software created a line pattern for WT based on zero-thickness gyroid model for each slice, while for MT, it created areas for melting based on 200 µm thickness gyroid model. Depending on the parameter settings, thin structures made with Wafer Theme can have a certain amount of through-holes. However, it is clear that application of the Wafer Theme for EBM of sheet-based lightweight and porous structures can bring significant improvement of the mechanical properties of the resulting components.

Specimens manufactured using Melt Theme and Wafer Theme are further referred to as MT and WT correspondingly. Examples of design structures and manufactured EBM WT and MT samples are presented in Figure 1.

In the build file, WT samples were placed directly on the base plate and were oriented with the side walls normal to the build direction. MT models were oriented in such a way that their cross-section diagonal was aligned with the build direction.

After manufacturing, the specimens were removed from the powder bed and separated from the base plate in a standard ARCAM Powder Recovery System. All specimens were subjected to compressed airflow for more than 10 min, thereby ensuring the removal of the powder particles loosely connected to the surfaces.

WT specimens were manufactured with turned off Multibeam^{TM} contour mode (continuous beam path), beam current of I = 5 mA, and scanning velocity of v = 1000 mm/s. Three specimens were manufactured for each of tension and compression tests. The fixation heads for all tension specimens were manufactured using Melt Theme with default parameter settings, Figure 1f, h. Equivalent parameters of the manufactured compression specimens are given in Table 1.

Table 1. Specimen parameters.

Specimen Parameters	Melt Theme	Wafer Theme
Mass m, g	3.4 ± 0.1	2.3 ± 0.2
Volume V, cm^3	3.4 ± 0.1	3.2 ± 0.1
Density ρ, g/cm^3	1.07 ± 0.13	0.74 ± 0.15
$\rho/\rho_{0\ (\rho 0\,=\,4.43\ g/cm3)}$	0.24 ± 0.3	0.17 ± 0.2
Porosity p, %	76 ± 3	86 ± 5

The surface morphology was studied by scanning electron microscopy Quanta 200 3D (FEI, Eindhoven, The Netherlands).

The outer dimensions of the fabricated scaffold samples were measured by a Vernier caliper; all specimens were weighted on an Acculab ALC-210d4 (Sartorius AG, Göttingen, Germany) scale. Their calculated solid volume and measured weight were used to determine the apparent density ρ of the structure (presented in Table 1). Assuming the density of solid Ti-6Al-4V is equal to ρ_{0} = 4.43 g/cm^3, the porosity P of the scaffolds in % was obtained by:

$$P = 1 - \frac{\rho}{\rho_0} \tag{2}$$

2.4. X-ray Computed Tomography

The X-ray computed tomography (XCT) measurements were performed at BAM using an XCT scanner, developed together with the company Sauerwein Systemtechnik (today RayScan Technologies GmbH, Meersburg, Germany). A microfocus X-ray tube XWT-225-SE (maximum voltage 225 kV) from X-RAY WorX GmbH (Garbsen, Germany) was used as a source. An XRD1620 (CsI scintillator, 2048 × 2048 pixel) detector from PerkinElmer Inc. (Waltham, MA, USA), with in-house built enclosure and cooling system was used. A tube voltage of 120 kV and a tube current of 120 μA were used during the scans. The voxel size was 15.3 μm. The reconstruction of 3D volumes from 2D projections was made by the software developed in BAM using a filtered back-projection algorithm [33]. The obtained raw data files were analyzed using VGStudio MAX 3.3 by Volume Graphics, Heidelberg, Germany. The STL files of the 3D models and the XCT-based reconstructions were used to conduct Nominal/Actual comparison. The wall thickness was evaluated by the sphere method.

2.5. Mechanical Tests

Quasi-static uniaxial compression and tension tests were performed using a universal testing machine INSTRON 3369 and INSTRON 5582, (Instron Deutschland GmbH, Darmstadt, Germany), with a 50 kN load cell. Tests were conducted at 20 °C according to ISO 13314:2011 [34], and using a displacement rate of 0.5 mm/min. Through the measurement of the applied load, we calculated the stress dividing the load by the effective area of the lattice structures. The failure strain was set at 50% of the specimen height. Results for the quasi-elastic gradient [35], compressive offset stress, first maximum compressive strength, energy absorption at 50% strain (Equation (3)), and specific energy absorption (Equation (4)) were calculated following ISO 1331 standard [34].

The ISO 13314 was devoted to describing the mechanical behavior of beam-based structures. TPMS sheet-based scaffolds represent structures with more complex shapes than beam-based structures. Indeed, the term "unit cell" is obvious for beam-based structures, but not for sheet-based structure. We therefore took ISO 13314 as a reference at the stage of designing the scaffold but did not strictly follow it. In fact, our goal was to qualitatively assess the behavior of our scaffolds, not to qualify them for production. Indeed, there are many scientific articles not strictly following the requirements of standard test or production methods [5,35–38].

The quasi-elastic gradient E_{qe} [4,39–43] of the porous samples is the gradient of the straight line determined within the linear deformation region at the beginning of the compressive stress-strain curve, i.e., this value is defined similarly to Young's modulus E for bulk material. Additionally, the compressive offset stress and the first maximum compressive strength for porous specimens are defined similarly to the yield stress σ_y and the compressive strength σ_{max} for bulk specimens. Yield strain was defined as 0.2% strain, and the compressive offset stress was determined accordingly. Quasi-elastic gradient E_{qe}, yield stress σ_y, and ultimate tensile strength σ_{max} were estimated for tensile specimens. The plateau stress σ_{pl} is the arithmetical mean of the stress values between 20% and 40% compressive strain. The point in the stress-strain curve at which the stress is 1.3 times the plateau stress is defined as the plateau end. It can be used for the determination of energy absorption and energy absorption efficiency:

$$W = \frac{1}{100}\int_{0}^{e_0} \sigma de \tag{3}$$

where W is energy absorption per unit volume (MJ/m^3), σ is the compressive stress (MPa), e_0 is the upper limit of the compressive strain. The energy absorption per unit volume was calculated from the area under the stress-strain curve up to 50% strain.

The crashworthiness of a material can be expressed in terms of its specific energy absorption. The specific energy absorption (Equation (3)) ψ is defined as the work W performed per unit weight when the material is compressed in a uniaxial manner up to a specific strain. The strain of 50% [41,44] was chosen for the evaluation of specific energy absorption:

$$\psi = \frac{W}{\rho}; W = \int_{0}^{\varepsilon} \sigma d\varepsilon \tag{4}$$

where ρ denotes the mass density, σ the axial stress, and ε the work-conjugate axial strain [45].

3. Results

3.1. Scanning Electron Microscopy

The gyroid structures present both partially melted powder particles on the surfaces (Figure 2a,b) and the stair effect (Figure 2c,d), typically observed in AM parts (due to the layer-wise manufacturing) [36–48]. While the minimum electron beam spot size of EBM machine is about 200 µm, the melt pool is commonly wider, and the smallest possible sheet thickness that can be resolved in EBM is about 200 µm [28]. The wall thickness of the WT specimens was smaller than that of MT, as can be observed in Figure 2a,b.

WT structures have quite small sheet thickness, as the beam energy used to manufacture them is rather low. The production of such thin structures is at the limits of the machine capabilities, since the powder particle size is between 75 and 125 µm. Consequently, some holes were present in the sheets (Figure 2b,d). Thicker sheets of the MT structures have visibly higher roughness but no through-holes. WT samples have the holes predominantly in the areas where the surface is parallel to the build plane. This is well-known in PBF techniques, where such problems occur in overhanging elements and thin structures normal to the build direction [48].

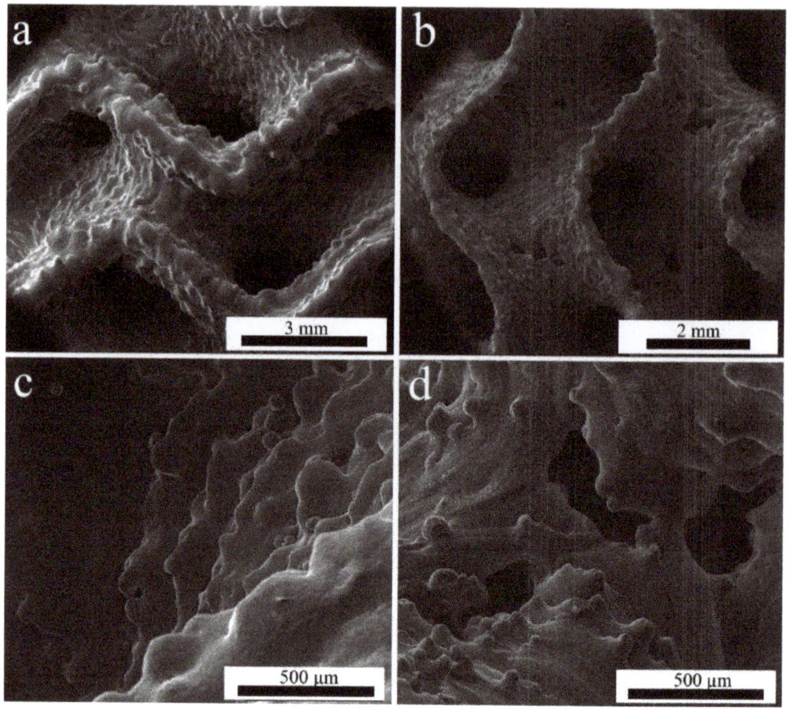

Figure 2. SEM images of (**a**,**c**)—MT specimen; (**b**,**d**)—WT specimen.

3.2. X-ray Computed Tomography

XCT was used to describe inner structure, evaluate wall thickness, and reveal the difference between the designed structure and the manufactured specimen. The wall thickness of WT and MT EBM structures evaluated from XCT reconstructions is presented in Table 2. The wall thickness distribution in the two specimens is presented in Figure 3a. Note that for both specimens, the targeted wall thickness was 0.2 mm. The actual mean wall thickness was about 0.4 mm and 0.25 mm for MT specimen (bulk-melt mode with contours) and WT specimen (Wafer-mode), respectively. It is clear that for thin-walled structures, the contour-enabled bulk melt mode is not optimal and leads to a larger than desired wall thickness.

Table 2. Summary of the results of the quantitative image analysis of XCT data. (Measurement errors cannot be estimated; the error intervals represent the variance of all the measured values).

Reconstructed Specimen Parameters	Melt Theme	Wafer Theme
Mean wall thickness, mm	0.38 ± 0.07	0.25 ± 0.06
Max. wall thickness, mm	0.56	0.44
Defect volume ratio, % (Micro-pores to bulk volume)	0.4	0.3

The reason for the difference between thicknesses of the designed and MT manufactured structures lies in additional thickness caused by two contours and, probably, wider than expected melt pool. Since the surface of the gyroid is curved, it is impossible to evaluate roughness by traditional methods. It is known that arithmetic roughness (R_a) for vertical struts of the EBM-manufactured structures is about 40 µm, while the mean value of the maximum height of the surface profiles of vertical struts (R_t) is 212 µm [28]. Comparison of a designed 3D model and as-manufactured samples performed with the standard

VGStudio function named 'Nominal/Actual Comparison' characterizes the manufacturing accuracy and supplements the wall thickness analysis, Figure 3b. Nominal/Actual Comparison is also an alternative way to describe roughness of the manufactured specimens in qualitative terms. The value of average surface roughness is comparable with the sheet thickness that is quite typical for EBM-manufactured porous structures [49].

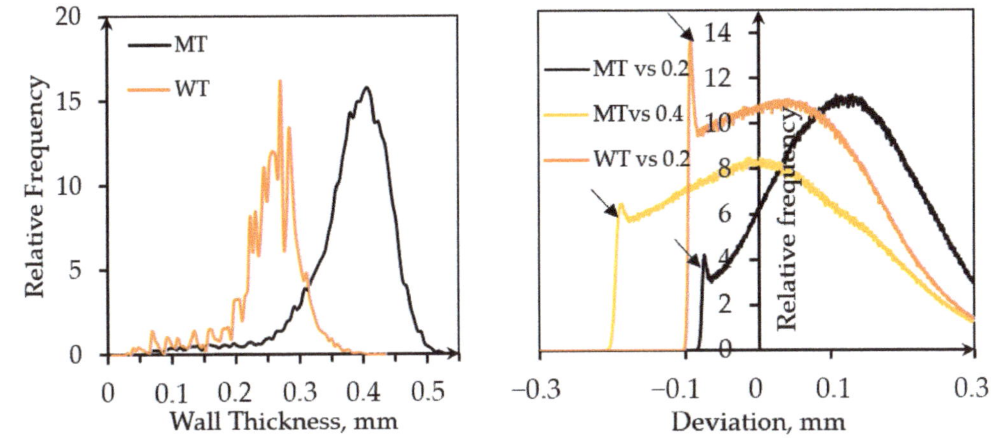

Figure 3. Evaluation of the parameters based on the XCT data analysis: (**a**) Wall thickness distribution; (**b**) Deviation distribution obtained from Nominal/Actual Comparison analysis.

It makes no sense comparing zero-thickness model used for designing WT samples and 0.25 mm thick reconstruction. Therefore, a model with the desired thickness was used for the Nominal/Actual Comparison. The average wall thickness of the MT specimen based on the 0.2 mm model was about 0.4 mm. A comparison with initial model and, additionally, with 0.4 mm thick model was performed to evaluate surface roughness more precisely.

The deviation of the structure was estimated from both sides of the walls. The searching distance was 0.3 mm, which is quite large in comparison with the average wall thickness. The positive deviation is caused by the presence of contours in MT samples, melt pool 'swelling' into the surrounding powder bed, and the presence of powder particles partially merged with the surface. There are sharp peaks of the relative frequency at the negative deviations followed by rapid drops (indicated by arrows). They may be attributed to the surface elements detected from the opposite side of the wall, and this will be typical for porous lattices with any unit cell design.

Figure 4 illustrates the Nominal/Actual analysis of the samples. White arrows in Figure 4a indicate stalactite-like structures on the horizontally oriented parts of the MT walls (areas parallel to the layers, purple color). This effect is quite common in the PBF-manufactured specimens [50]. Interestingly, it is much less pronounced for the WT specimens due to the smaller beam energy used. In the MT samples, clusters of the partially fused powder particles are also present. The white circles highlight the partially melted powder particles attached to the surface of the vertical areas. This effect was not found for WT specimens, presumably because of the smaller electron beam energy input, Figure 4b. The purple area on the lower part of WT specimen indicates larger dimensional deviation in the first layers. This is a known effect due to the incompletely stabilized temperatures in the semi-sintered powder surrounding the melt pool at the early stages of the build, and uneven compensation of the expansion of powder placed under the start plate. Uneven compensation of the powder expansion leads to the situation when during some of the first layers, EBM rake brings no powder for some parts of the working area. Moreover, sample material starts to be deposited only after a few nominal layers leading to the distortions of the samples that are due to start from the base plate. Additionally, the material adjacent to the start plate can be distorted and have some different microstructure due to diffusion of metal ions from base plate [28].

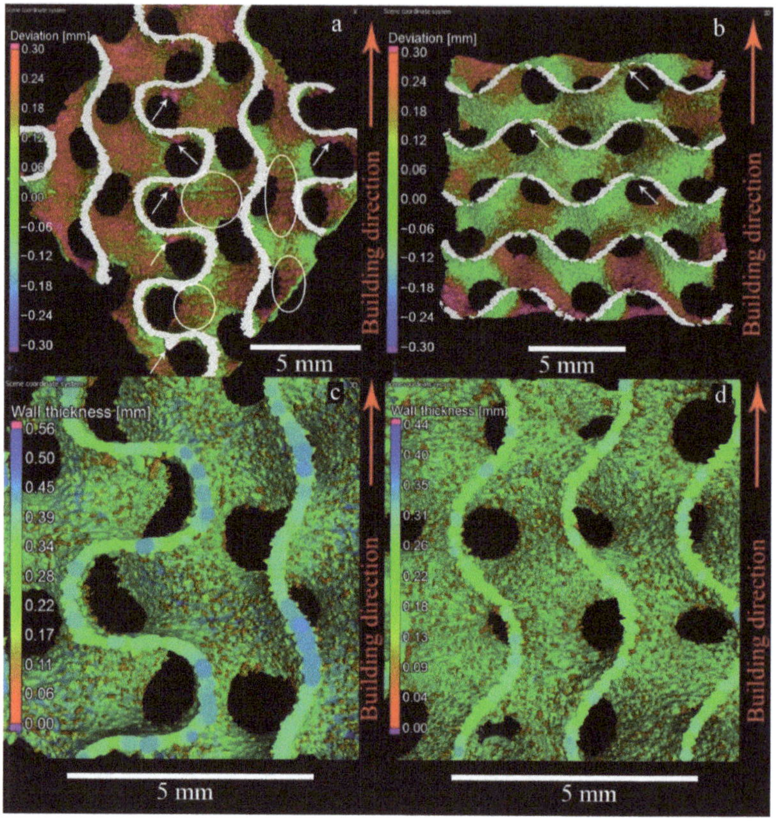

Figure 4. 3D rendering of the CT reconstructions for the Nominal/Actual Comparison of the MT gyroid (**a**) and WT gyroid (**b**). Visualization of the wall thickness for the MT gyroid (**c**) and WT gyroid (**d**). White arrows indicate stalactite-like structures on the horizontally oriented parts of the MT walls (areas parallel to the layers, purple color). The white circles highlight the partially melted powder particles attached to the surface of the vertical areas.

The through-holes are present nearly in each surface of WT samples parallel to base plate, Figure 5. The shapes of the pores are irregular; in some cases, holes are interconnected, Figure 5a,c. Red arrows in Figure 5b indicate the holes visible in the vertical cross-section of the sample. Careful investigation of the wall surfaces shows the tendency of the up-facing walls, Figure 5c, to have lower roughness than the down-facing ones, Figure 5a.

3.3. Mechanical Properties

3.3.1. Compression Tests

Stress and energy absorption vs. strain curves for the compression samples are presented in Figure 6. Characteristics of such curves for the WT samples show a larger scatter than those for the MT ones. Most probably this can be explained by the statistical variation of the shape of the hole-type defects in the WT structures (Figure 2a,b), but the intrinsic variation of the wall thickness also contributes to the scatter. The compressive strength, for instance, can vary by over 30% among samples. The first maximum compressive strength of the WT samples is about twice lower than that of MT samples.

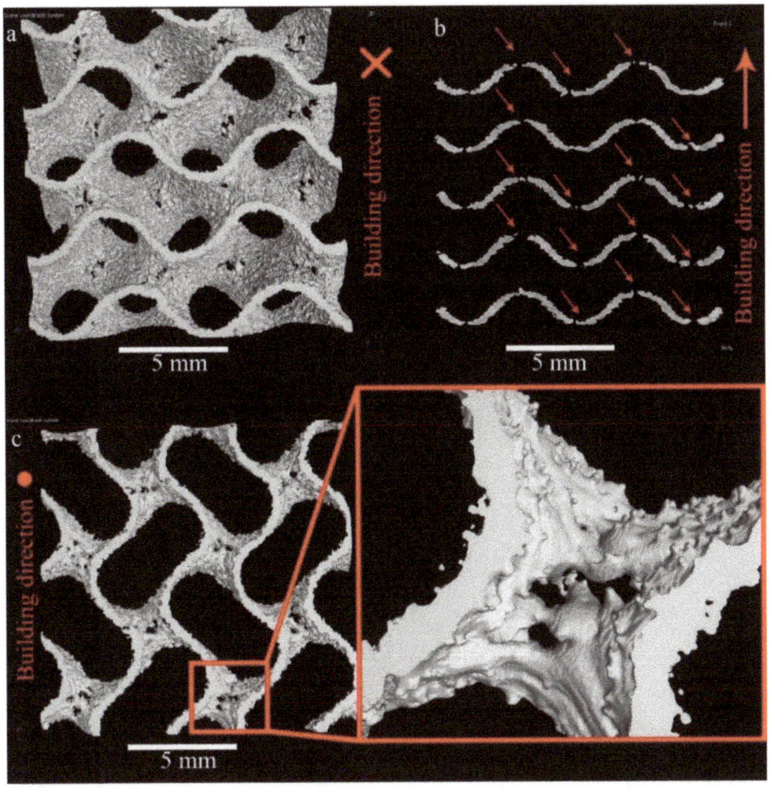

Figure 5. 3D rendering of CT-reconstruction of WT gyroid vertical (**a**,**c**) and horizontal (**b**) views (**a**—bottom view; **c**—top view; **b**—side view).

Conventional cellular structures with uniform density exhibit three deformation regimes during compressive testing [46]: a linear elastic compression, a plateau with approximately constant stress, and a final densification with steeply increasing stress. Sheet-based TPMS structures except gyroid ones are reported to have fluctuations of the curve in the plateau regime [4,6]. The mechanical behavior of our specimens does not fully follow this description. For MT specimens, a single drop and recovery in the strength in the plateau region can be observed, Figure 6a,b. The stress-strain curve of the WT shows a drop in the strength of the structure right after the peak strength is reached. This may be a result of sudden fracture of the wall element in one layer. The following fluctuations in the plateau region can be observed. According to Al-Ketan et al. [6], the fluctuations in strength can take place due to collapse or fracturing events of cell layers, while the recovery is due to local densification of the collapsed layer where the load is transmitted directly to the next layer of cells. The values of the plateau stress were calculated based on the ISO 13314 [34] (see Section 2.5). The plateau stress of WT gyroid is about 15 MPa, whereas it is about 49 MPa for MT gyroids.

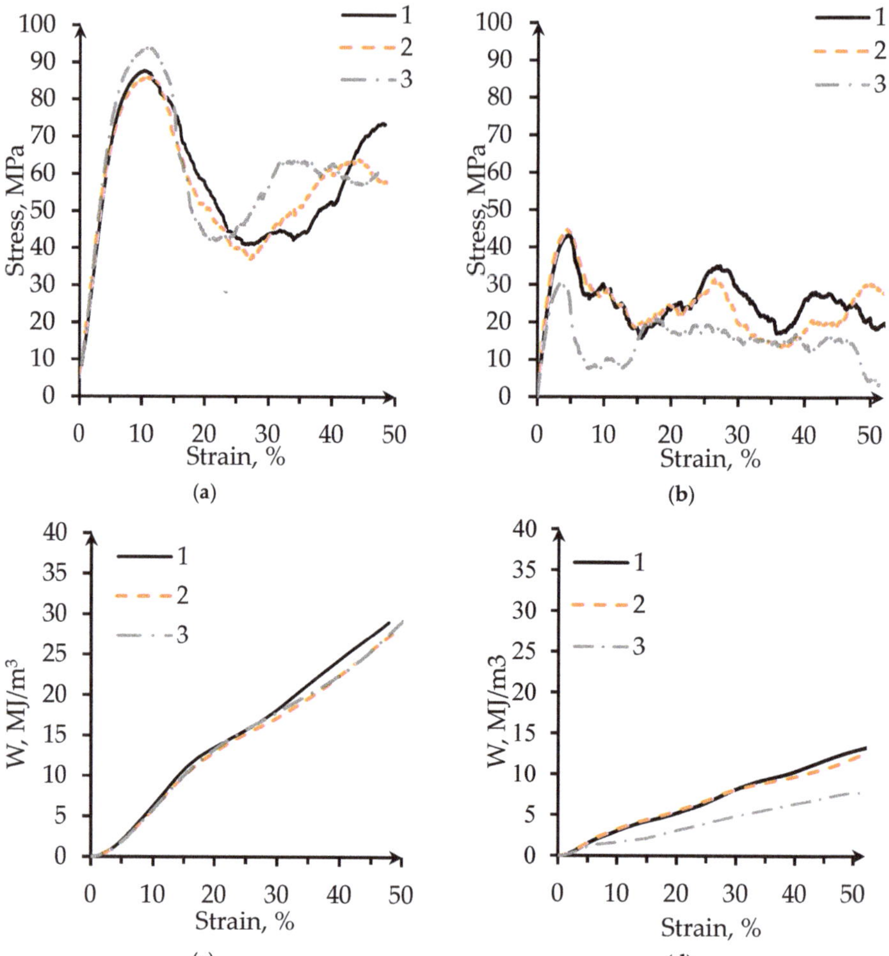

Figure 6. Compressive stress-strain curves for the gyroid samples manufactured in (**a**) MT; (**b**) WT. (Note the different y-axis scales). Energy absorption per unit volume versus strain curves for lattice samples of (**c**) MT; (**d**) WT. (Note the different y-axis scales).

Figure 7 presents photographs of the WT and MT structures at different deformation stages, corresponding to 6%, 18%, 24%, 36%, and 50% of the overall strain. Both WT and MT samples mainly display layer-by-layer deformation behavior. Plastic deformation is also visible in both sample types. Such deformation is more evident in the vicinity of the collapsing layers.

After the first maximum of compressive strength, a deep fall of the stress can be observed (Figure 6a,b). For the MT structure, this fall starts around 15% strain and corresponds to the collapse of the first layer (Figure 7b). The WT structures display several stress maxima and minima because the individual cell walls come into contact with each other after collapsing of a layer. In this case, the first minimum around 7–15% strain corresponds to the collapse of the first layer (Figure 7a). At larger strains, the stress grows again; this region is described by some authors as strain-hardening [46]. The structure becomes stronger because of the densification of the crushed layer. While losing its dimensions, the specimens recover a part of the initial crushing strength before reaching 50% strain [5].

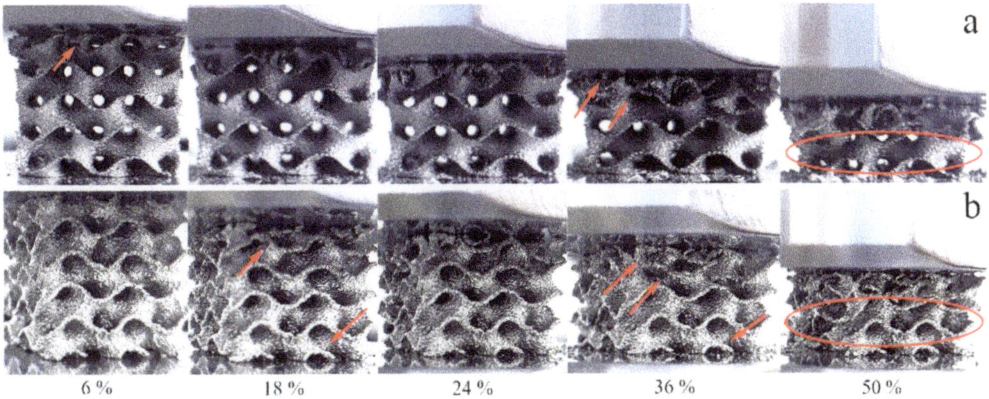

Figure 7. Steps of mechanical deformation during compression: (**a**) WT; (**b**) MT. Red arrows indicate places where the structure lost integrity during compression. The encircled layers keep integrity even at the 50% strain.

It is worth mentioning that WT specimens reach first maximum compression stress at 5% of strain, whereas MT specimens reach it at 10%. The yield strain, the compressive offset stress, and the first maximum compressive strength of the MT specimen are twice as large as that for WT specimen. Specific energy absorption at 30% strain for MT porous structure is 1.5 times higher than that for WT structure, see Figure 6c,d and Table 3. However, specific energy absorption at 50% strain for the two structures is quite similar.

Table 3. Mechanical properties of the specimens.

Parameters	Compression		Tension	
	MT	WT	MT	WT
Porosity, %	76	85	76	85
Quasi-elastic gradient E_{qe}, GPa	1.5 ± 0.1	1.5 ± 0.1	1.2 ± 0.1	1.2 ± 0.2
Compressive offset stress/Yield strength σ_y, MPa	65 ± 1	30 ± 5	37 ± 5	5 ± 0.6
Yield strain, %	4.6	2.3	3.5	0.7
First maximum compressive strength/σ_e, MPa	88 ± 2	40 ± 3		
UTS, MPa			76 ± 0.3	24 ± 0.6
Plateau stress $\sigma_{pl\,20\text{–}40}$, MPa	49 ± 2	15 ± 3	–	–
Energy absorption W_{50}, MJ/m^3	29 ± 0	11 ± 2	–	–
Specific energy absorption ψ (50%), J/g	27 ± 0	27 ± 1	–	–

3.3.2. Tensile Tests

Figure 8 shows the tensile stress-strain curves for MT and WT specimens. The tensile behavior of the samples manufactured using different Themes are similar. Ultimate tension stress (UTS) of the MT specimens is three times higher than that for WT specimens; however, their quasi-elastic gradients for both themes are equal to 1.2 GPa. The tensile failure strain of this type of specimen is about 10%.

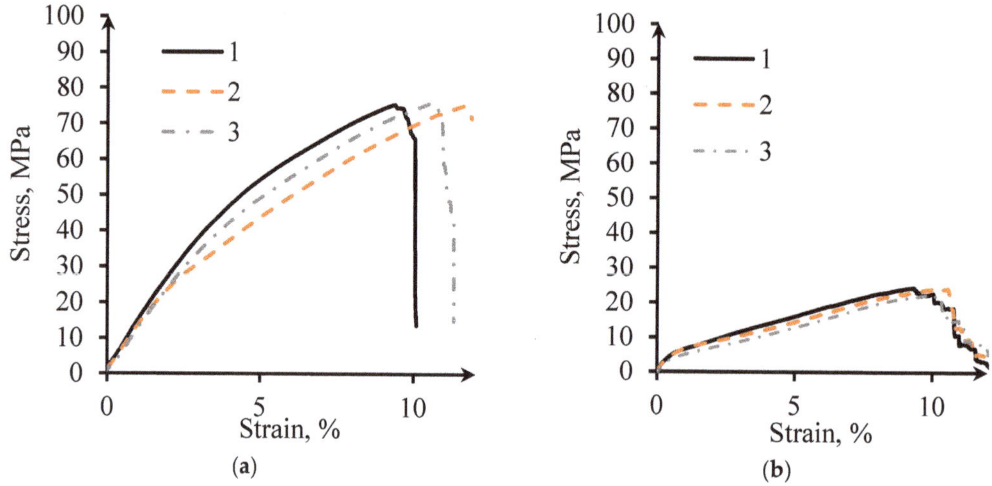

Figure 8. Tension stress-strain curves: (**a**) MT; (**b**) WT.

3.4. Finite Element Analysis

To analyze the elastic deformation process and the stress distribution during tensile tests, FE simulations were employed. The simulated stress distributions in the elastic region of a tensile test of both MT and WT models are displayed in Figure 9. The simulated stress distributions during a compression test (elastic range) were similar to tensile ones and are not presented here. In this research, a fine mesh has been used and the dependency of the results to the mesh size has been checked.

Interestingly, during elastic deformation of the structure, only the vertical elements of the gyroid wall are affected by mechanical deformation. These elements together have spiral shapes due to the specific design of the gyroid. They are stress-sensitive and reminiscent of elastic springs. This phenomenon matches the description of inclination angles influencing the manufacturability of TPMS described by Yang et al. [38]. As reported, the most frequent surface orientation within the experimental gyroid structure had an inclination angle of 55° to the build plane, while the least frequent possess around 0° inclination angle. The yielding process of the metal initiates in the vertical areas located parallel to the load direction and more actively continues in the diagonally (55°) oriented parts. The yielding process finally reaches horizontally aligned saddle points. Therefore, the holes appeared in these areas because of tessellation discrepancy influence on the mechanical behavior only after the whole specimen reaches the yield strain, and they do not affect the quasi-elastic gradient.

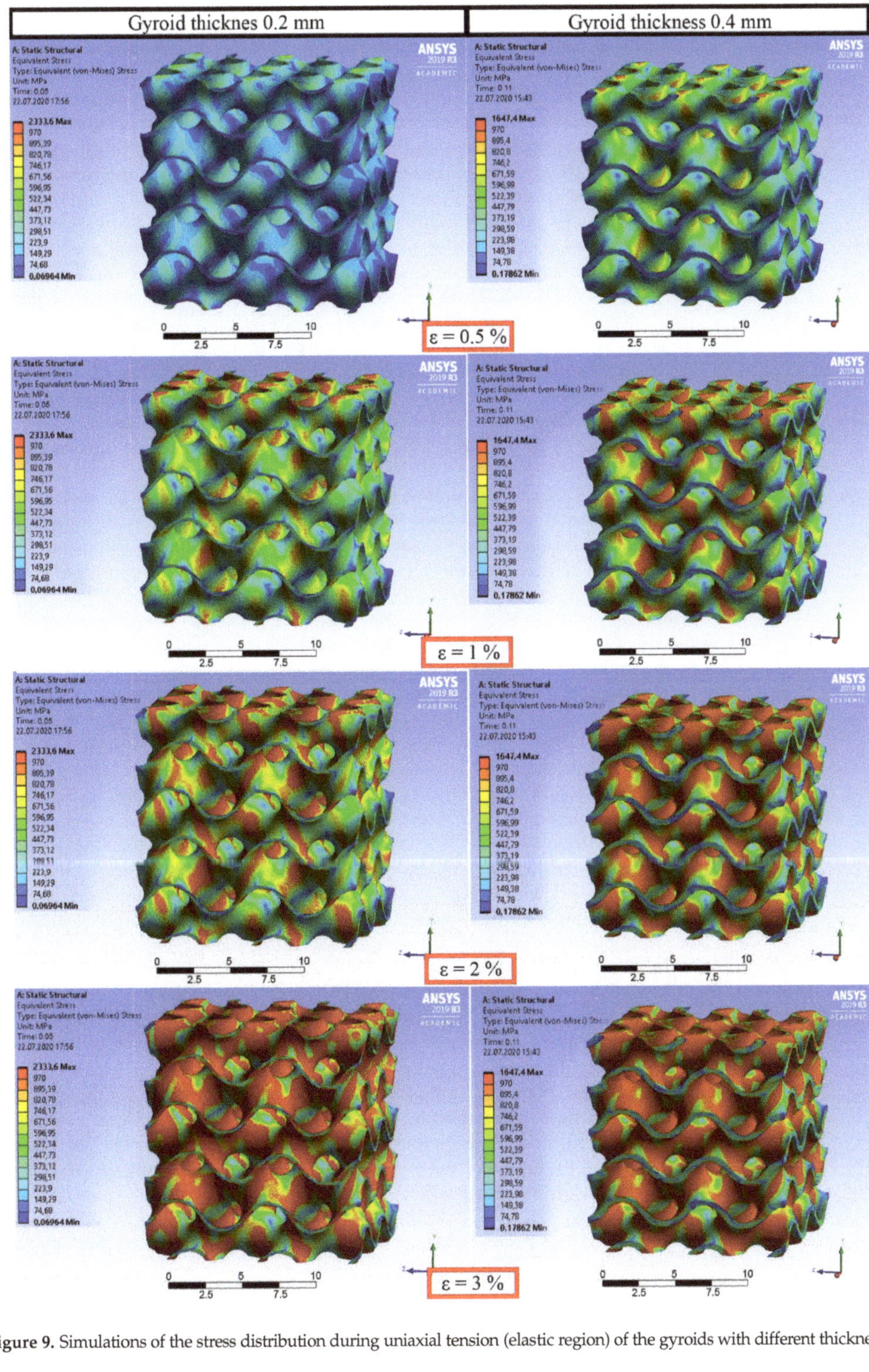

Figure 9. Simulations of the stress distribution during uniaxial tension (elastic region) of the gyroids with different thicknesses.

4. Discussion

The investigation results show that the sheet-based gyroid structures present an interesting combination of morphological properties and quasi-static mechanical behavior. Thin-wall gyroid structures obtained in EBM-manufacturing using different Themes were expected to have identical geometrical and mechanical properties due to initial settings. In the current research, the complex shape of the final gyroid samples was largely affected by the choice of the Themes, and fused powder particle distribution influencing the surface roughness and effective wall thickness. Such rough beaded surfaces are typical for AM lattice structures. Suard et al. investigated the influence of the fabrication angle on the surface roughness of EBM struts [49]. They reported that for a vertical strut, the roughness does not significantly fluctuate, while the roughness of oblique and horizontal struts increases on the down-facing surfaces. This was assumed to be due to the partial over-melting and reduced cooling, which lead to the 'leakage' of the melt and formation of the stalactite-like micro-columns and blobs (in our case indicated by white arrows in Figure 4a). It was established that for significant overhanging angles (more than 45° from horizontal), a greater number of partially molten powder particles are attached to the downward surfaces [51]. Moreover, Yang et al. demonstrated that sheet-based Ti-6Al-4V gyroids manufactured by LPBF have different surface morphology between up- and down-facing surface areas [38].

For the MT specimens, the beads are formed mainly on surfaces parallel to the building direction. Thus, the vertical walls had effective thickness up to 0.5 mm, while the oblique and horizontal walls had a thickness of 0.25 mm, Figure 4c. Typically, in the beam-like structures, there is an increase of the surface irregularities in struts produced by EBM with direction diverging from vertical [49]. However, in sheet-based structures, vertical walls are thicker, as it seems that the contours of the Melt Theme strongly influences resulting thickness of the oblique and horizontal walls. WT specimens possess quite uniform thickness for oblique and vertical wall areas, Figure 4d. Since holes are systematically present in the horizontal walls, no thickening in the WT was observed.

For porous samples, three types of compression failure are known [5]. The first type is layer-by-layer failure. This type is characterized by the collapse of cells in planes perpendicular to the manufacturing and loading direction: each layer collapses into the one below. The second type is brittle fracture of the cell walls and propagation of cracks through the lattice. Commonly, the fracture starts at a pre-existing defect, such as an internal pore of a surface irregularity. A crack can fork perpendicularly to its direction of propagation through the walls of the cells, implying that crack propagation through the structure is likely tortuous. The third type is diagonal shear. In some cases, combined diagonal shear and layer-by-layer failure occurs, for example, for sheet-based gyroid manufactured by SLS from maraging steel [6]. The compressive failure mode of the gyroid lattice is related to the size of its constituent cells.

The different stress values of the MT and WT specimens can be explained by the different relative density, while the different shape of the stress-strain curves is most probably due to the influence of the through-holes and smaller wall thickness in the WT samples, favoring Euler instability of the walls under compressive load. Such instability has a statistical character, as it occurs in single walls separately (or in groups of walls favorably oriented with respect to the load axis).

Interestingly, compression tests revealed that despite the difference in wall thickness and material volume, the two types of specimens have similar quasi-elastic gradient (around 1.5 GPa). Following a Gibson-Ashby law [52,53]:

$$E = E_0 \times (1 - P)^n \tag{5}$$

One can calculate the shape factor n using the known porosity values P and the Young's modulus of Ti6Al4V (E_0 = 110 GPa). Corresponding values are $n \sim 2.3$ for WT and $n \sim 3.2$ for MT samples. Since $n = 2$ corresponds to a cellular structure of compacted overlapping pores and $n = 3$ to a cellular structure of compacted non-overlapping pores [52], we can observe that thinning the walls not only causes an increase of porosity but also makes elastic behavior of the samples more similar to a strut-like cellular structure. The MT structure even tends to the elastic behavior of a cellular structure of overlapping spheres ($n = 4$).

Importantly, the values of the quasi-elastic gradient satisfy the limits of the elastic moduli for human cortical bone [54]. Moreover, the specific energy absorption at 50% strain for the two structures is quite similar, showing that EBM WT structures are as strain tolerant as MT.

It was demonstrated by Yang et al. that L-PBF Ti-6Al-4V gyroid sheet-based structures under compression load behave as bending-dominated structures [38], while Al-Ketan et al. [6] and Kelly et al. [35] showed stretching dominated deformation for the sheet-based gyroids made of maraging steel and Ti-6Al-4V, respectively. The linear elastic regime is driven by the bending of the inclined cell walls or by the stretching of the vertical cell walls.

Mechanical properties such as compressive offset stress, yield strain, first maximum compressive strength and energy absorption are different for sheet-based gyroid manufactured in Melt Theme and Wafer Theme. However, the quasi-elastic gradients and specific energy absorption at 50% strain of these structures are equal. Thus, WT structures are as strain tolerant as MT ones.

Comparison of the quasi-elastic gradient values (in compression) of the studied gyroid sheet-based samples to the ones reported in the literature are presented in Table 4.

Table 4. Ti-6Al-4V sheet-based gyroid properties.

Porosity, %	Wall Thickness, mm	E_{qe}, GPa	Method	Reference
70	0.25	4.0	L-PBF	[4]
50	0.33	6.0	L-PBF	[4]
78	0.25	5.3	L-PBF	[35]
85	0.25	3.0	L-PBF	[35]
87	0.25	3.4	L-PBF	[35]
86	0.25 (WT)	1.5	EBM	Current
76	0.38 (MT)	1.5	EBM	Current

L-PBF-manufactured sheet-based gyroids have quasi-elastic gradient significantly larger than EBM-manufactured ones. This can be attributed to surface roughness of the EBM-produced specimens. The highest value of quasi-elastic gradient in the Table 4 belongs to the specimen with the 50% porosity.

Thin walls allow decreasing the distance between neighboring walls, keeping the ability to effectively remove powder from the structure. In fact, it is essential that porous samples used for the mechanical and flow tests be very thoroughly cleaned from residual powder. The semi-sintered powder remaining inside the structure would influence its permeability and mechanical behavior. Additionally, thorough cleaning is critical for the applications where even slow release of loose powder during the component life is completely unacceptable, as in the case of implants. All TPMS structures used in this study were successfully cleaned with conventional powder recovery system (see e.g., [55]).

Using FE simulation, it was shown that relatively small number of through-holes occurring in thin-walled structures does not influence the elastic behavior of the lattice. In case of the WT structures, through-holes are wider and more abundant in the surface sections having small inclination to the layer plane (Figure 10). Two reasons could be responsible for that: first one, related to technology, and second one, to the design process. Technologically, overhanging elements without supports are always problematic in PBF AM. From the point of view of design, since the 3D model (CAD) for WT samples has zero thickness, the part of the STL file corresponding to this problem area cannot be registered by slicer software. The tangent point represents, therefore, a blind spot, where the electron beam could be turned off (Figure 10a). Further studies into this phenomenon are needed, and corresponding corrections of the zero-thickness design model should be incorporated in the future.

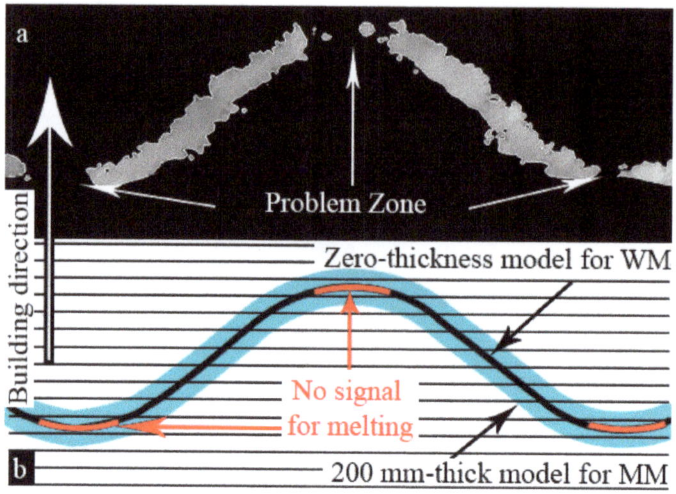

Figure 10. Assumption on the through-holes' appearance: (**a**) CT in vertical cross-sectional view; (**b**) Scheme of manufacturing process in vertical cross-sectional view.

It is interesting that the MT specimens, with porosity of 76% and wall thickness about 0.4 mm, have UTS of about 76 MPa, while L-PBF gyroid with the same porosity and wall thickness of 0.5 mm demonstrates UTS of 60 MPa [35]. Sheet-based gyroid manufactured by L-PBF from Ti-6Al-4V with wall thickness of 0.25 mm and porosity varying from 78 to 87% can possess quasi-elastic gradient from 1.9 to 5 GPa, depending on the unit cell size [35]. The UTS for a gyroid with $6 \times 6 \times 6$ mm unit cell and porosity of 87% was found to be equal to 24 MPa [35], equal to the results of the tensile testing of the WT gyroid (Table 3). However, the tensile strain at failure for EBM gyroid is approximately 10% while tensile failure strain of L-PBF gyroid with wall thickness of 0.25 mm ranges from 1.4% to 2.1%.

Achieving minimal thickness of the TPMS sheet-based lattices would yield many advantages: the structure would be lighter; it would be easier to remove entrapped powder; the cell size could be decreased, thereby allowing larger gradients. At the end of powder bed fusion processes, residual powder is always trapped in the final lattice. In the case of implants, if the powder will not be removed, inflammation and blockage of blood vessels may happen. To ensure efficiency of the dense porous structures in terms of bone cells' ingrowth and mechanical properties, an adequate method for the trapped powder removal is still to be found. We assume that in the case if the gyroid sheet-based structures will be produced using Wafer Theme, the size of the unit cell can be decreased while maintaining the ability of effective powder removal from the structure. Moreover, the Wafer Theme can be implemented for manufacturing functionally graded porous structures with optimized pore size. Finally, such structures satisfy the requirements for medical implants to avoid stress-shielding effect.

In conclusion, TPMS WT structures present a few advantages (high strain tolerance in both tension and compression, relatively high quasi-elastic gradient, relatively low surface roughness), and a few caveats (systematic presence of holes at the saddle points of the structure), but prove to be well adapted to mimic human cortical bone. Therefore, such structures certainly merit more attention and further investigations as potential design geometries for load-tolerant lightweight structures used in biomedicine (implants) and technology.

5. Conclusions

TPMS gyroid sheet-based structures were, for the first time, successfully manufactured by EBM using Melt and Wafer Themes (MT and WT, respectively). To produce specimens in MT, a 0.2 mm thick model was used, while to produce specimens in WT, a zero-thickness model was used. It was initially assumed that in using proper input design parameters, corresponding EBM-manufactured MT and WT samples would have equivalent geometrical and mechanical properties. However, we observed that:

- The minimum mean wall thickness, which can be achieved using standard Melt Theme in ARCAM EBM A2 machine, is around 380 μm, while the minimum mean wall thickness with Wafer Theme is 250 μm.
- Despite the difference in thickness, quasi-elastic gradient and specific energy absorption at 50% strain are approximately the same. Thus, MT and WT structures behave identically at small strains up to 5% (in the elastic range) and have similar strain tolerance.
- WT gyroids exhibit through-hole defects in the surface sections perpendicular to the building direction. They supposedly appear in each horizontal saddle point because the areas of zero-thickness 3D model are not detected by slicing software and, therefore, are not processed by the beam. Through-holes connect two separate void regions which TPMS consist of, thus, enabling better fluid transport, tissue ingrowth and differentiation.
- FEA simulation revealed that the yielding of the metal initiates in the vertical areas located parallel to the load direction and continues in the diagonally oriented surfaces. The yielding process reaches horizontally aligned saddle points only at a later stage. Therefore, the through-holes influence the mechanical behavior only in the plastic region.
- Thus, the Wafer Theme EBM-manufacturing is a promising method for TPMS-based structures.

Author Contributions: Conceptualization, D.K., S.E., T.M., R.S., M.S., A.K., and G.B.; methodology, D.K., M.K., S.E., T.M., W.S., and A.K.; CT measurements, D.M.; software D.K., S.E., T.M., D.M., K.M., and D.C., investigation, D.K., M.K., S.E., T.M. and D.M.; writing—original draft preparation, D.K.; writing—review and editing, D.K., S.E., T.M., R.S., M.S., A.K., A.P., D.C., and G.B. All authors have read and agreed to the published version of the manuscript.

Funding: The research was performed at Tomsk Polytechnic University within the framework of the Tomsk Polytechnic University Competitiveness Enhancement Program grant. The work was supported by the Ministry of Science and Higher Education of the Russian Federation (State Project "Science" No. WSWW-2020-0011). Sample manufacturing was supported by the EU Regional Development Fund and Swedish Agency for Economic and Regional Growth Grant No. 20202610. D. Khrapov would like to thank Kirill Evdokimov for support at modeling. The research at BAM was supported by the G-RISC program funded by the German Federal Foreign Office via the German Academic Exchange Service (DAAD) (Funding Decision No. M-2020a-6).

Institutional Review Board Statement: Not applicable.

Informed Consent Statement: Not applicable.

Data Availability Statement: The data presented in this study are available on request from the corresponding author. The data are not publicly available as they also form part of an ongoing study.

Conflicts of Interest: The authors declare no conflict of interest.

References

1. Niinomi, M.; Nakai, M. Titanium-Based Biomaterials for Preventing Stress Shielding between Implant Devices and Bone. *Int. J. Biomater.* **2011**, *2011*, 10. [CrossRef]
2. Ahmadi, S.M.; Yavari, S.A.; Wauthle, R.; Pouran, B.; Schrooten, J.; Weinans, H.; Zadpoor, A.A. Additively Manufactured Open-Cell Porous Biomaterials Made from Six Different Space-Filling Unit Cells: The Mechanical and Morphological Properties. *Materials* **2015**, *8*, 1871–1896. [CrossRef] [PubMed]
3. Zadpoor, A.A. Bone Tissue Regeneration: The Role of Scaffold Geometry. *Biomater. Sci.* **2014**, *3*, 231–245. [CrossRef]

4. Bobbert, F.S.L.; Lietaert, K.; Eftekhari, A.A.; Pouran, B.; Ahmadi, S.M.; Weinans, H.; Zadpoor, A.A. Additively Manufactured Metallic Porous Biomaterials Based on Minimal Surfaces: A Unique Combination of Topological, Mechanical, and Mass Transport Properties. *Acta Biomater.* **2017**, *53*, 572–584. [CrossRef] [PubMed]
5. Maskery, I.; Aboulkhair, N.T.; Aremu, A.O.; Tuck, C.J.; Ashcroft, I.A. Compressive Failure Modes and Energy Absorption in Additively Manufactured Double Gyroid Lattices. *Addit. Manuf.* **2017**, *16*, 24–29. [CrossRef]
6. Al-ketan, O.; Rowshan, R.; Al-Rub, R.K.A. Topology-Mechanical Property Relationship of 3D Printed Strut, Skeletal, and Sheet Based Periodic Metallic Cellular Materials. *Addit. Manuf.* **2018**, *19*, 167–183. [CrossRef]
7. Liu, F.; Zhang, D.Z.; Zhang, P.; Zhao, M.; Jafar, S. Mechanical Properties of Optimized Diamond Lattice Structure for Bone Scaffolds Fabricated via Selective. *Materials* **2018**, *11*, 374. [CrossRef] [PubMed]
8. Melchels, F.P.W.; Bertoldi, K.; Gabbrielli, R.; Velders, A.H.; Feijen, J.; Grijpma, D.W. Mathematically Defined Tissue Engineering Scaffold Architectures Prepared by Stereolithography. *Biomaterials* **2010**, *31*, 6909–6916. [CrossRef] [PubMed]
9. Luca, A.; Di Longoni, A.; Criscenti, G.; Mather, M.L.; Morgan, S.P.; White, L.J. Surface Curvature in Triply-Periodic Minimal Surface Architectures as a Distinct Design Parameter in Preparing Advanced Tissue Engineering Scaffolds. *Biofabracation* **2017**, *9*, 12.
10. Maskery, I.; Sturm, L.; Aremu, A.O.; Panesar, A.; Williams, C.B.; Tuck, C.J.; Wildman, R.D.; Ashcroft, I.A.; Hague, R.J.M. Insights into the Mechanical Properties of Several Triply Periodic Minimal Surface Lattice Structures Made by Polymer Additive Manufacturing. *Polymer* **2018**, *152*, 62–71. [CrossRef]
11. Yan, C.; Hao, L.; Hussein, A.; Raymont, D. Evaluations of Cellular Lattice Structures Manufactured Using Selective Laser Melting. *Int. J. Mach. Tools Manuf.* **2012**, *62*, 32–38. [CrossRef]
12. Yánez, A.; Cuadrado, A.; Martel, O.; Afonso, H.; Monopoli, D. Gyroid Porous Titanium Structures: A Versatile Solution to Be Used as Scaffolds in Bone Defect Reconstruction. *Mater. Des.* **2018**, *140*, 21–29. [CrossRef]
13. Yánez, A.; Herrera, A.; Martel, O.; Monopoli, D.; Afonso, H. Compressive Behaviour of Gyroid Lattice Structures for Human Cancellous Bone Implant Applications. *Mater. Sci. Eng. C* **2016**, *68*, 445–448. [CrossRef]
14. Ataee, A.; Li, Y.; Fraser, D.; Song, G.; Wen, C. Anisotropic Ti-6Al-4V Gyroid Scaffolds Manufactured by Electron Beam Melting (EBM) for Bone Implant Applications. *Mater. Des.* **2018**, *137*, 345–354. [CrossRef]
15. Abueidda, D.W.; Elhebeary, M.; Shiang, C.A.; Pang, S.; Al-Rub, R.K.A.; Jasiuk, I.M. Mechanical Properties of 3D Printed Polymeric Gyroid Cellular Structures: Experimental and Finite Element Study. *Mater. Des.* **2019**, *165*, 107597. [CrossRef]
16. Dalaq, A.S.; Abueidda, D.W.; Al-Rub, R.K.A.; Jasiuk, I.M. Finite Element Prediction of Effective Elastic Properties of Interpenetrating Phase Composites with Architectured 3D Sheet Reinforcements. *Int. J. Solids Struct.* **2016**, *83*, 169–182. [CrossRef]
17. Kapfer, S.C.; Hyde, S.T.; Mecke, K.; Arns, C.H.; Schröder-Turk, G.E. Minimal Surface Scaffold Designs for Tissue Engineering. *Biomaterials* **2011**, *32*, 6875–6882. [CrossRef] [PubMed]
18. Aremu, A.; Maskery, I.; Tuck, C.; Ashcroft, I.; Wildman, R.; Hague, R. A comparative finite element study of cubic unit cells for selective laser melting. In Proceedings of the 25th Annual International Solid Freeform Fabrication Symposium (SFF Symp 2014), Austin, TX, USA, 4–6 August 2014; pp. 1238–1249.
19. Zhao, X.; Li, S.; Zhang, M.; Liu, Y.; Sercombe, T.B.; Wang, S.; Hao, Y.; Yang, R.; Murr, L.E. Comparison of the Microstructures and Mechanical Properties of Ti–6Al–4V Fabricated by Selective Laser Melting and Electron Beam Melting. *Mater. Des.* **2016**, *95*, 21–31. [CrossRef]
20. Wang, P.; Nai, M.; Tan, X.; Vastola, G.; Raghavan, S.; Sin, W.J.; Tor, S.B.; Pei, Q.X.; Wei, J. Recent Progress of Additive Manufactured Ti-6Al-4V by Electron Beam Melting. In Proceedings of the 26th Annual International Solid Freeform Fabrication Symposium (SFF Symp 2016), Austin, TX, USA, 8–10 August 2016; pp. 691–704.
21. Negi, S. Review on Electron Beam Based Additive Manufacturing. *Rapid Prototyp. J.* **2020**, *26*, 485–498. [CrossRef]
22. Zhang, L.; Liu, Y.; Li, S.; Hao, Y. Additive Manufacturing of Titanium Alloys by Electron Beam Melting: A Review. *Adv. Eng. Mater.* **2017**, *20*, 1700842. [CrossRef]
23. Wong, H. Bitmap Generation from Computer-Aided Design for Potential Layer-Quality Evaluation in Electron Beam Additive Manufacturing. *Rapid Prototyp. J.* **2020**, *26*, 941–950. [CrossRef]
24. Pal, S.; Lojen, G.; Gubeljak, N.; Kokol, V.; Drstvensek, I. Melting, Fusion and Solidification Behaviors of Ti-6Al-4V Alloy in Selective Laser Melting at Different Scanning Speeds. *Rapid Prototyp. J.* **2020**, *26*, 2–9. [CrossRef]
25. Zäh, M.F.; Lutzmann, S. Modelling and Simulation of Electron Beam Melting. *Prod. Eng.* **2010**, *4*, 15–23. [CrossRef]
26. Srivastava, S.; Garg, R.K.; Alba-baena, N.G. Multi-Physics Continuum Modelling Approaches for Metal Powder Additive Manufacturing: A Review. *Rapid Prototyp. J.* **2020**, *26*, 737–764. [CrossRef]
27. Xiang, D.; Tan, X. Comparison of Wear Properties of Ti6Al4V Fabricated by Wrought and Electron Beam Melting Processes in Simulated Body Fluids. *Rapid Prototyp. J.* **2020**, *26*, 959–969. [CrossRef]
28. Suard, M. Characterization and Optimization of Lattice Structures Made by Electron Beam Melting. Ph.D. Thesis, Université Grenoble Alpes , Grenoble, France, 2015.
29. Wolfram Research. *Wolfram Mathematica 12*; Wolfram Research, Inc.: Champaign, IL, USA, 2020.
30. Cignoni, P.; Callieri, M.; Corsini, M.; Dellepiane, M.; Ganovelli, F.; Ranzuglia, G. MeshLab: An Open-Source Mesh Processing Tool. In Proceedings of the Sixth Eurographics Italian Chapter Conference, Salerno, Italy, 2–4 July 2008; pp. 129–136.
31. Riegel, J.; Mayer, W.; Van Havre, Y. FreeCAD, (Open-Source Software, Version 0.18.4). Available online: https://www.freecadweb.org (accessed on 15 May 2019).

32. Stedfeld, R.L. *ASM Handbook—Properties and Selection: Nonferrous Alloys and Special-Purpose Materials*; ASM International: Russel Township, OH, USA, 1998; Volume 2.
33. Weitkamp, T.; Haas, D.; Wegrzynek, D.; Rack, A. ANKA Phase: Software for Single-Distance Phase Retrieval from Inline X-Ray Phase-Contrast Radiographs. *J. Synchrotron Radiat.* **2011**, *18*, 617–629. [CrossRef]
34. International Organization for Standartization. *Mechanical Testing of Metals. Ductility Testing. Compression Test for Porous and Cellular Metals*; International Organization for Standartization: Geneva, Switzerland, 2011.
35. Kelly, C.N.; Francovich, J.; Julmi, S.; Safranski, D.; Robert, E.; Maier, H.J.; Gall, K. Fatigue Behavior of As-Built Selective Laser Melted Titanium Scaffolds with Sheet-Based Gyroid Microarchitecture for Bone Tissue Engineering. *Acta Biomater.* **2019**, *94*, 610–626. [CrossRef]
36. Plessis, A.; Razavi, S.M.J.; Berto, F. The Effects of Microporosity in Struts of Gyroid Lattice Structures Produced by Laser Powder Bed Fusion The Effects of Microporosity in Struts of Gyroid Lattice Structures Produced by Laser Powder Bed Fusion. *Mater. Des.* **2020**, *194*, 108899. [CrossRef]
37. Ma, S.; Tang, Q.; Hanb, X.; Feng, Q.; Song, J.; Setchic, R.; Liu, Y.; Liu, Y.; Tsee, Y.; Zhend, N.; et al. Manufacturability, Mechanical Properties, Mass-Transport Properties and Biocompatibility of TPMS Scaffolds Fabricated by Selective Laser Melting. *Mater. Des.* **2020**, *195*, 109034. [CrossRef]
38. Yang, E.; Leary, M.; Lozanovski, B.; Downing, D.; Mazur, M.; Sarker, A.; Khorasani, A.; Jones, A.; Maconachie, T.; Bateman, S.; et al. Effect of Geometry on the Mechanical Properties of Ti-6Al-4V Gyroid Structures Fabricated via SLM: A Numerical Study. *Mater. Des.* **2019**, *184*, 108165. [CrossRef]
39. Van Hooreweder, B.; Apers, Y.; Lietaert, K.; Kruth, J.P. Improving the Fatigue Performance of Porous Metallic Biomaterials Produced by Selective Laser Melting. *Acta Biomater.* **2017**, *47*, 193–202. [CrossRef]
40. Wauthle, R.; Van Der Stok, A.; Humbeeck, J.; Van Kruth, J.; Abbas, A.; Weinans, H.; Mulier, M.; Schrooten, J. Additively Manufactured Porous Tantalum Implants. *Acta Biomater.* **2015**, *14*, 217–225. [CrossRef]
41. Choy, S.Y.; Sun, C.N.; Leong, K.F.; Wei, J. Compressive Properties of Functionally Graded Lattice Structures Manufactured by Selective Laser Melting. *Mater. Des.* **2017**, *131*, 112–120. [CrossRef]
42. Burton, H.E.; Eisenstein, N.M.; Lawless, B.M.; Jamshidi, P.; Segarra, M.A.; Addison, O.; Shepherd, D.E.T.; Attallah, M.M.; Grover, L.M.; Cox, S.C. The Design of Additively Manufactured Lattices to Increase the Functionality of Medical Implants. *Mater. Sci. Eng. C* **2019**, *94*, 901–908. [CrossRef]
43. Harris, J.A.; Winter, R.E.; McShane, G.J. Impact Response of Additively Manufactured Metallic Hybrid Lattice Materials. *Int. J. Impact Eng.* **2017**, *104*, 177–191. [CrossRef]
44. Tancogne-Dejean, T.; Spierings, A.B.; Mohr, D. Additively-Manufactured Metallic Micro-Lattice Materials for High Specific Energy Absorption under Static and Dynamic Loading. *Acta Mater.* **2016**, *116*, 14–28. [CrossRef]
45. Lei, Y.; Yan, C.; Song, B.; Wen, S. Compression-Compression Fatigue Behaviour of Gyroid-Type Triply Periodic Minimal Surface Porous Structures Fabricated by Selective Laser Melting. *Acta Mater.* **2019**, *181*, 40.
46. Yang, L.; Mertens, R.; Ferrucci, M.; Yan, C.; Shi, Y.; Yang, S. Continuous Graded Gyroid Cellular Structures Fabricated by Selective Laser Melting: Design, Manufacturing and Mechanical Properties. *Mater. Des.* **2019**, *162*, 394–404. [CrossRef]
47. Guo, N.; Leu, M.C. Additive Manufacturing: Technology, Applications and Research Needs. *Front. Mech. Eng.* **2013**, *8*, 215–243. [CrossRef]
48. Wauthle, R.; Vrancken, B.; Beynaerts, B.; Jorissen, K.; Schrooten, J.; Kruth, J.P.; Van Humbeeck, J. Effects of Build Orientation and Heat Treatment on the Microstructure and Mechanical Properties of Selective Laser Melted Ti6Al4V Lattice Structures. *Addit. Manuf.* **2015**, *5*, 77–84. [CrossRef]
49. Suard, M.; Martin, G.; Lhuissier, P.; Dendievel, R.; Vignat, F.; Blandin, J.J.; Villeneuve, F. Mechanical Equivalent Diameter of Single Struts for the Stiffness Prediction of Lattice Structures Produced by Electron Beam Melting. *Addit. Manuf.* **2015**, *8*, 124–131. [CrossRef]
50. Leary, M.; Mazur, M.; Elambasseril, J.; Mcmillan, M.; Chirent, T.; Sun, Y.; Qian, M.; Easton, M.; Brandt, M. Selective Laser Melting (SLM) of AlSi12Mg Lattice Structures. *Mater. Des.* **2016**, *98*, 344–357. [CrossRef]
51. Pyka, G.; Kerckhofs, G.; Papantoniou, I.; Speirs, M.; Schrooten, J.; Wevers, M. Surface Roughness and Morphology Customization of Additive Manufactured Open Porous Ti6Al4V Structures. *Materials* **2013**, *6*, 4737–4757. [CrossRef]
52. Gibson, L.J.; Ashby, M.F. The Mechanics of Three-Dimensional Cellular Materials. *Proc. R. Soc. A Math. Phys. Eng. Sci.* **1982**, *382*, 43–59.
53. Bruno, G.; Efremov, A.M.; Levandovskyi, A.N.; Clausen, B. Connecting the Macro- and Microstrain Responses in Technical Porous Ceramics: Modeling and Experimental Validations. *J. Mater. Sci.* **2011**, *46*, 161–173. [CrossRef]
54. Katsamanis, F.; Raftopoulos, D.D. Determination of Mechanical Properties of Human Femoral Cortical Bone by the Hopkinson Bar Stress Technique. *J. Biomech.* **1990**, *23*, 1173–1184. [CrossRef]
55. Khrapov, D.; Koptyug, A.; Manabaev, K.; Léonard, F.; Mishurova, T.; Bruno, G.; Cheneler, D.; Loza, K.; Epple, M.; Surmenev, R.; et al. The Impact of Post Manufacturing Treatment of Functionally Graded Ti6Al4V Scaffolds on Their Surface Morphology and Mechanical Strength. *J. Mater. Res. Technol.* **2019**, *9*, 1866–1881. [CrossRef]

materials

MDPI

Article

Microstructural Analysis and Mechanical Properties of $TiMo_{20}Zr_7Ta_{15}Si_x$ Alloys as Biomaterials

Adriana Savin [1,*], Mihail Liviu Craus [1,2,*], Alina Bruma [3], František Novy [4], Sylvie Malo [5], Milan Chlada [6], Rozina Steigmann [1], Petrica Vizureanu [7], Christelle Harnois [5], Vitalii Turchenko [2] and Zdenek Prevorovsky [6]

1 Nondestructive Testing Department, National Institute for Research and Development for Technical Physics, 700050 Iasi, Romania; steigmann@phys-iasi.ro
2 Frank Laboratory for Neutron Physics, Joint Institute for Nuclear Research, Dubna 141980, Russia; turchenko@jinr.ru
3 National Institute of Standards and Technology, Gaithersburg, MD 20899, USA; bruma.alina@outlook.com
4 Department of Materials Engineering, University of Zilina, 010 26 Zilina, Slovak Republic; frantisek.novy@fstroj.uniza.sk
5 Normandie Université, ENSICAEN, UNICAEN, CNRS, CRISMAT, 14000 Caen, France; sylvie.malo@ensicaen.fr (S.M.); christelle.harnois@ensicaen.fr (C.H.)
6 Institute of Thermomechanics, Academy of Sciences of the Czech Republic, 182 00 Prague, Czech Republic; chlada@it.cas.cz (M.C.); zp@it.cas.cz (Z.P.)
7 Faculty of Materials Science and Engineering, Technical University Gheorghe Asachi, 700050 Iasi, Romania; peviz2000@yahoo.com
* Correspondence: asavin@phys-iasi.ro (A.S.); crausmihailliviu@gmail.com (M.L.C.); Tel.: +40-232-430680 (A.S.)

Received: 24 September 2020; Accepted: 23 October 2020; Published: 28 October 2020

Abstract: TiMoZrTaSi alloys appertain to a new generation of metallic biomaterials, labeled high-entropy alloys, that assure both biocompatibility as well as improved mechanical properties required by further medical applications. This paper presents the use of nondestructive evaluation techniques for new type of alloys, $TiMo_{20}Zr_7Ta_{15}Si_x$, with x = 0; 0.5; 0.75; 1.0, which were obtained by vacuum melting. In Ti alloys, the addition of Mo improves tensile creep strength, Si improves both the creep and oxidation properties, Zr leads to an α crystalline structure, which increases the mechanical strength and assures a good electrochemical behavior, and Ta is a β stabilizer sustaining the formation of solid β-phases and contributes to tensile strength improvement and Young modulus decreasing. The effects of Si content on the mechanical properties of the studied alloys and the effect of the addition of Ta and Zr under the presence of Si on the evolution of crystallographic structure was studied. The influence of composition on fracture behavior and strength was evaluated using X-ray diffraction, resonant ultrasound spectroscopy (RUS) analyses, SEM with energy dispersive X-ray spectroscopy, and acoustic emission (AE) within compression tests. The β-type $TiMo_{20}Zr_7Ta_{15}Si_x$ alloys had a good compression strength of over 800 MPa, lower Young modulus (69.11–89.03 GPa) and shear modulus (24.70–31.87 GPa), all offering advantages for use in medical applications.

Keywords: titanium alloys; microstructure; mechanical properties; resonant ultrasound spectroscopy; acoustic emission; inhomogeneities

1. Introduction

Implanted prosthetics are regulated by the FDA [1] as class III devices after demonstrating reasonable assurance of safety and efficacy. All metal devices, regardless of alloying elements, are subjected to a variety of non-clinical tests (bench and/or animal) followed by specific FDA documentation or national/international standards.

The clinical success of modern medical implants depends on the integration with the bone to resist functional loading [2]. Their mechanical properties depend on the manufacturing process or thermal and mechanical treatments, processes that can change the microstructure [3]. The stainless steels (iron-based alloys), Co and Ti alloys are very useful for medical implants. As an alternative to those metals, commercially pure titanium (cpTi) and its alloys were promoted due to their physical and mechanical features being of interest in prosthodontics or for endosseous implants, as well as plates and screws for bones fixtures. Recent research in material science and continuous technological developments have stimulated innovations in design and the diversification of alloys used in implant fabrication [4].

Used in physiological conditions, alloys are exposed to mechanical and biological factors that can affect long-term use. Some materials for implants contain metals that are normally found in small quantities in the human body, having an essential role in biological functioning (Cu, Mn, Co, Zn, Ca, etc.), having a critical role for structure, and/or proper functioning. Disruption of biological functions in the presence of implants can occur through associated signs or symptoms, when both essential and non-essential metals are present in concentrations which are either too high or too low. These biological events can be avoided by identifying and evaluating all factors that influence biocompatibility [5] according to ISO standards [6].

Titanium presents few crystallographic phases in biocompatible alloys due the high corrosion strength and adequate hardness/weight ratio [7].

In pure Ti, the modification of the crystalline β-phase cannot be obtained at ambient temperature, even by quenching with very high cooling rates; the β-phase passes into the α′-hexagonal form through a martensitic type of transformation. The difference between implant and bone stiffness can lead to the β-phase based implant weakening [8]. The contemporary and complex micro-alloyed systems research has been focused on reducing the rigidity and the effect of stress shielding. For titanium alloys, α, α + β, and β categories are known according to their equilibrium constitution, function of the type of alloying elements, and ratios [9].

Ti and Mo are compatible elements that, when combined, can form binary alloys leading to stable crystal structures, optimal for biomedical integration. Part of the elements used for titanium alloys ensure β stabilization, such as Mo (which improves the tensile strength of creep) [10] or Si (which improves the creep and oxidation properties). The β-phase proportion in Ti alloys affects the mechanical and corrosive properties [11,12], therefore, a compromise needs to be achieved between strength and plasticity, by alloying the TiMoSi ternary alloy with stabilizing elements for medical prostheses [13]. Previously published studies [14,15] focused on determining the properties of $TiMo_ySi_x$ alloys with x = 0; 0.5; 0.75; 1.0 and y = 15; 20; 25, and showed that Mo assures the stability of β-Ti (spatial group Im–3m) by lowering the transition temperature of α-Ti to β-Ti to ambient temperature [16], and ensuring a fine structure of titanium alloy. TiMoZrTaSi takes part from the new generation of metallic biomaterials labeled high-entropy alloys (HEAs) [17], that assure both biocompatibility as well as improved mechanical properties required by further medical applications. Zr and Ta act as consolidation elements: the alloying with Zr leads to an α crystalline structure which increases the mechanical strength and assures a good electrochemical behavior; Ta is a β stabilizer sustaining the formation of solid β-phases in the titanium alloy. The addition of Ta contributes to the improving of tensile strength and the decrease in the Young modulus.

The studied samples are from $TiMo_{20}Si_x$ alloys with x = 0; 0.5; 0.75; 1.0 and doped with 7 wt.% Zr and 15 wt.% Ta. Selection of alloying elements was made according to [18], referring to bond order (B_o) values and metal d-orbital energy level (M_d) on the phase stability diagram.

Combining characterization techniques such as energy-dispersive spectroscopy (EDS) and X-ray diffraction (XRD) analysis as well as nondestructive evaluation (NDE) with ultrasound and resonant ultrasound spectroscopy (RUS), the physical properties of these compounds were evaluated. After that, acoustic emission (AE) was carried out during the compression testing.

This paper presents few nondestructive testing methods for selected biocompatible alloys $TiMo_ySi_x$ when Mo percent is 20 wt.% and Si varies from 0 to 1.0 wt.% in the presence of Ta and Zr doping. The mechanical properties were noninvasively evaluated. The studies have been completed with the analysis of crystal structure of samples and microstructural characterization based on their dependence by microstructure, phase composition and the alloying elements content, knowing that the evolution of Ti alloy microstructures during the production has a special importance [8].

2. Materials and Methods

2.1. Samples Preparation

In [15], the elastic modulus of $TiMo_{15}Zr_7Ta_5$ and $TiMo_{15}Si_{0.5}$ are compared. It has shown that, for a lower elastic modulus comparable with the human bone, a combination of ratios of alloying elements in TiMoZrTa is required, so that the value should reach that of cortical bone.

Using raw materials with high purity as starting materials, mini-ingots were prepared, with the composition empirically designed as to reduce elastic modulus and increase the compatibility of $TiMo_{20}Zr_7Ta_{15}Si_x$ with elastic properties of the cortical bone. The procedure of alloying was developed by the Faculty of Materials Science and Engineering, Gh. Asachi Technical University, Iasi, Romania [16]. In [19], a modeling approach is presented for understanding the hardening mechanisms of Ti alloys, seeking to evaluate the contribution of the effect of solute elements to the hardening mechanism. For binary and ternary β metastable titanium alloys (Ti-Mo; Ti-Mo-Si), the neutral alloying additions as Ta and Zr assure the formation of microstructures in β-phase and improve tensile strength while decreasing the Young modulus [20], and improving the strength at fatigue crack propagation and deformation. Doping with Zr and Ta has a minor influence on the β-transition temperature, but together with Si qualitatively influences the stabilization of the β-phase. Si has low solubility in both the α- and β-phases, contributing to hardening and decreased ductility, but can improve creep resistance at high temperatures, increase corrosion resistance and increase fluidity during casting. The stabilizing effect strongly depends on the wt.% of the other alloying elements in the compound, Ta being known to cause solid solution hardening [21], however the experimental results cannot be discernible due to the transition phases and changes in the deformation mechanisms [22].

The mini-ingot quaternary TiMoZrTa alloys investigated were obtained from raw precursor elements, melted in a vacuum arc remelting VAR MRF ABJ 900 furnace (Allenstown, NH, USA) in high purity Ar atmosphere and a vacuum (4×10^{-3} mbar); which assured temperature control. The remelting was carried out with minimal overheating to reduce the segregation [23]. The mini-ingots were obtained by melting (at temperatures higher than 3000 °C), then remelted between 2 to 7 times to assure their homogeneity.

2.2. Surface Analysis; Microstructure

The samples were fixed in EpoFix Resin (Struers, Rotherham, UK) and their surface was polished and etched to reveal the microstructure. The microstructure was observed on sections, after mechanical processing, by optical microscopy using AxioCam MRc5-Zeiss (Jena, Germany) and by electron microscopy using SEM JEOL JSM7200F (JEOL Ltd., Tokyo, Japan), equipped with a Bruker EDS analyzer (BRUKER AXS Inc., Madison, WI, USA).

After that, the samples were re-polished and the Vickers micro hardness HV0.5 was measured. Few methods for material characterization (i.e., phase identification via X-ray diffraction (XRD) and optical microscopy) were employed. Tests to identify the phase composition and the type of crystal structure of crystallographic phase were carried out by XRD—Xpert Pro MPD PANalytical diffractometer (Phillips, New York, NY, USA) with CoKα radiation, and a Bruker AXS D8 Advance diffractometer (Karlsruhe, Germany) with Cu-Kα radiation, respectively [16].

2.3. Ultrasound Measurements

The longitudinal (C_l) and transverse (C_t) ultrasound speeds of waves through the samples were determined using the method described in [14] using two sets of wide-band transducers: for the longitudinal wave sensor, G5 KB GE [14]; and for the transversal wave sensor, MB4Y GE. The PR 5077 pulse/receiver system was used for emission and reception of the signals. The timing of flight and digitizing of the signal were carried out with digital oscilloscope Wave Runner 64Xi, LeCroy (LeCroy Corporation, Chestnut Ridge, NY, USA). Elastic properties (E, G and ν) were calculated from the ultrasound (US) time of flight.

2.4. Mechanical Properties

2.4.1. Resonant Ultrasound Spectroscopy (RUS)

RUS was employed to evaluate the quality of the studied samples. RUS implies the exploration of the resonant structure of a compact sample on the basis of resonance frequency modification according to ASTM E2001-18 [24], for the detection and assessment of variations and mechanical properties of a test object. RUS techniques can employ small samples, where the evaluation is made from the entire volume; the method can also be applied for phase transition studies [25]. The resonance frequencies depend on mechanical properties (size, shape, density) and on the sample's elasticity. Most resonance frequencies of a cube do not respond to analytical solutions, even if it is isotropic. However, there are analytical solutions offered by Mindlin-Lamé [26] and Damarest, whose formulations offer the possibility to manage anisotropy. For small samples of the order of centimeters, RUS is applicable for allowing the connection of low frequency stress strain methods and ultrasound time delay. Each frequency corresponds to a propagating wave beam, the maximum of resonance modes having 1/f period. (f is the resonance frequency). RUS is frequently combined with Finite Element Analysis (FEA) to identify the resonance frequency, f_r. Typically, in conventional RUS, the sample is fixed between the sensors by "edge-to edge" contact [27], assuring a weak coupling with the sensors. An emission transducer (US Em) and a receiver transducer (US Re) type P111.O.06P3.1 with high bandwidth were used to carry out the experiment. The experimental assembly included an Agilent 4395A Network/Spectrum/Analyzer generating a frequency sweep in 1 kHz steps between 120–200 kHz, as shown in Figure 1.

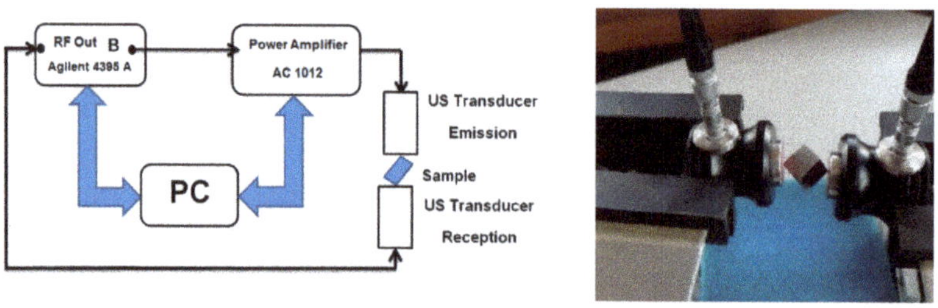

(a) (b)

Figure 1. Resonant ultrasound spectroscopy (RUS) experimental set-up: (**a**) basic diagram; (**b**) detail of sample fixture.

2.4.2. Acoustic Emission (AE)

AE is a highly accurate method for detecting active microscopic events in samples and for detecting the initiation/propagation of a crack [14]. In order to test the mechanical performances of $TiMo_{20}Si_xZr_7Ta_{15}$ alloys designed and prepared in the form of specimens with different chemical composition, AE measurements were carried out during compression loading. Structural modifications in titanium alloys during compression loading were accompanied by different frequencies noises; AE analysis method showed information useful for the identification of failed mechanisms.

The samples were tested by AE during compressive tests on an INSTRON 1195 with 0.2 mm/min displacement speed. The compression tests were carried out up to an 80 kN load (the limit of the equipment was 100 kN). Crack and microcrack initiation and propagation during compressive deformation generated AE events, with the elastic waves propagating through the sample towards the piezoelectric transducers and converting to electric signals. These signals were amplified and distinguished after root-mean-square (RMS) calculations, the number of thresholds was counted (N_c), count rate ($C_x = dN_c/dt$), intensity parameters, frequency spectrum, signal waveform, amplitude, etc., [27–30].

C_x was evaluated as the number of AE signal crossings over the voltage level x, allowing the suppression of extraneous noise and the enhancement of higher energy processes. AE activity depends on damage and transformation processes during the compression of samples.

$$RMS = \left(\frac{1}{T} \int_0^T a^2(t)dt \right)^{\frac{1}{2}} \quad (1)$$

The RMS value is proportional to the emitted acoustic energy, which mostly characterizes running fracture processes such as micro-crack jumps in the sample.

The strength of materials can be predicted by recording the AE events from the initiation of microcracks [15,31–33]. The AE sensors (two DAEL IDK09) could not be applied directly to the sample, due its small dimensions, thus, they were glued onto support plates, acting as waveguides (Figure 2).

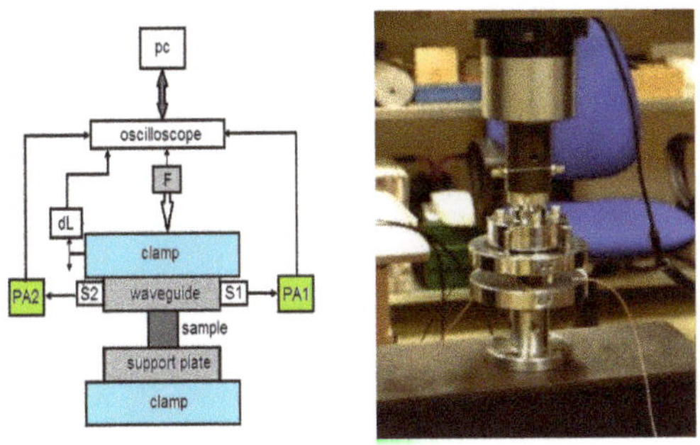

(a) (b)

Figure 2. Experimental set-up of acoustic emission (AE): (**a**) basic diagram; (**b**) positioning of the sensors on the support plate ensuring a waveguide.

Signals from the sensors were amplified by 20 and 40 dB, respectively, in two preamplifiers. Continuous AE signals recording with 14-bit amplitude resolution, and 2.5 MHz sampling by USB oscilloscope were employed.

Standard microindentation hardness (HV) was measured according to ASTM E384 [34] with a ZwickRoell ZHμ Vickers micro-hardness tester with a loading of 500 gf/10 s, calculating the average of 6 measurements at 1 mm distance for each sample.

3. Results and Discussions

3.1. Characterization of the Samples

3.1.1. Mechanical Properties of Studied Alloys

The mechanical properties determined by the ultrasound method are presented in Table 1.

Table 1. Mechanical properties, dependent on Si content (x) for $TiMo_{20}Zr_7Ta_{15}Si_x$.

Sample	Composition	Density ρ [g/cm^3]	Young Modulus E [GPa]	Shear Modulus G [GPa]	Poisson Ratio ν	C_l [m/s]	C_t [m/s]	HV0.5
#1	$TiMo_{20}$	5.126 ± 0.004	109.23 ± 1.037	40.15 ± 0.329	0.36	5993 ± 0.7	2799 ± 0.5	278 ± 1.2
#2	$TiMo_{20}Si_{0.75}$	5.078 ± 0.006	111.51 ± 0.898	41.30 ± 0.304	0.35	5993 ± 0.5	2852 ± 0.4	339 ± 0.8
#3	$TiMo_{20}Zr_7Ta_{15}$	7.049 ± 0.009	91.18 ± 0.248	32.61 ± 0.692	0.39	5222 ± 0.4	2151 ± 0.5	388 ± 1.7
#4	$TiMo_{20}Zr_7Ta_{15}Si_{0.5}$	6.845 ± 0.009	89.03 ± 0.456	31.87 ± 0.640	0.39	5210 ± 0.4	2158 ± 0.3	337 ± 1.1
#5	*$TiMo_{20}Zr_7Ta_{15}Si_{0.75}$	5.404 ± 0.001	69.11 ± 0.478	24.70 ± 0.260	0.39	5215 ± 0.4	2138 ± 0.4	354.8 ± 1.9
#6	$TiMo_{20}Zr_7Ta_{15}Si_{1.0}$	6.868 ± 0.005	78.78 ± 0.297	27.91 ± 0.276	0.41	5191 ± 0.5	2016 ± 0.4	356 ± 0.9

OBS * it is possible that after increasing the Si content over 0.75 wt.%, a supersaturation and distortions of the lattice appeared in the matrix, modifying the distance between atoms, and leading to modification of the elastic modulus.

The results also include the binary alloy $TiMo_{20}$ in order to emphasize the influence of doping with Ta and Zr on the density and Young modulus, as well as the influence of Si on the mechanical properties.

Table 1 shows that, compared to the binary/ternary alloy (sample #1 and #2), the presence of doping with 15 wt.% Ta (sample #3–#6) made the Young modulus decrease, thus the alloy was more elastic; a result that is in agreement with [35]. It is observed that the presence of Zr and Ta increased the density of the binary/ternary alloy, due to the large difference in density of Zr [36], while Si in both binary and ternary structures resulted in a decrease in density.

3.1.2. Microstructure of Studied Alloys

XRD is used to investigate the surface layer up to 10 μm depth. Using XRD analysis, the crystal structures and lattice parameters of the constituent phases were determined.

Table 2 shows the Si concentration (x), average size of crystalline blocks (D), the value of microstrains (ε), lattice constants observed (a) and calculated (a_{calc}), and unit cell volume (V) [16]. Here, B_o represents the covalent bond strength and M_d represents metal d-orbital energy level for $TiMo_{20}Zr_7Ta_{15}Si_x$ (x = 0.0; 0.5; 0.75; 1.0) alloy samples at room temperature. Table 3 presents the atomic concentration of alloying elements, and lattice constants and densities calculated using PowderCell and/or FullProf programs. The lattice constants were calculated from the number of atoms on the main diagonal of the elementary cell.

Table 2. Dependence of crystallographic structure on the Si concentration of samples of $TiMo_{20}Zr_7Ta_{15}Si_x$.

Sample	(x) Mass Concentration	a = b= c * (Å)	V (Å3)	D (Å) **	[ε ***]	B_o	M_d (eV)	ρ *** (g/cm³)
#3	0.0	3.2785	35.239	204	0.001427	2.918	2.396	5.964
#4	0.50	3.2804	35.300	339	0.000844	2.917	2.395	5.931
#5	0.75	3.2818	35.346	790	0.000630	2.917	2.395	5.894
#6	1.00	3.2663	34.847	437	0.000967	2.916	2.394	5.952

* Lattice constants observed and determined using PowderCell. ** The average crystalline length and size of the microstrains were determined by means of the PowderCell program. *** Densities were calculated using the nominal compositions and experimentally determined constant lattices (PowderCell).

Table 3. Atomic concentration of elements $\left(C_i^{Ti}, C_i^{Mo}, C_i^{Zr}, C_i^{Ta}, C_i^{Si}\right)$, calculated lattice constants (a_{calc} *) and calculated densities (ρ **).

Sample	$C_i^{Ti}(\%)$	$C_i^{Mo}(\%)$	$C_i^{Zr}(\%)$	$C_i^{Ta}(\%)$	$C_i^{Si}(\%)$	a_{calc} * (Å)	ρ ** (g/cm³)
#3	76.70	13.20	4.90	5.20	0	3.2714	6.003
#4	75.69	13.14	4.84	5.22	1.12	3.26620	6.008
#5	74.68	13.07	4.81	5.20	2.23	3.2551	6.041
#6	73.69	13.01	4.79	5.18	3.33	3.2477	6.055

* Lattice constants calculated by using the atomic radii of Ti, Si, Mo, Zr and Ta. ** Calculated densities with nominal compositions and calculated lattice constants (PowderCell).

The main phase of the TiMoSi alloys is a Ti solid solution, being characterized by space group Im–3m. Increasing Si content leads to a decrease in the lattice constant when only one phase is present at the substitution of Ti with Si, Mo, Zr and Ta. Here, however, a maximum of the lattice constant was observed for x = 0.75. The calculated lattice value was obtained using the following equation:

$$a_i^{calc} = \frac{4}{\sqrt{3}}(C_i^{Ti} * 1.40 + C_i^{Mo} * 1.45 + C_i^{Zr} * 1.55 + C_i^{Ta} * 1.45 + C_i^{Si} * 1.10) \quad (2)$$

The interpretation of the lattice constant values shown in Table 2, up to a Si concentration of 0.75 wt.% can be seen, as Si did not substitute with Ti, and probably formed some Si-based compounds. This supposition is justified by the presence of some small and broad maxima on the diffractograms.

The calculated density from the experimental lattice constants was very close to the calculated density of the theoretical values of lattice constants (Table 3). We assumed that the weight of the unit cell was given by the nominal formula of the alloys. The design of titanium alloys and prediction of β stability and deformation mechanisms were suggested in [37] and is based on the calculations of electron structures. The electron parameters which predict the alloy stability are B_o and M_d, in relation with the electric properties and metallic radius of atoms [19,38–42].

For an alloy, the average values $\overline{B}_o$ and $\overline{M}_d$ are given by compositional averages [41]

$$\overline{B}_o = \sum_{i=1}^{5} x_i B_{oi} \tag{3}$$

and, respectively,

$$\overline{M}_d = \sum_{i=1}^{5} x_i M_{di} \tag{4}$$

where x_i represents the atomic weight/atomic mass, and the B_{oi}, M_{di} values are of each ith element in the alloy composition. Mo, Ta and Zr will contribute to the obtaining of high values for B_{oi}, and therefore are added to the design of a good β-Ti alloy, with improved properties suitable for biomedical applications, such as reducing the elasticity modulus. According to [38], the values of the average B_o and the average M_d indicate that all casted alloys are in the stable β region. That is in agreement with the phase composition of the present alloys, which contain practically only the β-phase, with small concentrations of extrinsic phases (Figure 3).

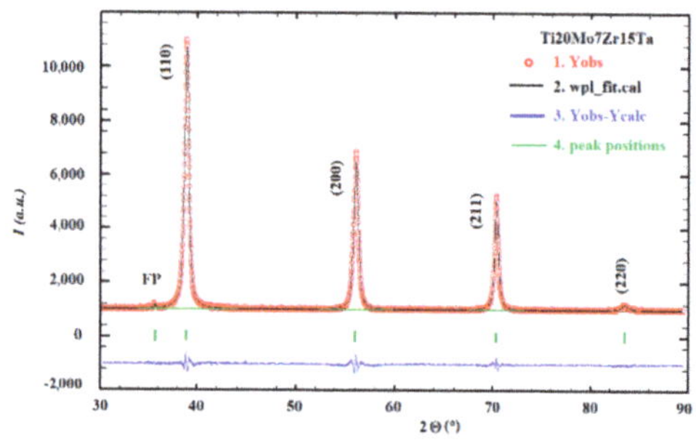

(**a**)

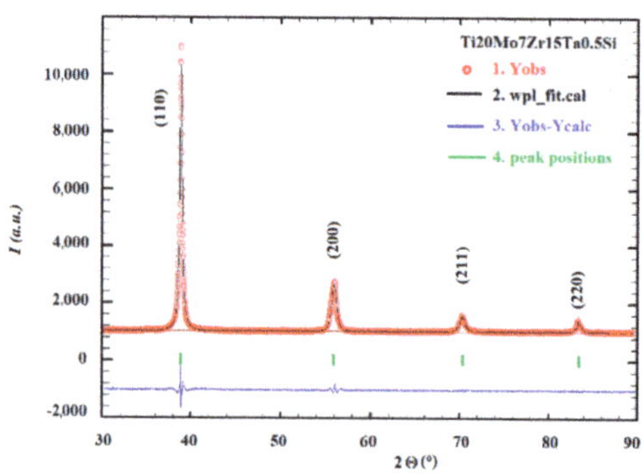

(**b**)

Figure 3. *Cont.*

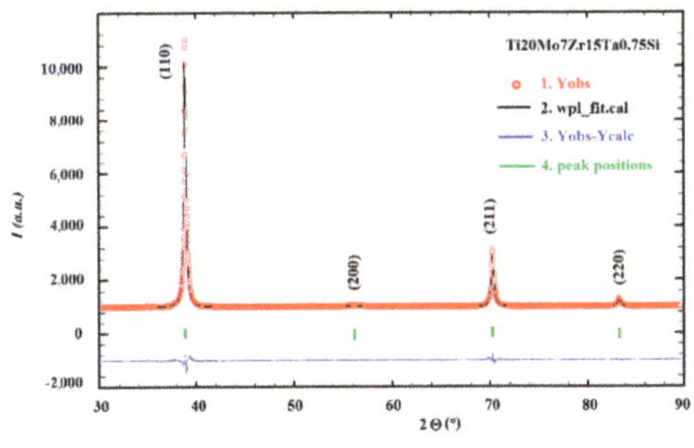

(c)

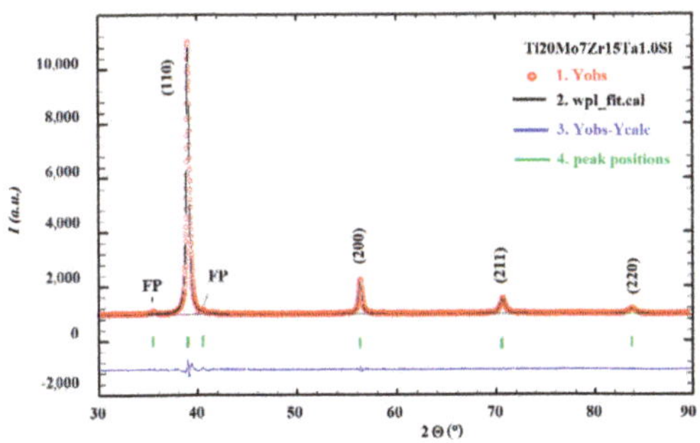

(d)

Figure 3. The observed (red), calculated (black) and the difference between the observed and calculated (blue) diffractograms of $TiMo_{20}Zr_7Ta_{15}Si_x$ corresponding to: (**a**) x = 0.0; (**b**) x = 0.5; (**c**) x = 0.75; (**d**) x = 1.0.

It is possible that some maxima, due to some compounds with Si, cannot be identified due to the small dimensions of the mosaic blocks or very small concentrations.

The results emphasize that all the samples contained a low number of extrinsic phases; these phases are represented by two/one maxima, at the limit of the intensity errors. We considered a maximum when its intensity was at least three times larger than the intensity error.

Figure 4 presents the microstructure and semi-quantitative composition spectrum for the $TiMo_{20}Zr_7Ta_{15}$ sample, showing a dendritic structure, irregular grain boundaries with several segregations, and large grains from dendritic colonies. The SEM and EDS analyses were performed at 15 kV, with a working distance of 9.9 mm.

The dendrite structures in Ti samples obtained by casting can be observed in Figure 5. A redistribution of the constituent elements during the solidification resulted in the formation of a dendritic structure.

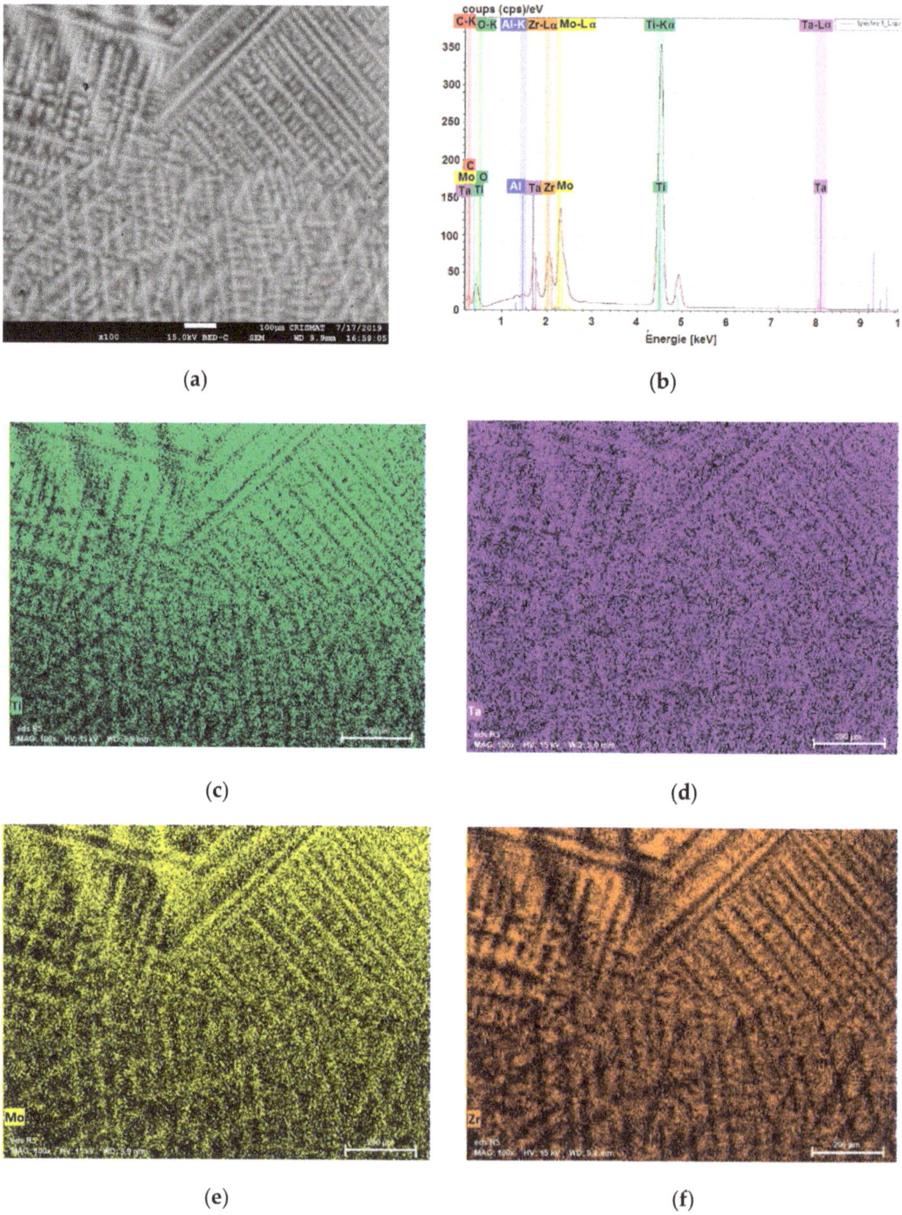

Figure 4. Microstructure and semi-quantitative composition spectrum for the sample #3: (**a**) SEM image; (**b**) EDS spectrum; (**c**–**f**) EDS maps of Ti, Ta, Mo, Zr—the scale is 200 µm.

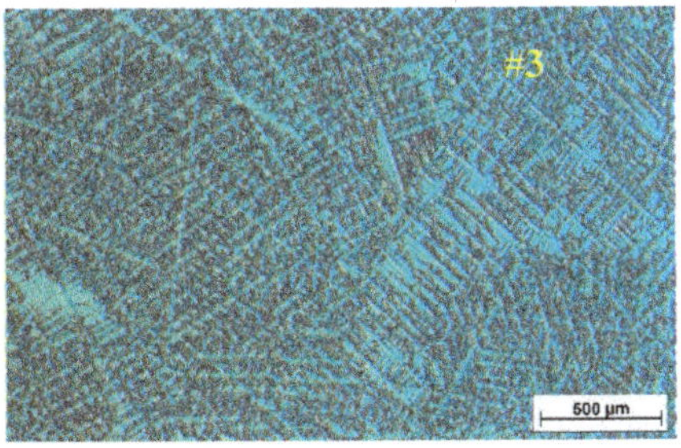

(a)

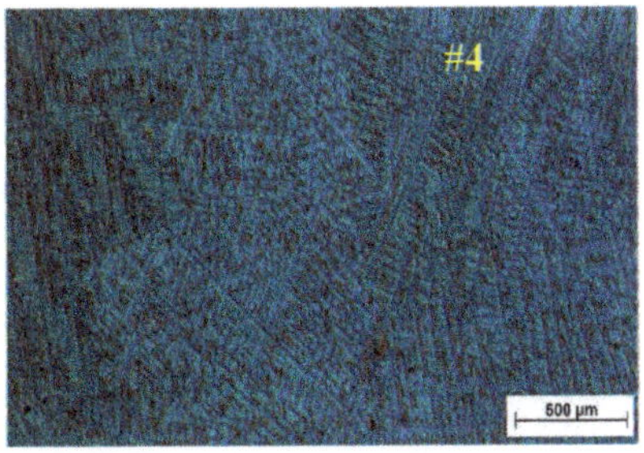

(b)

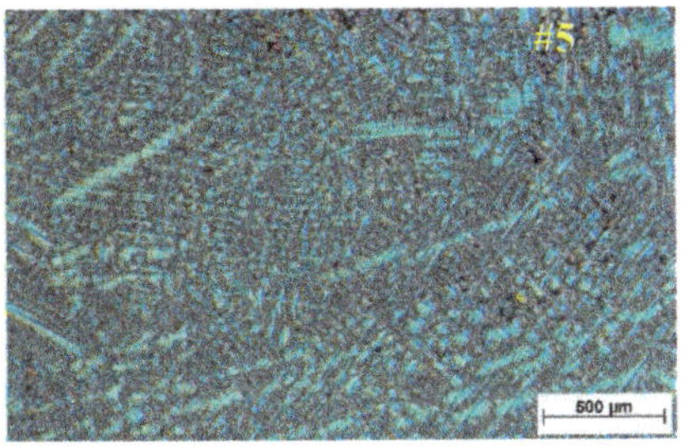

(c)

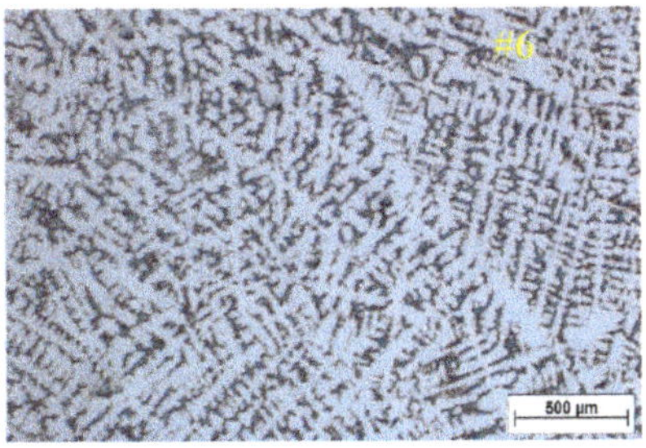

(d)

Figure 5. Optical microstructures for $TiMo_{20}Zr_7Ta_{15}Si_x$ samples with: (**a**) x = 0; (**b**) x = 0.5; (**c**) x = 0.75; (**d**) x = 1.0.

A non-uniaxial fine dendrite structure of $TiMo_{20}Zr_7Ta_{15}Si_x$ samples is depicted in Figure 5. Due to the small dimensions of cast ingots, those dendrite structures were independent of the sample position and orientation. A tendency of evolution of the distance of the main dendrites axis dependent on Si content can be observed, but cannot be quantified using standard metallographic methods due to the random dendrite orientation.

It is also observed that, in addition to the structural body of the main dendrite with a white contrast, these structures appear with an inhomogeneous light gray contrast near the black regions visible in the Figure 4 EDS maps. These trends of microstructures can be explained by the tendency of Ti and Zr to be abundant in the inter-dendritic region [43], (as well as the tendency of Ta and Mo to be more abundant in dendrites).

3.2. RUS for Elasticity Measurements

RUS is used to determine the elastic properties of compact samples with regular geometries based on resonance frequencies [44,45]. According to [46] the elastic constants are connected to resonance frequency. In order to find resonant modes, a frequency sweep was made with a frequency generator in a range established for each sample by simulation. The fundamental mode's frequency [46] is $f = \frac{m}{2h}\sqrt{\frac{G}{\rho}}$ with h as the side of the cube, G as the shear modulus, ρ as the density and m as an integer. To use this method, it is useful to determine the sample's eigen frequencies and then to obtain an FEM model equivalent of eigen frequencies calculations, and finally, minimizing the object function (F) [47]:

$$F = \sum_i w_i \left(f_i^{(p)} - f_i^{(m)} \right)^2 \quad (5)$$

where $f^{(p)}$ and $f^{(m)}$ are the computed and measured frequencies, respectively, and w_i is the weight showing the reliability of the measurements. RUS is not entirely an experimental method, it presents an inverse problem [48], which most often implies a conjugate gradient method [49] minimizing the object function (F) using the determined frequencies. The symmetry produces more resonant modes at the same resonance frequency (f_r). The presence of inhomogeneities can be detected by the modification of resonance spectra or by the variation of resonance frequencies (f_r), and the splitting of the modes may be correlated with the dimensions of inhomogeneities [50].

Figure 6 presents the resonance spectra for samples #1; #3; #5 and #6 in the range from 120 to 220 kHz.

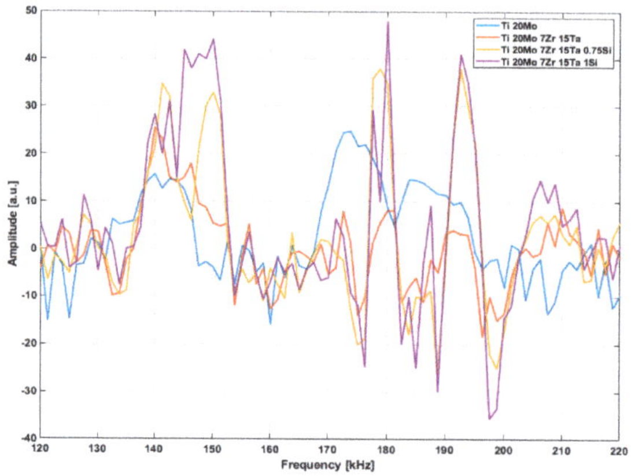

Figure 6. Resonance ultrasound spectra for samples shows the frequency shift, peaks splitting, and magnitude increasing.

The proposed method locates the fundamental frequency, knowing that the resonant modes for metals are sharp. Figure 6 shows that the samples had a different response in the amplitude and range of eigen frequencies, which can be correlated with the elastic properties of the samples, dimensions of grains and the crystalline structures (Table 2). The intervals of eigen frequencies present slight displacement (Table 2), and the resonant frequencies in the spectrum do not appear as separated peaks.

It can be observed that with the increase in Si content of over 0.75 wt.%, the frequency spectrum modifies its amplitude and shape and sharp peaks appear, which is specific to Si present in a Ti matrix, which is in good correlation with the values presented in Table 1. In addition, this denotes a modification of grain structure that seems to slightly decrease with the increase in Si content and leads to increasing hardness (Table 2). This modification of the structure, smaller than that of pure Ti, may be due to the hardening of the solid solution induced by the addition of Mo. The results show that the mechanical properties for the alloys were improved when the addition of Si was up to 0.75 wt.%.

The resonance frequencies were predicted by FEM simulation in SolidWorks 2019, a mesh of 28,264 nodes and 19,287 total elements. The typical response for the two modes obtained by simulation, extensional and flexural, for the studied samples #3 and #5 in a 120–250 kHz frequency range, was determined according the physical properties of samples [16].

From the RUS spectrum, it can be observed that, for sample #5, both extensional f = 189 kHz and flexural f = 151 kHz modes are shifted according to mechanical parameters from Table 1, as well as with the crystallographic structure from Table 2.

Figure 7 presents the elastic properties according to Table 1, for samples studied, showing that these values reached a minimum for Si content of 0.75 wt.%, as ν = 0.39 and G = 24.70 GPa.

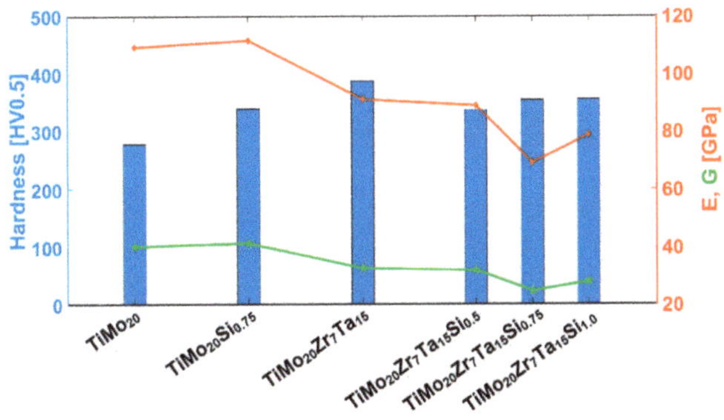

Figure 7. Mechanical properties of samples as a function of Si content.

The results of RUS and US measurements can be in good concordance for homogeneous samples with high density and without voids and pores. Both methods assure the analysis in the volume of samples.

3.3. AE Response During Compressive Test

AE is used for monitoring the possible damage during compression tests, and assessing the influence of the microstructure over the final mechanical properties of titanium alloys. From data recorded during AE compression tests, the threshold crossing, RMS/s, and the level of stress and strain available at loading machine analog outputs are calculated [14].

To illustrate the spectral content of burst AE signals, an approach has been developed (described below), solving the problem of the extraction of AE hits, which can be relatively noisy and time-overlapping. As a simple tool, the thresholding of the signal envelope was performed (Figure 8). The continuous signal envelope was calculated as the convolution of the originally sampled signal with the Hamming window of an appropriately chosen length. AE hit impact was then defined as a signal part (light-green colored in Figure 8) corresponding to the multiple of time period T, in which the signal envelope was higher than detection level (set slightly above the noise). Multiplication coefficient, k, was set around the value $k = 1.3$ to capture the AE hit impact decay as well.

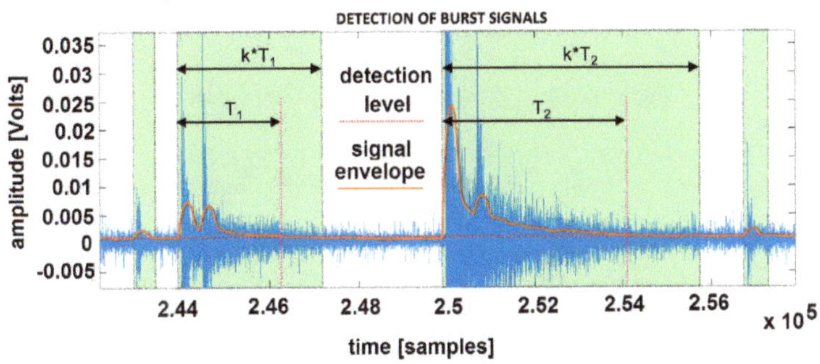

Figure 8. HD spectrogram computation.

After the extraction of AE hits, the "correlation" method of spectral analysis was applied, considering the unsuitability of classical Fast Fourier Transform (FFT), namely the variable length of signals and resulting problems with common frequency resolutions. Analogically, compared to Fourier

Transform, it is possible to compute the measure p_{HD} representing the inherence of frequency $\underline{f_{\mathrm{rel}}}$ in signal *s*:

$$p_{HD}(f_{rel}) = \left|\sum_{n=0}^{N-1} s(n)e^{i\Pi n f_{rel}}\right|^2 \tag{6}$$

where arbitrary chosen frequencies $f_{rel} \in (0,1)$ are relative to the Nyquist frequency, Nq. All computations were made accordingly with the selected frequency resolution 1 kHz, while Nq = 500 kHz. Correlation values p_{HD} for those 500 frequencies represented the estimate of a particular AE hit periodogram. To show the global changes of the spectral content of AE hits, partial periodograms of extracted hits detected within each second of the experiment were summed into corresponding column vectors of HD Power Spectral Distribution (PSD), and figured out as an alternative to classical spectrograms (Figure 8). By the above-mentioned method, only the parts of continuous signal with significant AE bursts were taken into account. In Figure 9, the spectral analysis of AE activities of Ti alloys for sample #5 and #6 are presented.

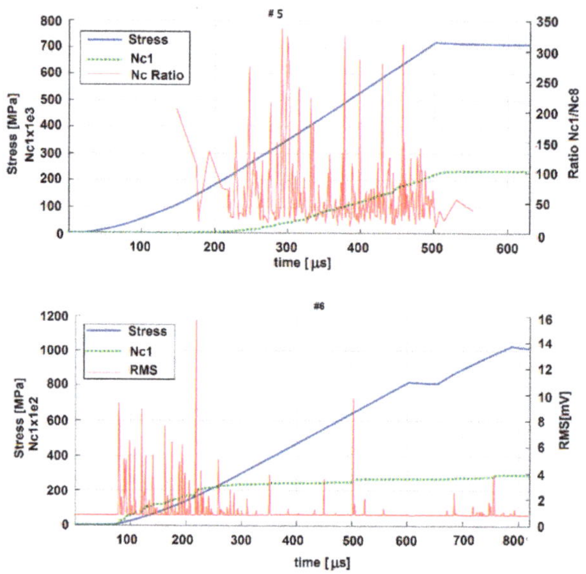

Figure 9. Acoustic emission results—spectral analysis.

Performed compression tests of samples have shown a major influence of chemical composition on their fracture behavior and resulting strength due to different damage mechanisms. The formation of microcracks and their coalescence can be well documented by AE activity during the loading. Internal cracking manifested by the accompanying AE activity resulted in different fractures.

The compression curves have the characteristic S shape for sample #5 and sample #6. The AE signal (Figure 9) is very intense during the start of compression, close to the compressive failure point (sample #6); high amplitude AE events and a maximum at the yield point are presented. The AE activity is connected to the inflection points [51], and distinct differences in the compression behavior occur in relation to the Si content.

The AE signal in sample #6 can be attributed to the twining of the sample, which also has a higher modulus of elasticity. Tests have shown that the compressive strength of a Ti alloy is high enough (about 800 MPa), however internal cracking shown by the AE signal appeared at a stress of about 300 MPa, with rapid degradation to 600 MPa.

The decrease in AE activity in the region of plastic deformation of the sample could be due to the dynamics of structural modifications. After stopping the compression (maximum peaks of the stress curves) the material began to relax, producing only weak and apparently decreased secondary AEs.

The significantly higher AE signal in the case of alloys for which x = 1.0 compared to that for x = 0.75 can be explained by the differences between the internal microstructures. Most events take place around the possible dislocations where the crack initiates, forming microstructures, and their coalescence propagates towards the external surface in the compression direction, leading to high AE events during loading. The microcracks develop into the breaking of small particles, leading to the failure of the sample.

4. Conclusions

The changes of the mechanical properties and crystallographic structures of the TiMoZrTa quaternary alloy with variable doping by Si were investigated using various non-destructive evaluation (NDE) methods. The Si content in the $TiMo_{20}Zr_7Ta_{15}Si_x$ alloy was varied with x = [0.0; 0.5; 0.75; 1.0]. The main NDE investigation methods were RUS (resonant ultrasound spectroscopy), AE (acoustic emission) and X-ray diffraction. Those methods revealed the importance of the Ta and Zr content in the TiMo binary alloy, that together with NDE methods highlighted the influence of the Si addition on the mechanical properties and crystallographic structure. The changes in crystallographic structure showed that all the samples contained a small foreign phase in the error limit. The ascertained values of the average bond order (B_o) and the averaged orbital energy level (M_d) indicated that all casted alloys were in the stable β-region. RUS and compression tests showed an influence of the composition on the strength and fracture behaviors with various damage mechanisms.

The test results verified that $TiMo_{20}Zr_7Ta_{15}Si_x$ alloys have good mechanical behavior, with a Young modulus in the range of 69.11 to 89.03 GPa, and shear modulus from 24.70 to 31.87 GPa. All the tested samples were slightly affected by the Si content through their moduli with non-linear trends. The compressive strength of the studied Ti alloy, highlighted by AE, turned out to be over 800 MPa. Despite this, the internal cracking, which appeared in AE signals at 300 MPa stress, was followed by rapid failure at 600 MPa. The decrease in AE activity in the region of plastic deformation could be due to the structural modification dynamics. Although Si improves the creep strength of Ti alloys, AE test activity has shown that the plasticity is affected at a Si content higher than 0.75 wt.%.

The dependence of crystallographic structure on Si concentration is probably due to modification of the sample's resonance spectrum symmetry, demonstrated by the splitting of the peaks at certain resonance frequencies in RUS.

Structural modifications of samples caused by different Si doping ratios could easily be monitored by RUS, emphasizing the changing of eigen frequencies. This led to the splitting of some spectral peaks, increasing their amplitudes and displacement of the RUS spectra from their initial position. The shape of the RUS spectrum and the resonance frequencies were also modified by the lower content of alloying elements and smaller elastic properties than those prescribed.

The analysis of the alloy microstructure dependence on the $TiMo_{20}Zr_7Ta_{15}Si_x$ components, as well as on the constitutive phases will be reported in detail in another paper.

Author Contributions: Conceptualization: A.S., F.N. and A.B.; Data curation: A.S., M.L.C., C.H. and Z.P.; Formal analysis: A.S., S.M., M.L.C., M.C., A.B., R.S. and F.N.; Funding acquisition: A.S.; Investigation: A.S., V.T., C.H., M.L.C., P.V. and S.M.; Methodology: A.B., F.N., M.C., R.S., P.V. and Z.P.; Project administration: A.S.; Resources: A.S., P.V., V.T., F.N. and S.M.; Software: M.L.C., M.C., R.S. and C.H.; Supervision: A.S., S.M., M.L.C. and Z.P.; Validation: A.S., F.N., M.L.C., A.B., C.H. and Z.P.; Visualization A.B., F.N., R.S. and C.H.; Roles/Writing—original draft: A.S., F.N., A.B., Z.P., R.S. and M.L.C.; Writing—review and editing: All authors. All authors have read and agreed to the published version of the manuscript.

Funding: This work was supported by the Ministry of Education and Research, Romania under CCCDI-UEFISCDI, project number PNCDI III-P1-1.2-PCCDI-207-03239/60PCCDI 2018—MedicalMetMat.

Conflicts of Interest: The authors declare no conflict of interest.

References

1. Food Drug Administration. *Biological Responses to Metal Implants*; Food Drug Administration: Silver Spring, MD, USA, 2019.
2. Ryu, D.J.; Ban, H.Y.; Jung, E.Y.; Sonn, C.H.; Hong, D.H.; Ahmad, S.; Gweon, B.; Lim, D.; Wang, J.H. Osteo-Compatibility of 3D Titanium Porous Coating Applied by Direct Energy Deposition (DED) for a Cementless Total Knee Arthroplasty Implant: In Vitro and In Vivo Study. *J. Clin. Med.* **2020**, *9*, 478. [CrossRef] [PubMed]
3. Bai, W.; Xu, G.; Tan, M.; Yang, Z.; Zeng, L.; Wu, D.; Liu, L.; Zhang, L. Diffusivities and atomic mobilities in bcc Ti-Mo-Zr alloys. *Materials* **2018**, *11*, 1909. [CrossRef] [PubMed]
4. Dryburgh, P.; Patel, R.; Pieris, D.M.; Hirsch, M.; Li, W.; Sharples, S.D.; Smith, R.J.; Clare, A.T.; Clark, M. Spatially resolved acoustic spectroscopy for texture imaging in powder bed fusion nickel superalloys. In Proceedings of the 45th Annual Review of Progress in Quantitative Nondestructive Evaluation, AIP Conference Proceedings, Burlington, VT, USA, 15–19 July 2018; AIP Publishing LLC: Melville, NY, USA, 2019; Volume 38.
5. Cambiaghi, A. *Biological Evaluation of Medical Devices as an Essential Part of the Risk Management Process: Updates and Challenges of ISO 10993-1*; Eurofins Group: Luxembourg, 2018.
6. ISO 10993-1. *Biological Evaluation of Medical Devices—Part 1: Evaluation and Testing within a Risk Management Process*; International Organization for Standardization: Geneva, Switzerland, 2018.
7. Hegyeli, R.J. Limitations of current techniques for the evaluation of the biohazards and biocompatibility of new candidate materials. *J. Biomed. Mater. Res.* **1971**, *5*, 1–14. [PubMed]
8. Rack, H.J.; Qazi, J.I. Titanium alloys for biomedical applications. *Mater. Sci. Eng. C* **2006**, *26*, 1269–1277. [CrossRef]
9. Villars, P.; Prince, A.; Okamoto, H. *Handbook of Ternary Alloy Phase Diagrams*; ASM International: Almere, The Netherlands, 1995; Volume 10, p. 13621.
10. Mao, Y.; Li, S.; Zhang, J.; Peng, J.; Zou, D.; Zhong, Z. Microstructure and tensile properties of orthorhombic Ti–Al–Nb–Ta alloys. *Intermetallics* **2000**, *8*, 659–662. [CrossRef]
11. Zhang, L.C.; Chen, L.Y. A review on biomedical titanium alloys: Recent progress and prospect. *Adv. Eng. Mater.* **2019**, *21*, 1801215. [CrossRef]
12. Li, Y.; Yang, C.; Zhao, H.; Qu, S.; Li, X.; Li, Y. New developments of Ti-based alloys for biomedical applications. *Materials* **2014**, *7*, 1709–1800. [CrossRef]
13. Geetha, M.; Singh, A.K.; Asokamani, R.; Gogia, A.K. Ti based biomaterials, the ultimate choice for orthopaedic implants–A review. *Prog. Mater. Sci.* **2009**, *54*, 397–425. [CrossRef]
14. Savin, A.; Vizureanu, P.; Prevorovsky, Z.; Chlada, M.; Krofta, J.; Baltatu, M.S.; Istrate, B.; Steigmann, R. Noninvasive Evaluation of Special Alloys for Prostheses Using Complementary Methods. *MS&E* **2018**, *374*, 012030.
15. Sandu, A.V.; Baltatu, M.S.; Nabialek, M.; Savin, A.; Vizureanu, P. Characterization and mechanical properties of new TiMo alloys used for medical applications. *Materials* **2019**, *12*, 2973. [CrossRef] [PubMed]
16. Savin, A.; Craus, M.L.; Bruma, A.; Novy, F.; Minarik, P.; Jabor, M.; Malo, S.; Steigmann, R.; Vizureanu, P.; Harnois, C.; et al. Material properties mapping using complementary methods in titanium alloys TiMoSi used in medical application. In *Low-Dimensional Materials and Devices*; International Society for Optics and Photonics: Bellingham, WA, USA, 2020; p. 114650Z.
17. Trincă, L.C.; Mareci, D.; Solcan, C.; Fântânariu, M.; Burtan, L.; Vulpe, V.; Hriţcu, L.D.; Souto, R.M. In vitro corrosion resistance and in vivo osseointegration testing of new multifunctional beta-type quaternary TiMoZrTa alloys. *Mat. Sci. Eng. C* **2019**, *108*, 110485. [CrossRef]
18. Kuroda, D.; Niinomi, M.; Morinaga, M.; Kato, Y.; Yashiro, T. Design and mechanical properties of new β type titanium alloys for implant materials. *Mat. Sci. Eng. A* **1998**, *243*, 244–249. [CrossRef]
19. Zhao, G.H.; Liang, X.Z.; Kim, B.; Rivera-Díaz-del-Castillo, P.E.J. Modelling strengthening mechanisms in beta-type Ti alloys. *Mat. Sci. Eng. A* **2019**, *756*, 156–160. [CrossRef]
20. Inaekyan, K.; Brailovski, V.; Prokoshkin, S.; Pushin, V.; Dubinskiy, S.; Sheremetyev, V. Comparative study of structure formation and mechanical behavior of age-hardened Ti–Nb–Zr and Ti–Nb–Ta shape memory alloys. *Mater. Charact.* **2015**, *103*, 65–74. [CrossRef]

21. Kudrman, J.; Fousek, J.; Brezina, V.; Míková, R.; Vesely, J. Titanium alloys for implants in medicine. *Kov. Mater.* **2007**, *45*, 199.
22. Correa, D.R.N.; Vicente, F.B.; Araújo, R.O.; Lourenço, M.L.; Kuroda, P.A.B.; Buzalaf, M.A.R.; Grandini, C.R. Effect of the substitutional elements on the microstructure of the Ti-15Mo-Zr and Ti-15Zr-Mo systems alloys. *J. Mater. Res. Tech.* **2015**, *4*, 180–185. [CrossRef]
23. Karimi-Sibaki, E.; Kharicha, A.; Wu, M.; Ludwig, A.; Bohacek, J. A Parametric Study of the Vacuum Arc Remelting (VAR) Process: Effects of Arc Radius, Side-Arcing, and Gas Cooling. *Metall. Mater. Trans. B* **2020**, *51*, 222–235. [CrossRef]
24. ASTM E2001-18. *Guide for Resonant Ultrasound Spectroscopy for Defect Detection in Both Metallic and Non-Metallic Parts*; ASTM International: West Conshohocken, PA, USA, 2018.
25. Stráský, J.; Harcuba, P.; Václavová, K.; Horváth, K.; Landa, M.; Srba, O.; Janeček, M. Increasing strength of a biomedical Ti-Nb-Ta-Zr alloy by alloying with Fe, Si and O. *J. Mech. Behav. Biomed.* **2012**, *71*, 329–336. [CrossRef]
26. Liu, G.; Maynard, J.D. Measuring elastic constants of arbitrarily shaped samples using resonant ultrasound spectroscopy. *J. Acoust. Soc. Am.* **2012**, *131*, 2068–2078. [CrossRef]
27. BS EN 13554. *Non-Destructive Testing-Acoustic Emission Testing-General Principles*; British Standard Institution Group: London, UK, 2011.
28. BS EN 1330-9. *Non-Destructive Testing. Terminology. Terms Used in Acoustic Emission Testing*; British Standard Institution Group: London, UK, 2017.
29. ASTM E1316-18a. *Standard Terminology for Nondestructive Examinations*; ASTM International: West Conshohocken, PA, USA, 2018.
30. ASTM E1932-12. *Standard Guide for Acoustic Emission Examination of Small Parts*; ASTM International: West Conshohocken, PA, USA, 2017.
31. Ereifej, N.; Silikas, N.; Watts, D.C. Initial versus final fracture of metal-free crowns, analyzed via acoustic emission. *Dent. Mater.* **2008**, *24*, 1289–1295. [CrossRef]
32. Mavrogordato, M.; Taylor, M.; Taylor, A.; Browne, M. Real time monitoring of progressive damage during loading of a simplified total hip stem construct using embedded acoustic emission sensors. *Med. Eng. Phys.* **2011**, *33*, 395–406. [CrossRef]
33. Cormier, J.M.; Manoogian, S.J.; Bisplinghoff, J.; McNally, C.; Duma, S. The use of acoustic emission in facial fracture detection. *Biomed. Sci. Instrum.* **2008**, *44*, 147–152. [PubMed]
34. ASTM. *E384: Standard Test Method for Knoop and Vickers Hardness of Materials*; ASTM International: West Conshohocken, PA, USA, 2012; pp. 1–43.
35. Sakaguchi, N.; Niinomi, M.; Akahori, T.; Takeda, J.; Toda, H. Effect of Ta content on mechanical properties of Ti–30Nb–XTa–5Zr. *Mater. Sci. Eng. C* **2005**, *25*, 370–376. [CrossRef]
36. Lide, D.R. (Ed.) *CRC Handbook of Chemistry and Physics*; CRC Press: Boca Raton, FL, USA, 2004; Volume 85.
37. Morinaga, M.; Kato, M.; Kamimura, T.; Fukumoto, M.; Harada, I.; Kubo, K. Theoretical Design of Beta-Type Titanium Alloys. In *Titanium'92: Science and Technology*; The Minerals, Metals & Materials Society: Pittsburgh, PA, USA, 1993; pp. 217–224.
38. Biesiekierski, A.; Wang, J.; Gepreel, M.A.H.; Wen, C. A new look at biomedical Ti-based shape memory alloys. *Acta Biomater.* **2012**, *8*, 1661–1669. [CrossRef] [PubMed]
39. Ehtemam-Haghighi, S.; Cao, G.; Zhang, L.C. Nanoindentation study of mechanical properties of Ti based alloys with Fe and Ta additions. *J. Alloys Compd.* **2017**, *692*, 892–897. [CrossRef]
40. Ehtemam-Haghighi, S.; Liu, Y.; Cao, G.; Zhang, L.C. Influence of Nb on the $\beta \rightarrow \alpha$ martensitic phase transformation and properties of the newly designed Ti–Fe–Nb alloys. *Mater. Sci. Eng. C* **2016**, *60*, 503–510. [CrossRef] [PubMed]
41. Li, C.; Lee, D.G.; Mi, X.; Ye, W.; Hui, S.; Lee, Y. Phase transformation and age hardening behavior of new Ti–9.2 Mo–2Fe alloy. *J. Alloys Compd.* **2013**, *549*, 152–157. [CrossRef]
42. Laheurte, P.; Prima, F.; Eberhardt, A.; Gloriant, T.; Wary, M.; Patoor, E. Mechanical properties of low modulus β titanium alloys designed from the electronic approach. *J. Mech. Behav. Biomed.* **2010**, *3*, 565–573. [CrossRef]
43. Nagase, T.; Todai, M.; Hori, T.; Nakano, T. Microstructure of equiatomic and non-equiatomic Ti-Nb-Ta-Zr-Mo high-entropy alloys for metallic biomaterials. *J. Alloys Compd.* **2018**, *753*, 412–421. [CrossRef]
44. Visscher, W.M.; Migliori, A.; Bell, T.M.; Reinert, R.A. On the normal modes of free vibration of inhomogeneous and anisotropic elastic objects. *J. Acoust. Soc. Am.* **1991**, *90*, 2154–2162. [CrossRef]

45. Migliori, A.; Maynard, J.D. Implementation of a modern resonant ultrasound spectroscopy system for the measurement of the elastic moduli of small solid specimens. *Rev. Sci. Instrum.* **2005**, *76*, 121301. [CrossRef]
46. Landau, L.D.; Lifshitz, E.M. *Theory of Elasticity*, 3rd ed.; Pergamon Press: London, UK, 1986.
47. Zadler, B.J.; Le Rousseau, J.H.; Scales, J.A.; Smith, M.L. Resonant ultrasound spectroscopy: Theory and application. *Geophys. J. Int.* **2004**, *156*, 154–169. [CrossRef]
48. Balakirev, F.F.; Ennaceur, S.M.; Migliori, R.J.; Maiorov, B.; Migliori, A. Resonant ultrasound spectroscopy: The essential toolbox. *Rev. Sci. Instrum.* **2019**, *90*, 121401. [CrossRef] [PubMed]
49. Leisure, R.G. *Ultrasonic Spectroscopy: Applications in Condensed Matter Physics and Materials Science*; Cambridge University Press: Cambridge, UK, 2017.
50. Flynn, K.; Radovic, M. Evaluation of defects in materials using resonant ultrasound spectroscopy. *J. Mater. Sci.* **2011**, *46*, 2548–2556. [CrossRef]
51. Drozdenko, D.; Bohlen, J.; Yi, S.; Minárik, P.; Chmelík, F.; Dobroň, P. Investigating a twinning–detwinning process in wrought Mg alloys by the acoustic emission technique. *Acta Mater.* **2016**, *110*, 103–113. [CrossRef]

materials

MDPI

Article

Pulsed Multifrequency Excitation and Spectrogram Eddy Current Testing (PMFES-ECT) for Nondestructive Evaluation of Conducting Materials

Jacek Michał Grochowalski * and Tomasz Chady *

Faculty of Electrical Engineering, West Pomeranian University of Technology, 70-313 Szczecin, Poland
* Correspondence: grochowalski.jacek@zut.edu.pl (J.M.G.); tchady@zut.edu.pl (T.C.); Tel.: +48-91-449-41-34 (T.C.)

Abstract: This paper presents a new method for nondestructive testing—a pulsed multifrequency excitation and spectrogram eddy current testing (PMFES-ECT), which is an extension of the multifrequency excitation and spectrogram eddy current testing. The new method uses excitation in the form of pulses repeated at a specified time, containing several periods of a waveform consisting of the sum of sinusoids with a selected frequency, amplitude and phase. This solution allows the maintenance of the advantages of multifrequency excitation and, at the same time, generates high energy pulses similar to those used in pulse eddy current testing (PECT). The effectiveness of the new method was confirmed by numerical simulations and the measurement of thin Inconel plates, consisting of notches manufactured by the electric-discharge method.

Keywords: multifrequency excitation and spectrogram eddy current testing; pulsed eddy current testing; nondestructive testing

Citation: Grochowalski, J.M.; Chady, T. Pulsed Multifrequency Excitation and Spectrogram Eddy Current Testing (PMFES-ECT) for Nondestructive Evaluation of Conducting Materials. *Materials* **2021**, *14*, 5311. https://doi.org/10.3390/ma14185311

Academic Editor: Giovanni Bruno

Received: 21 July 2021
Accepted: 13 September 2021
Published: 15 September 2021

1. Introduction

NDT (nondestructive testing) refers to the many techniques for assessing components or systems' technical conditions and properties without causing damage. Standard NDT methods are X-ray testing, ultrasonic testing, magnetic flux leakage testing and eddy current testing. Hidden defects, cracks and stresses are inherent in the production and use of structural elements, and therefore the detection and assessment of this damage is essential.

Eddy current testing (ETC) has been used as a nondestructive testing method in the aerospace [1], petrochemical [2] and shipbuilding industries. It is used for surface inspection [3], quality inspection [4] and thickness measurements [5], among other things. Reasonable sensitivity, high speed of measurement and affordability are its main advantages.

The earliest version of ETC is the single frequency method using coils as the excitation element due to the simplicity of implementation [6]. The single-frequency eddy current technique with a phase analysis has limited potential for the inspection of surface defects in inhomogeneous ferromagnetic materials, as signals from cracks are hidden in noise caused by the lift-off of the probe or local changes to the tested specimen's permeability [7]. In the eddy current tests, the impedance plane plots show that the phase angles between the impedance planes change with the frequency.

The phase angles between lift-off, cracks and permeability variations may be at one set of values at one frequency and another set of values at a different frequency. Two excitation frequencies used simultaneously in the test coil allow the reduction of the unwanted influence of external structures (i.e., support elements). Dual-frequency EC testing is currently standard, but up to eight frequencies are processed and analyzed simultaneously in many systems. Two testing frequencies were applied to ensure good phase discrimination. One frequency is likely to be about ten times the other. The phase and impedance sensitivity for each frequency is then adjusted independently on the vector display. A dual-frequency ECT system for detecting cracks in welds was proposed in [7]. Changes in the amplitude ratio and phase difference were utilized to enhance defect detection. Dual-frequency systems [8] with the use of data fusion were discussed in [8]. The changes in impedance for two different frequencies were analyzed and the resulting images were combined. Dual-frequency systems create the possibility of detecting surface and hidden defects. In patent [9], an extended system using four different test frequencies sequentially applied to the transducer is proposed.

The development of digital measuring systems and signal processing allows the use of many testing frequencies. Adding more frequencies increases the sensitivity and resolution of the ECT systems [10].

In [11,12], a new multifrequency method (MFES-ECT) was proposed. In this system, measurements using many testing frequencies (more than 15) and spectrograms were utilized. This method creates a new opportunity to characterize the defects occurring in the tested materials accurately.

Another approach is to use a square wave pulse instead of a single/multifrequency signal. In this method, pulsed eddy current testing (PECT), the exciting coil is fed by a current pulse. The time-domain response of the probe is measured and further processed. An example of computer simulations of this method is given in [13] and its application to the detection and characterization of hidden cracks around fastener sites is presented in [1]. Additionally, PECT using pulses reduces energy use and thus can be used in portable measuring systems.

Instead of induction coils, a motion against the permanent magnets can be used to induce eddy currents. Moving permanent magnets makes it possible to obtain a high amplitude of eddy currents deep in the material. An example is the system shown in [14] and used to inspect ferromagnetic objects, such as pipelines and heat exchanger tubes, utilizing eddy currents induced by rotating permanent magnets. It contains permanent magnets mounted on the arms, rotating around a shaft placed inside a ferromagnetic pipe. Another transducer, shown in [15], is used to inspect planar conducting plates. It consists of a rotating head with permanent magnets used to generate variable magnetic fields and thus induce eddy currents in the tested material.

This paper presents a novel method of nondestructive testing—the pulsed multifrequency excitation and spectrogram eddy current testing (PMFES-ECT) extension of the multifrequency excitation and spectrogram eddy current testing. The new method uses excitation in the form of pulses repeated at a specified time. The pulses consist of several periods of a waveform containing the sum of sinusoids with a selected frequency, amplitude and phase. This solution allows the maintenance of the advantages of multifrequency excitation and at the same time generates high energy pulses similar to those used in pulse eddy current testing (PECT) and creates good conditions for the accurate detection of flaws located deeply under the surface of the conductive materials.

The concepts and working principles of PMFES-ECT are provided in Section 2. Section 3 shows the main components of the system for testing and validating the PMFES-ECT method, including the description of the sensor and the tested specimen. Based on this, computer simulations were carried out using COMSOL software (the methodology and the results are presented in Section 4), and experimental tests of the system were carried out, the results of which are presented in Section 5. The conclusions are provided in Section 6.

2. Pulsed Multifrequency Excitation and Spectrogram Eddy Current Testing

2.1. Pulsed Multifrequency Excitation and Spectrogram ECT Method

In the conventional technique, the alternating magnetic field generated by the coil/set of coils acts on the tested sample. Under the influence of this excitation, a secondary magnetic field is created from the eddy currents, which can be measured with appropriate sensors. There are two distinct methods of shaping the current in the excitation coil: the pulse method and the multifrequency method (including the single frequency method).

Each has its advantages; i.e., for the latter, a broad spectrum of frequencies makes it suitable for detecting hidden defects, and for the former, a much higher power output and power efficiency.

The pulsed multifrequency excitation and spectrogram eddy current testing (PMFES-ECT) is a novel extension of the MFES-ECT. This method combines the advantages of MFES-ECT and PECT.

This method uses excitation in the form of pulses. Each pulse consists of several periods of a waveform. The waveform is generated as a sum of sinusoids with selected frequencies, amplitudes and phases. The frequency spectrum of the pulse is identical to the continuous MFES-ECT. Also, a short pulse duration enables a large excitation current, thus detecting deep-lying defects without worrying about overheating the excitation coil. Figure 1 shows the comparison of the excitation signal for the MFES-ECT or PMFES-ECT method. The algorithm for generating the multifrequency pulse is described in the following subsection.

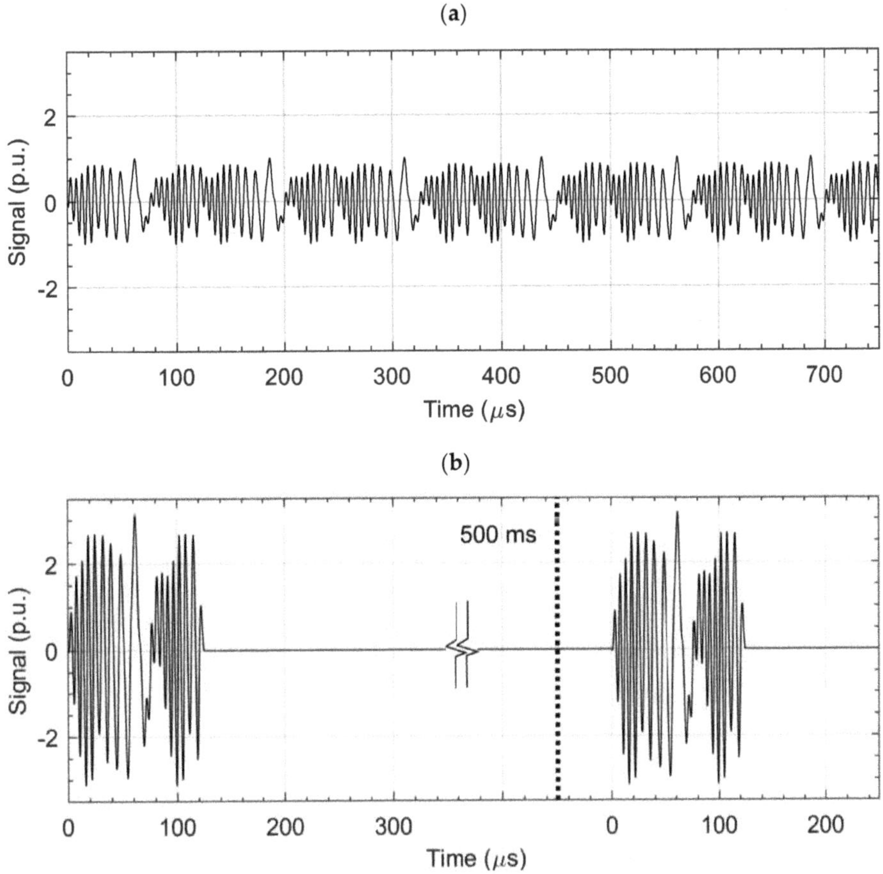

Figure 1. Comparison of excitation voltage for MFES-ECT (**a**) and PMFES-ECT (**b**). There is an interval of about 500 ms between successive pulses in the PMFES-ECT method. The pulse in PMFES-ECT has a power 10 dB more than in the MFES-ECT.

The spectrum of multifrequency pulse in the frequency range of interest is much richer than in the case of a rectangular pulse. For comparison, in Figure 2, the spectra for the PMFES/MFES-ECT and the PECT waveforms are presented. The rectangular pulse parameters (i.e., the duration and the amplitude of the pulse) used in PECT were selected so that the value of the fundamental (first) frequency amplitude in the spectrum was identical for each type of method.

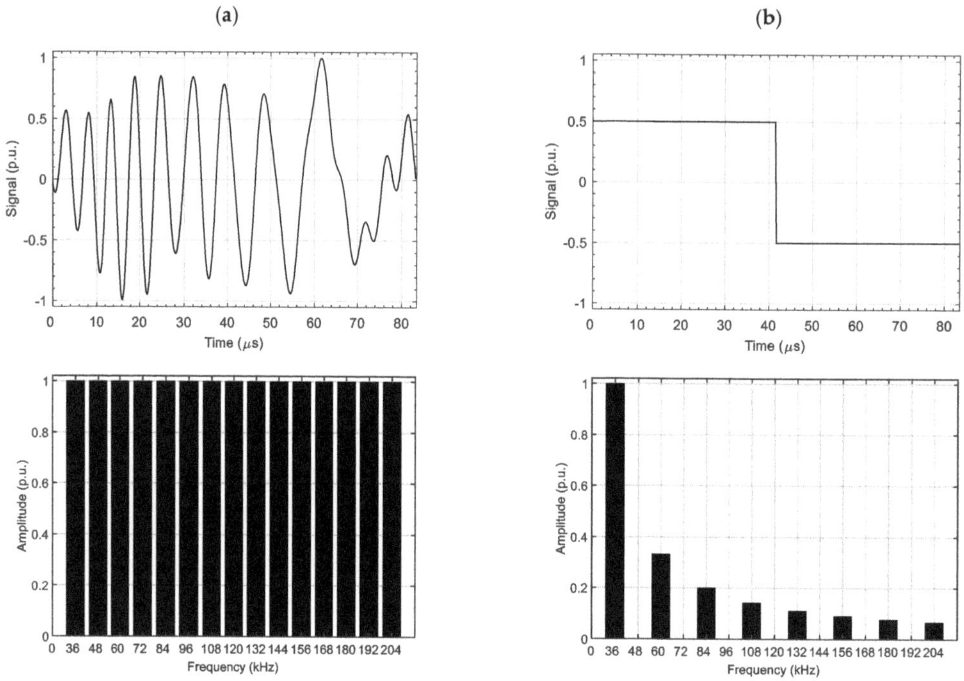

Figure 2. Comparison of the frequency spectrum for PMFES/MFES-ECT (**a**) and PECT (**b**). The signal is at the top and its spectrum (FFT) at the bottom. Signal parameters were selected to obtain the equal amplitude of the first harmonic for both signals.

In the case of the PECT method, successive frequency components have successively lower values. The MFES-ECT/PMFES-ECT methods allow a constant amplitude value in the entire frequency spectrum to be obtained.

2.2. Generation of the Multi-Frequency Signal

An exciting multifrequency signal is generated by combining multiple sinusoidal waveforms. The general formula of this signal is following:

$$u_{exct}(t) = \sum_{i=1}^{n} a_i U_i \sin(2\,\pi f_i + \phi_i) \tag{1}$$

where U_i is the amplitude of the i-th sinusoid, a_i is normalization factor f_i is frequency and ϕ_i is the phase angle of i-th sinusoid. The factor ϕ_i is calculated using Equation (2):

$$\phi_i = \pi \frac{i^2}{N} \tag{2}$$

where N is the total number of sinusoids. Setting the parameter ϕ_i reduces the crest factor of the signal, and thus improves the power delivery to excitations coils. The a_i factor is calculated using Equation (3) presented in the following section.

A complex excitation signal containing many sinusoids creates an opportunity to detect all kinds of defects. The frequency range has been selected so that it is possible to detect surface and deeply located defects in tested specimens. Their selection also considered the maximum RMS value of the current flowing in the coil (to avoid overheating). More specifically, the number of frequencies was $N = 15$ and the frequencies were in the range $f_i = 36, 48, \ldots, 204$ kHz. The signal amplitude at the beginning and end of a pulse rises and falls linearly to shorten the transition state.

Compared to the continuous wave method (MFES-ECT), the pulse method uses a pulse—a packet constituting a fragment of the excitation wave from the MFES-ECT method. The pulse length is selected empirically to eliminate the transient state in the response signal and obtain good frequency resolution in further analysis.

2.3. Maintaining a Constant Amplitude over the Entire Frequency Spectrum

In order to ensure a constant amplitude over the entire spectrum, the excitation signal adjustment/normalization procedure is performed. If the value of U_i from Equation (1) is equal for each frequency, the amplitudes of successive frequencies in the measured signal are not constant. This is due to the variable coils impedance for different frequencies, different signal attenuation, the inductance of connection cables and parasitic capacities.

To mitigate this, we first measured the response to the excitation signal with a constant and equal U_i value in a homogeneous region (without any defects). Figure 3a shows the harmonic amplitudes in the simulated measured signal with no normalization.

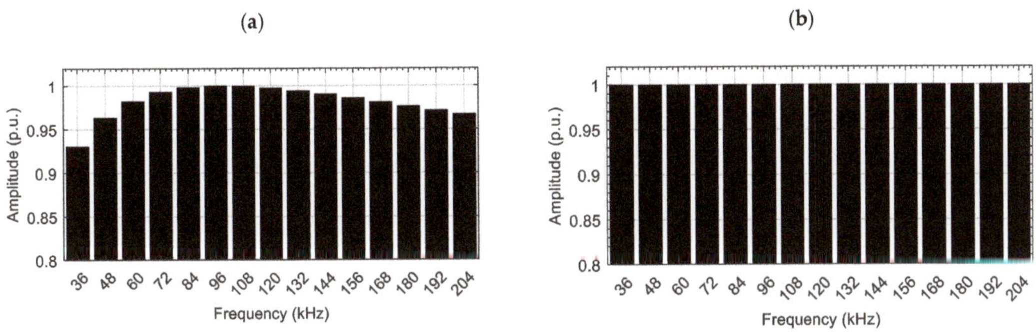

Figure 3. Spectrum comparison for non-normalized (**a**) and normalized (**b**) measured signals.

Then, we calculated the correction coefficients a_i for successive U_i values according to the formula:

$$a_i = \frac{U_{ref}}{U_{i\ measured}} \tag{3}$$

where U_{ref} is the reference value we want to acquire, U_i measured is the current value of the amplitude of successive components. The U_{ref} value is selected so as not to exceed the maximum sensor operating current.

The normalization factor is then used to generate a new excitation signal according to Equation (1). Figure 3b shows the spectrum of the measured signal after normalization in a homogenous area.

2.4. Information Extraction and Spectrogram Creation

The signal obtained from the pickup coil, after processing and analysis, is presented in the form of the spectrogram. The spectrogram is a plot of the amplitude of subsequent frequency components of a signal from the coil versus the sensor position. The presented amplitude for each frequency on the spectrogram (ΔU_{RMS}) is calculated as the difference between the actual amplitude measured at the measurement point and the amplitude measured at the uniform material location.

Selected parameters of spectrogram are defined as:

1. S_{MAX}—the maximum value of the spectrogram,
2. f_{MAX}—the frequency for which the spectrogram achieves the maximal value
3. $S(f)_{x=X_{MAX}}$—the frequency characteristic at the point $x = X_{MAX}$.

Figure 4 shows the sample spectrogram (in the two-dimensional view) with the above parameters marked. These parameters are utilized in the evaluation of the characteristics of the defect (i.e., position, depth). The estimation of these parameters based on the spectrogram is presented in [11,12].

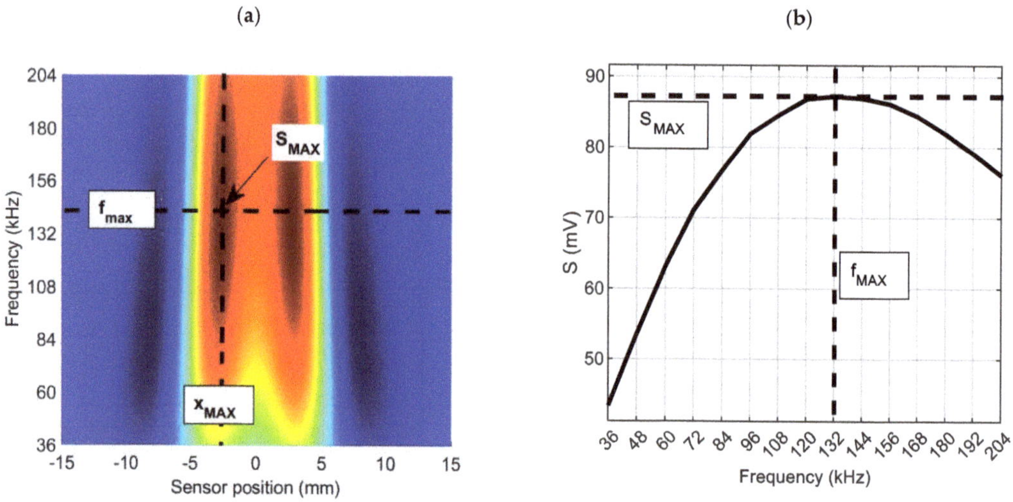

Figure 4. Spectrogram: the two-dimensional plot (**a**) and frequency characteristic for X_{MAX} point (**b**) with marked parameters (S_{MAX}, f_{MAX}) corresponding to the material defect.

The measurement procedure assumes that the RMS value of the signal is constant for a given measurement (one pulse). The pulse generator is synchronized with the A/D, the sampling frequency and the number of samples are selected so that a rectangular window can be used in the FFT analysis and there is no spectral leakage.

The signal for each measurement point was decomposed using the FFT algorithm and the amplitudes of successive frequencies were determined. In order to enhance the sensitivity and reduce noises or trends, additional signal processing algorithms can be applied.

Due to the pulse characteristic of the operation, the signal waveforms from the pickup coil contain a transient state. Only a fragment immediately after the end of the transient state was selected for further analysis (determination of the amplitudes of subsequent frequency components). Figure 5 shows an example of the recorded signal waveform and the indication of the part selected for further analysis.

As previously mentioned, an important aspect is to eliminate the transient state when carrying out further analysis. The influence of this state on the generated spectrograms is presented in Figure 6 below.

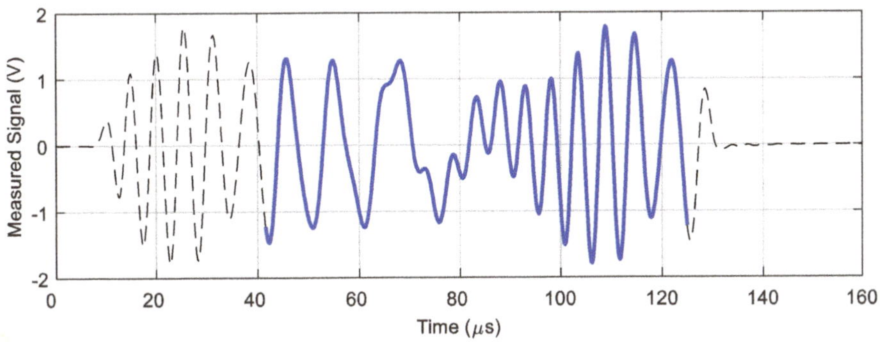

Figure 5. The measured signal on the pickup coil. The blue line marks the signal segment that is subject to further analysis.

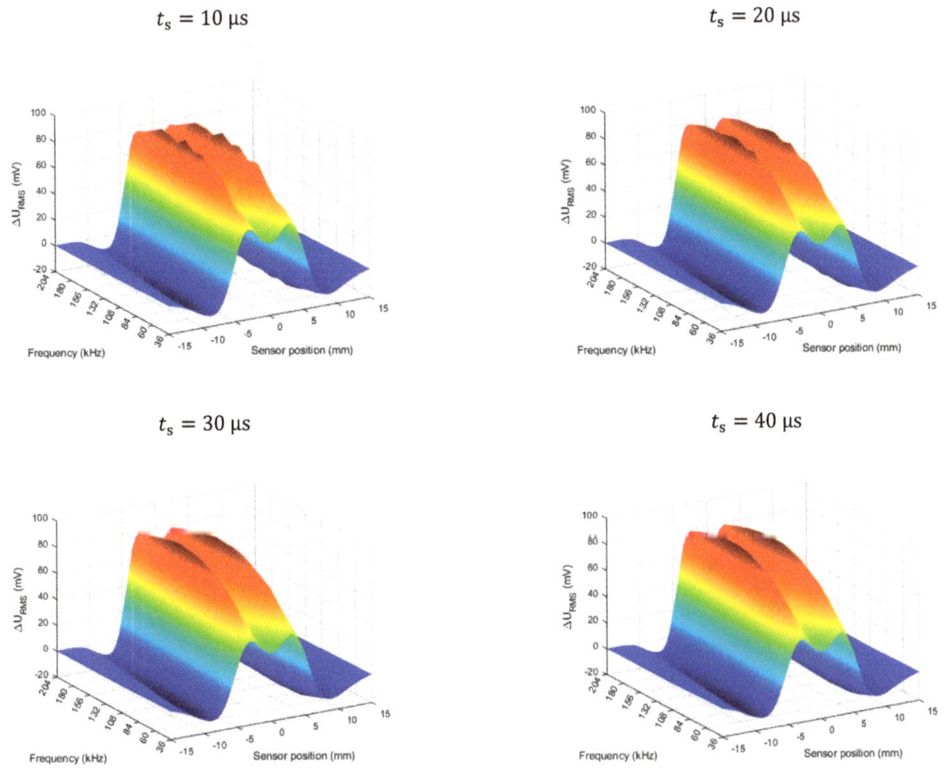

Figure 6. Effect of the transition state on the spectrogram. Flaw depth 60%. The t_s is the start time of the data analysis window (the length of the window is constant).

3. Experimental Setup

The essential components of the PMFES-ECT measurement system are an excitation signal generator (NI PXI 5422 manufactured by NI based in Austin, TX, USA, sample rate 200 MS/s, 80 MHz bandwidth, 16-Bit waveform generator), a power amplifier (HSA 4101, manufactured by NF Corporation based in Yokohama, Japan, frequency range DC to 10 MHz, slew rate 5000 V/µs, max., current 1.4 A, amplification gain 1–20) used as the excitation coils driver, an eddy current transducer, an A/D capture device (NI PXI 5922 manufactured by NI, Austin, TX, USA, A/D converter maximum sampling rate

15 MS/s, maximum resolution 24 bits) and a PC with dedicated software. A photo of the experimental setup is presented in Figure 7.

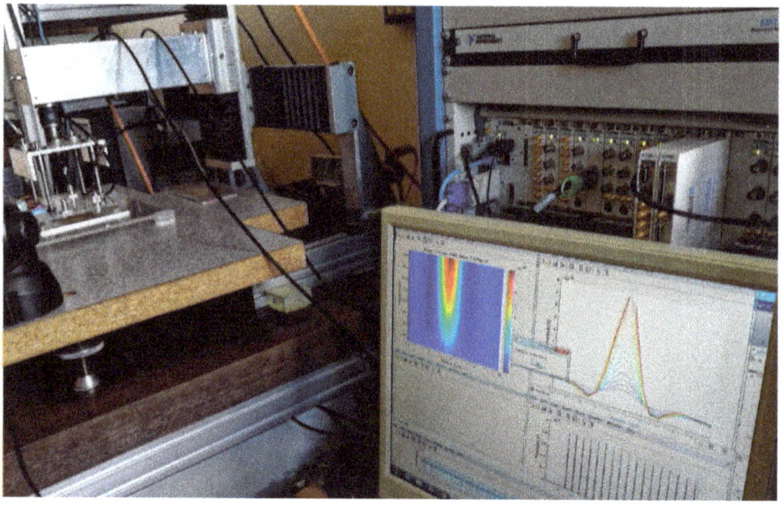

Figure 7. Photo of the experimental setup.

The waveform generator outputs pulsed signals to the power amplifier, which supplies the excitation coils in the transducer, thus inducing eddy currents in a tested specimen. The pickup coil measures the effective magnetic field, consisting of a field due to excitation and one caused by eddy currents in the inspected material. The signal from the pickup coil is captured by an A/D converter and saved on a PC for further analysis. Then, the transducer is moved to the next measuring point by the XY linear positioning unit by a step of 0.5 mm. The software on PC manages the whole system. The software also performs information extraction and spectrogram creation on a PC.

A plate specimen with a thickness of 1.25 mm made of INCONEL600 alloy (Nippon Steel Corporation, Tokyo, Japan) with the electrical conductivity σ = 1 MS/m is used in this experiment. The plate is 165 mm long, 85 mm wide and contains notches of different depths. The length of the notches is 5 mm and the width is 0.25 mm. The relative depth of the notches varies from 10% to 100% (through the entire thickness of the plate).

Pulsed multifrequency eddy current testing (PMFES-ECT) using an ECT transducer [11], presented in Figure 8, is employed to gather responses from flaws during the movement of the probe. For comparison, the standard MFES-ECT (continuous wave excitation) is also utilized.

The pickup coil (S) mounted on a center column of the 5-column ferrite core measures a differential flux generated by two oppositely oriented pairs of the excitation coils (E_A, E_B and E_C, E_D). Flux produced in the pickup coil by one pair of the excitation coils flows in the opposite direction to the one caused by another pair. The resulting flux in the pickup coil in the equilibrium state is about zero. In the event of a defect in the tested specimen, a signal appears on the measuring coil S.

The ferrite core has a relative permeability of $\mu_r = 1000$. The exciting coils are driven by a multifrequency signal generated from the arbitrary wave generator and amplified by a high-frequency power amplifier. Dimensions of the sensor are presented in Figure 8a,b. Detailed transducer characteristic is presented in Table 1.

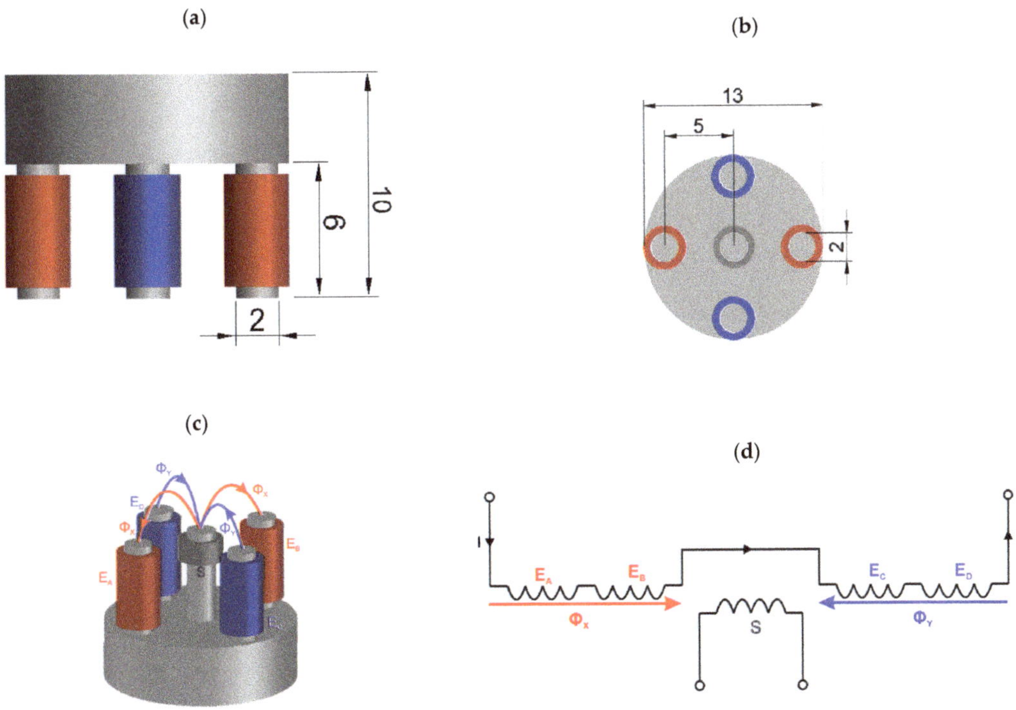

Figure 8. Views of the five columns of ECT transducer used in the experiments: (**a**) front view with dimensions, (**b**) bottom view with dimensions, (**c**) view with magnetic fluxes, the two groups of the excitation coils are marked with red and blue color, (**d**) the simplified electrical scheme of the probe.

Table 1. Transducer parameters.

Parameter	Value
Coil E_A, E_B, E_C, E_D winding turns	25 turns, Φ 0.14 mm
Coil S winding turns	100 turns, Φ 0.02 mm
Core relative permeability	$\mu_R = 1000$
Maximum working flux density	200 mT

The instantaneous waveforms of induced voltage on the pickup coil are acquired by an A/D converter and saved for further analysis.

As mentioned earlier, the plate made of INCONEL600 contains six notches manufactured by the electric-discharge method. The plate is inspected from the opposite side, i.e., opposite of the notches opening (except the 100% notch). The notches have the same length of 5 mm and width of 0.25 mm, while they differ in relative depths d = 10%, 20%, 40%, 60%, 80%, 100%. The view of the specimen is shown in Figure 9.

The transducer is placed over the tested specimen (INCONEL600 plate), touching its surface (as shown in Figure 10). Measurements are taken along the defect ranging from −15 mm to 15 mm from the center of each notch using the XY linear positioning unit in 0.5 mm increments.

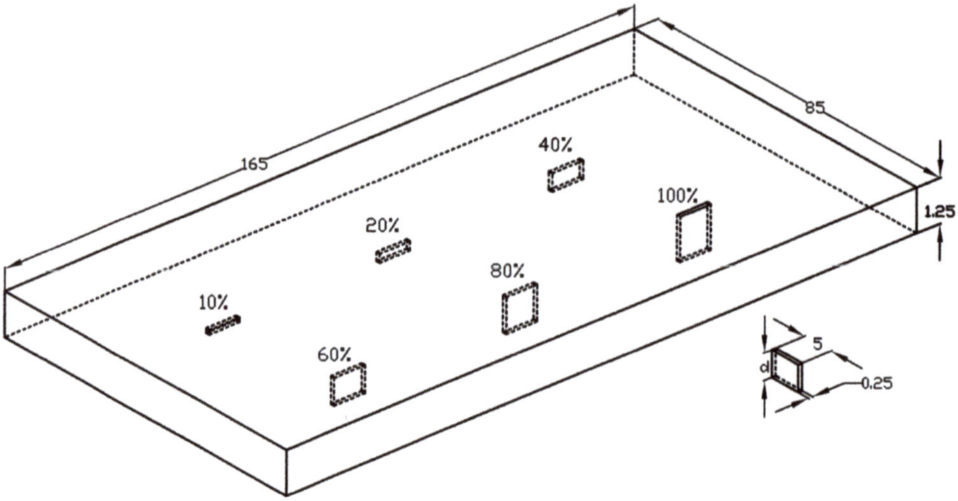

Figure 9. A simplified view of the specimen with artificial flaws.

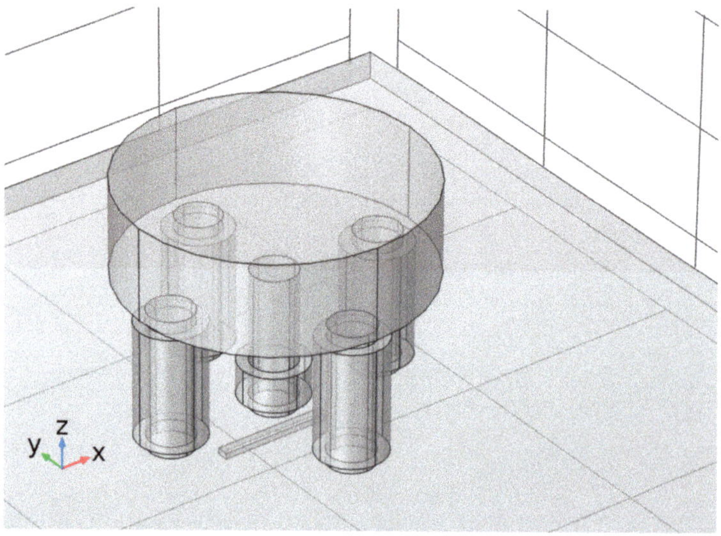

Figure 10. The transducer above the tested specimen with a flaw.

4. Simulation Analysis

4.1. FEM Model

In order to initially verify the method, FEM two-dimensional and three-dimensional models were made and simulations were carried out in the COMSOL Multiphysics environment. The "magnetic fields" model was used. The size of the mesh elements was selected so as to take into account the skin effect for the highest utilized frequency. Two types of simulations were carried out: "frequency domain" and "time dependent". Figure 11 shows the model geometry used to analyze the influence of the defect on the distribution of the electromagnetic field and, ultimately, the value of the voltage induced on the pickup coil.

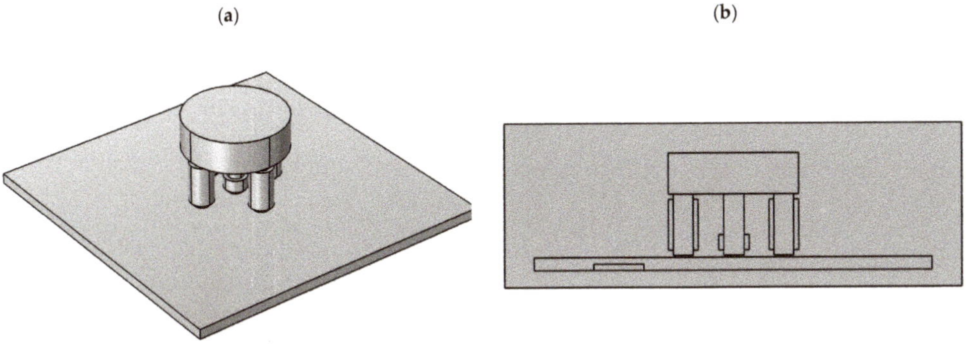

Figure 11. The geometry of the sensor and the specimen with flaw arrangements (the three-dimensional (**a**) and two-dimensional (**b**) configurations).

In the simulation experiment, the specimen with dimensions analogous to those in the experimental setup was used, i.e., 1.25 mm thick. The plate conductivity is 1 MS/m, and the relative permeability is 1. The conductivity of the ferrite core is $\sigma = 1$ S/m, while the relative magnetic permeability is $\mu_r = 1000$. The conductivity of the surrounding air is 0 S/m and the relative permeability is 1.

The defect dimensions are the same as those presented in Section 3, as are the core dimensions. The flaw length is 5 mm and the relative depth is in the range of 10–100%.

A series of simulations with a different location of the flaw in relation to the sensor was performed—at a distance of −15 mm to 0 mm (the defect was directly under the middle column of the sensor) with a step of 0.5 mm. In order to compare the MFES-ECT and PMFES-ECT methods, a series of simulations, "frequency sweep" and "time simulation", were performed for each flaw position.

4.2. Simulation Parameters and Settings

Frequency domain and time simulations were carried out for each measuring point in COMSOL software. In the frequency domain, simulations were carried out for the frequencies 36–204 kHz, with a step of 12 kHz and a constant amplitude. The simulation results in the form of amplitudes and phases calculated for successive frequencies were saved for further analysis.

In the time dependent simulation, the pulse described in Section 2.2 was used as the feeding voltage for excitation coils. The pulse duration was 125 μs, and the entire simulation time was 150 μs, with time step not larger than 0.20 μs. The voltage waveform on the pickup coil was saved for further analysis. An example of the signal achieved from the pickup coil is shown in Figure 12.

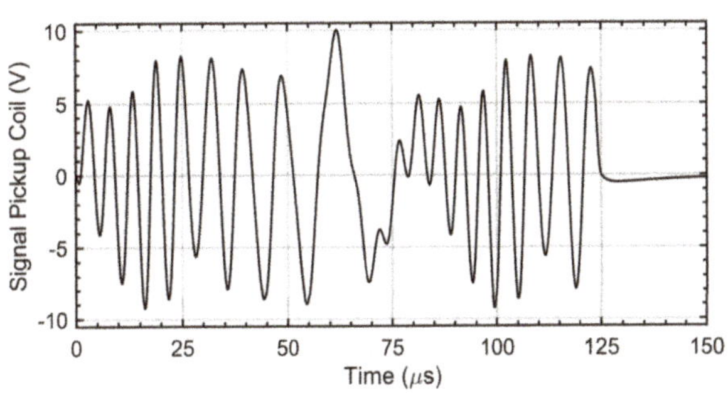

Figure 12. The pickup coil output signal obtained from time dependent simulations. The relative depth of the flaw is 60%.

4.3. Simulation Results

This section presents simulation results that allow us to compare both methods: PMFES-ECT and MFES-ECT. The simulations were performed for various depths of flaws, according to Section 3. Figure 13 shows the comparison of the spectrograms achieved from the MFES-ECT and PMFES-ECT methods for various flaw depths.

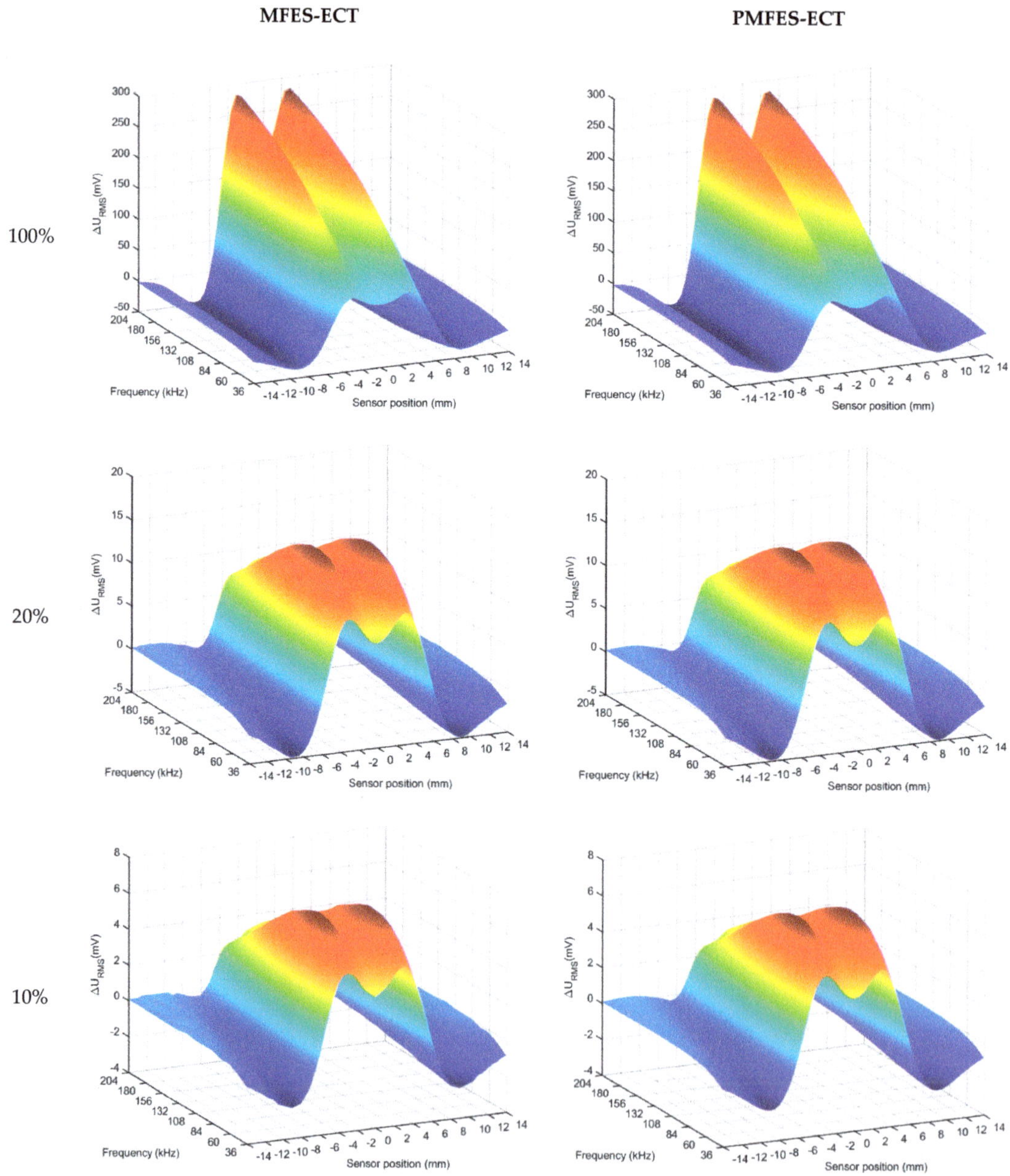

Figure 13. Comparison of spectrograms achieved from two methods MFES-ECT and PMFES-ECT, for various flaws depth.

For each defect depth for each method, the f_{MAX}, x_{MAX} and S_{MAX} parameters were almost identical. Both methods in the simulation environment provide the same results and enable precise estimation of the length and depth of a defect, and therefore can be used interchangeably.

5. Experimental Results

After the validation of the proposed method in computer simulations, physical tests were performed. In the experimental studies, INCONEL alloy samples with notches manufactured by the electric-discharge method were used. The RMS value of the excitation voltage was constant for each depth of the flaw. In the PMFES-ECT method, the RMS voltage value for one period (one pulse) was the same as for the MFES-ECT method.

5.1. Experimental Results for Flaws Having Depths in the Range of 10–100%

Figure 14 shows a comparison of the results of experiments with both methods. The results are presented for different depths of the flaws.

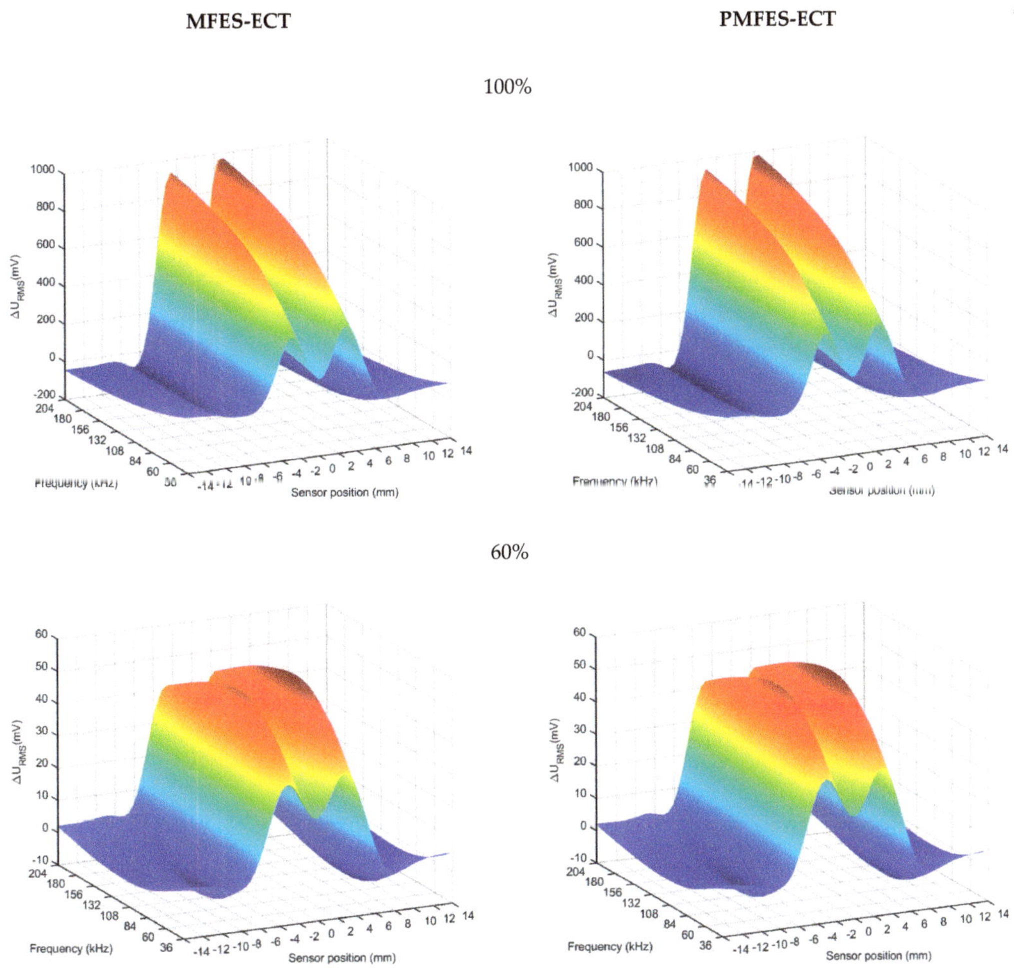

Figure 14. *Cont.*

20%

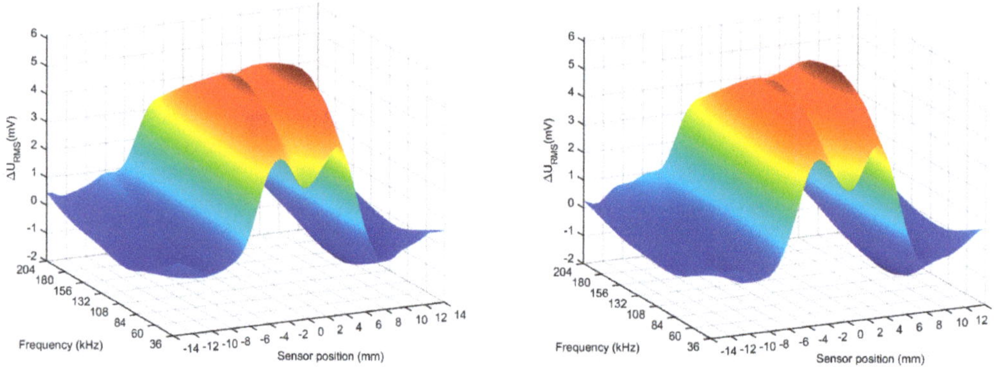

10%

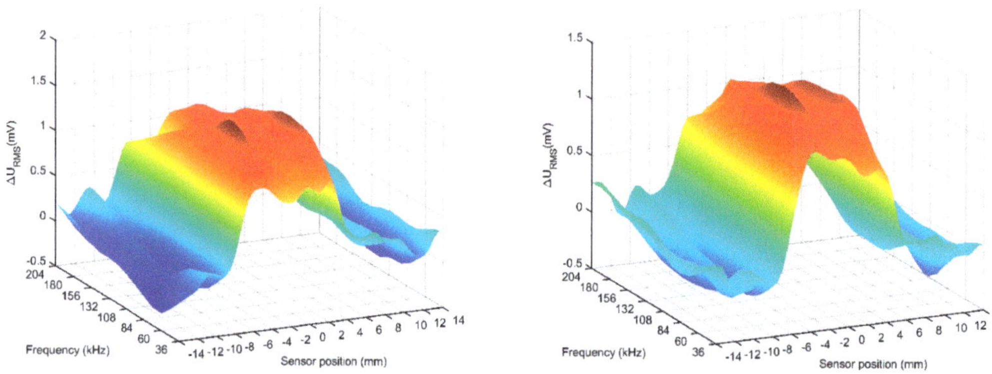

Figure 14. Comparison of the results obtained for the MFES-ECT and PMFES-ECT methods for various flaw depths in the range of 10–100%.

As in the case of simulation, the spectrogram parameters in the range of 20–100% defect depths are almost identical for the PMFES-ECT and MFES-ECT methods. The detection and characterization of the deepest defect (10%) is subject to high uncertainty due to the distortion of the generated spectrogram. This is due to the high amount of noise in the measured signal. The Table 2 below shows the signal-to-noise ratio (SNR) values for each defect depth.

Table 2. SNR (dB) values for different defect depths.

Depth	MFES-ECT	PMFES-ECT
100%	24.94	24.86
60%	26.53	26.55
20%	22.87	21.64
10%	14.08	17.55

The pulse method offers the opportunity to improve the SNR value by increasing the pulse power, as shown in the next subsection.

5.2. Experimental Results for a Defect with a Depth of 10% at a Higher Pulse Powers

In the case of the deepest laying flaw (relative depth of 10%), a series of measurements were performed with a different pulse power value. The 0 dB reference value has been established on the level of previous measurements for the 10–100% defect depth. Additionally, the results were compared without and with the applied filter. The filter type is low-pass Butterworth and was applied to the generated spectrogram in the x (position) and f (frequency) domains. The results are shown in Figure 15.

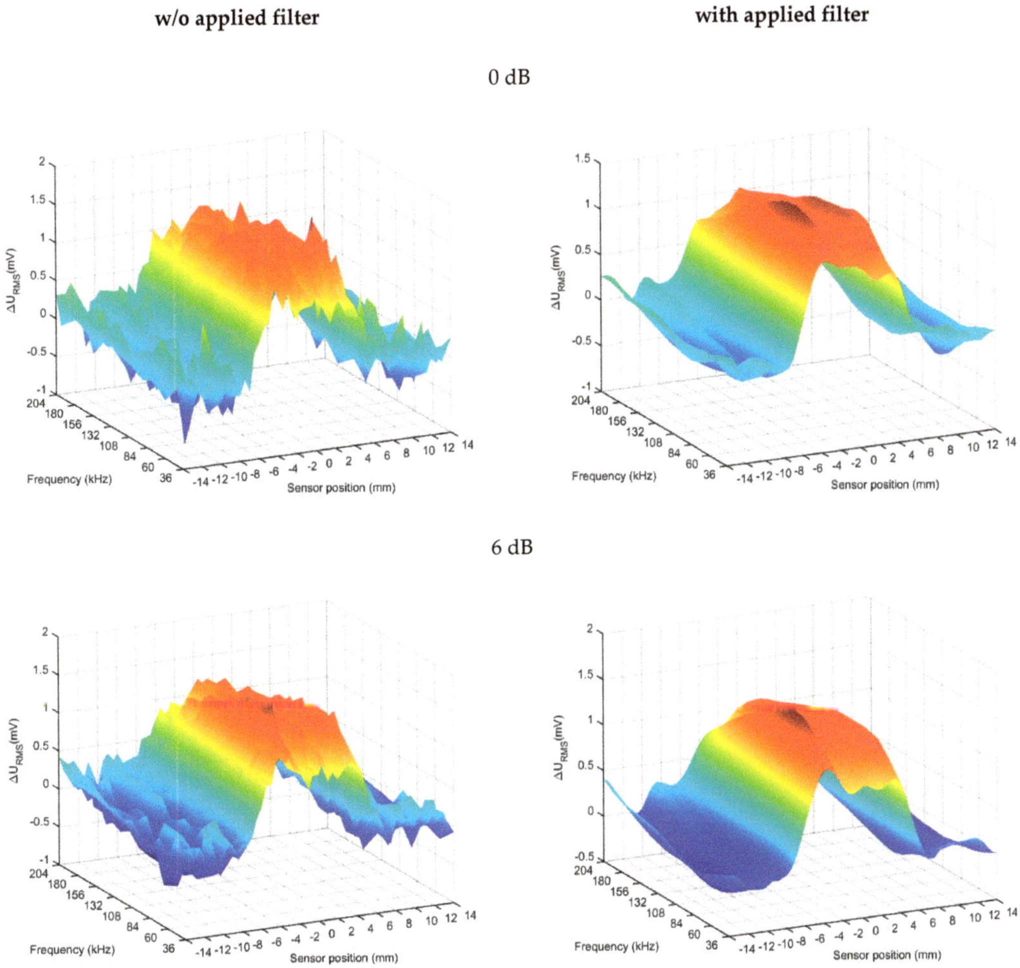

Figure 15. *Cont.*

10 dB

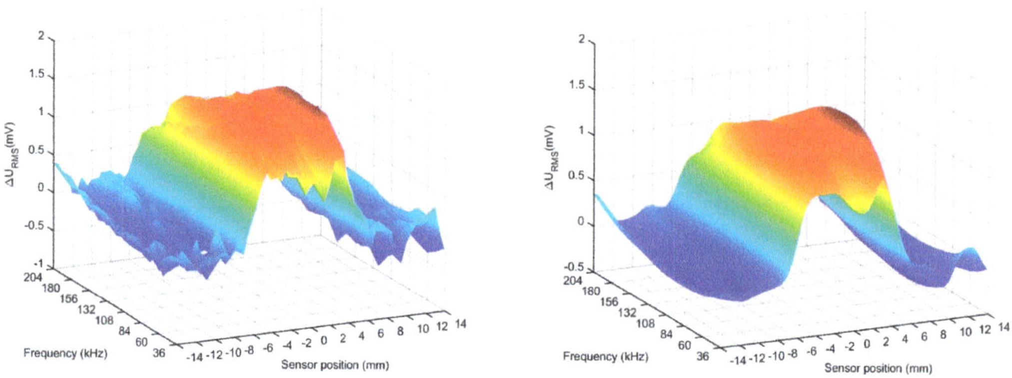

Figure 15. Comparison of the results for the 10% flaw depth for different excitation pulse powers. Results are shown with and without filtering.

The results obtained for a pulse with a reference power (0 dB) show a significant amount of noise. Increasing the pulse power improves the signal-to-noise ratio (SNR) and enables more accurate visualization, defect detection and identification. The use of filters allows you to significantly improve the SNR. The Table 3 below shows a comparison of the SNR values for different pulse powers.

Table 3. SNR (dB) values for different pulse powers with and without filtering for defect depth 10%.

Pulse Power	Without Filtering	With Filtering
0 dB	14.48	17.55
6 dB	16.16	16.50
8 dB	13.68	17.68
10 dB	13.97	19.29

The results of the experiments show that the use of the PMFES-ECT method with increased pulse power and the use of filtering significantly increases the SNR value and the possibility of detecting deeply lying defects.

In the case of a 10% defect, the gain for PMFES-ECT was as high as 5 dB compared to the MFES-ECT method.

6. Conclusions

This article presents the novel pulsed multi-frequency eddy current method for non-destructive testing (PMFES-ECT). The method was initially verified by the computer simulations and then experimentally validated.

The proposed measurement method can be utilized for the inspection of conducting materials with different thicknesses. It combines the advantages of the multifrequency and pulse ECT methods. A rich frequency spectrum enables accurate and detailed identification of flaws, and the pulsed operating principle increases the range and detection of deeply located flaws. Moreover, it also reduces energy consumption, which is essential in portable systems.

The spectrogram and peak frequency enable the determination of the depth of the defect and location in examined material, while the maximum amplitude of the spectrogram is correlated with the defect size.

In future work, this method could be used together with artificial intelligence methods to create an autonomous, portable system for detecting and identifying defects.

Author Contributions: Idea and concept of the PMFES-ECT method, T.C.; implementation of the method, J.M.G.; software preparation, J.M.G.; resources—transducer, samples and system elements preparation, T.C.; measurement, J.M.G.; data curation, J.M.G.; writing—original draft preparation, J.M.G.; writing—correction and extensive editing, T.C. and J.M.G.; visualization, J.M.G.; supervision, T.C.; All authors have read and agreed to the published version of the manuscript.

Funding: This research received no external funding.

Institutional Review Board Statement: Not applicable.

Informed Consent Statement: Not applicable.

Data Availability Statement: The data presented in this study are available on request from the corresponding author. The data are not publicly available due to a complicated structure that requires additional explanations.

Conflicts of Interest: The authors declare no conflict of interest. The funders had no role in the design of the study; in the collection, analyses, or interpretation of data; in the writing of the manuscript, or in the decision to publish the results.

References

1. Yang, G.; Tamburrino, A.; Udpa, L.; Udpa, S.S.; Zeng, Z.; Deng, Y.; Que, P. Pulsed eddy-current based giant magnetoresistive system for the inspection of aircraft structures. *IEEE Trans. Magn.* **2010**, *46*, 910–917. [CrossRef]
2. Camerini, C.; Rebello, J.M.A.; Braga, L.; Santos, R.; Chady, T.; Psuj, G.; Pereira, G. In-line inspection tool with eddy current instrumentation for fatigue crack detection. *Sensors* **2018**, *18*, 2161. [CrossRef] [PubMed]
3. Cartz, L. *Nondestructive Testing*; ASM International: Cleveland, OH, USA, 1995; ISBN 978-0-87170-517-4.
4. Kacprzak, D.; Taniguchi, T.; Nakamura, K.; Yamada, S.; Iwahara, M. Novel eddy current testing sensor for the inspection of printed circuit boards. *IEEE Trans. Magn.* **2001**, *37*, 2010–2012. [CrossRef]
5. Nonaka, Y. A double coil method for simultaneously measuring the resistivity, permeability, and thickness of a moving metal sheet. *IEEE Trans. Instrum. Meas.* **1996**, *45*, 478–482. [CrossRef]
6. Holler, P.; Becker, R.; Sharpe, R.S. *The Application of Eddy Currents in Weld Testing*; Le Soudage Dans Le Monde: Roissy, France, 1984; Volume 22, pp. 164–177. ISSN 0043-2288.
7. Gilstad, C.W.; Dersch, M.F.; Denale, R. Multi-Frequency Eddy Current Testing of Ferromagnetic Welds. In *Review of Progress in Quantitative Nondestructive Evaluation*; Thompson, D.O., Chimenti, D.E., Eds.; Springer: Boston, MA, USA, 1990. [CrossRef]
8. Mina, M.; Yim, J.; Udpa, S.S.; Udpa, L.; Lord, W.; Sun, K. Two Dimensional Multi-Frequency Eddy Current Data Fusion. In *Review of Progress in Quantitative Nondestructive Evaluation*; Thompson, D.O., Chimenti, D.E., Eds.; Springer: Boston, MA, USA, 1996. [CrossRef]
9. Davis, T.J.; Perry, C.B. Multi Frequency Eddy Current Test Apparatus and Method. European Patent Application EP 0 035 391 A2, 9 September 1981. Available online: https://data.epo.org/publication-server/document?iDocId=22264&iFormat=2 (accessed on 3 September 2021).
10. Schiller, D.; Meier, T.; Bical, D. Improvement of the reliability of fatigue crack detection on holes of typical aircraft structures by using multi-frequency EC-technique. In Proceedings of the ATA NDT FORUM NDT & "The Human Element", Albuquerque, NM, USA, 27 September–1 October 1998; Available online: https://www.ndt.net/article/v04n01/ata_dasa/ata_dasa.htm (accessed on 3 September 2021).
11. Chady, T.; Enokizono, M. Multi-frequency exciting and spectrogram-based ECT method. *J. Magn. Magn. Mater.* **2000**, *215*, 700–703. [CrossRef]
12. Chady, T. Inspection of Clad Materials Using Massive Multifrequency Excitation and Spectrogram Eddy Current Method. In Proceedings of the 19th World Conference on Nondestructive Testing 2016, Munich, Germany, 13–17 June 2016; Available online: https://www.ndt.net/article/wcndt2016/papers/p76.pdf (accessed on 20 July 2021).
13. Pavo, J. Numerical calculation method for pulsed eddy-current testing. *IEEE Trans. Magn.* **2002**, *38*, 1169–1172. [CrossRef]
14. Nestleroth, J.B.; Davis, R.J. Application of eddy currents induced by permanent magnets for pipeline inspection. *NDT E Int.* **2007**, *40*, 77–84. [CrossRef]
15. Chady, T.; Grochowalski, J.M. Eddy current transducer with rotating permanent magnets to test planar conducting plate. *Sensors* **2019**, *19*, 1408. [CrossRef] [PubMed]

materials

MDPI

Article

Examining Ferromagnetic Materials Subjected to a Static Stress Load Using the Magnetic Method

Tomasz Chady [1,*] and Ryszard Łukaszuk [2]

[1] Faculty of Electrical Engineering, West Pomeranian University of Technology in Szczecin, Sikorski St. 37, 70-313 Szczecin, Poland
[2] Independent Researcher, Sikorski St. 37, 70-313 Szczecin, Poland; tramwajarz.szczecinski@gmail.com
* Correspondence: tchady@zut.edu.pl; Tel.: +48-91-449-41-34

Abstract: This paper discusses the experimental examination of anisotropic steel-made samples subjected to a static stress load. A nondestructive testing (NDT) measurement system with a transducer, which enables observation of local hysteresis loops and detection of samples' inhomogeneity, is proposed. Local hysteresis loops are measured on two perpendicular axes, including one parallel to the rolling direction of the samples. The results confirm that the selected features of the local hysteresis loops provide important information about the conditions of ferromagnetic materials. Furthermore, it is shown that the selected parameters of the statistical analysis of the achieved measurements are beneficial for evaluating stress and fatigue changes induced in the material.

Keywords: nondestructive evaluation; magnetic methods of testing; NDT

Citation: Chady, T.; Łukaszuk, R. Examining Ferromagnetic Materials Subjected to a Static Stress Load Using the Magnetic Method. *Materials* **2021**, *14*, 3455. https://doi.org/10.3390/ma14133455

Academic Editor: Giovanni Bruno

Received: 19 May 2021
Accepted: 15 June 2021
Published: 22 June 2021

1. Introduction

Steel is susceptible to the harmful effects of certain external environmental factors. For this reason, it is necessary to subject steel products to examination at the stage of both production and operation. If the internal and external structure of the object must remain intact, nondestructive testing (NDT) is performed. Detection of small inhomogeneities in the material allows us to observe degradation at an early stage, reducing the possibility of a catastrophic failure and alternative repair costs. The good electrical conductivity and high permeability of steel create possibilities to detect discontinuities in its structure using electromagnetic methods of NDT. The electromagnetic methods have high sensitivity, so apart from detecting a defect, it is also possible to pinpoint its location and assess its dimensions. In the case of steel-made sheets rolled in the direction opposite to the grain orientation, magnetic anisotropy is a particular obstacle during an examination. In addition, anisotropy can also be induced by stress. This results in the need for the inspection to be carried out in two orthogonal directions of the material.

Ferromagnetic materials can be tested using several electromagnetic methods of NDT:

- the magnetic flux leakage method is based on observation of the magnetic flux distribution over the material surface [1]. The primary magnetic field source causes a magnetic flux in the material. A barrier to the secondary flux is any inhomogeneity in the material structure that has a significant reluctance value [2]. The flux leakage method allows us to assess the tested object's surface and subsurface inhomogeneity [3]. The main advantages are high sensitivity, easiness of signal acquisition, and the possibility of automation [2,4]. However, this method also has some disadvantages, including sensitivity to material impurities and the need to magnetize the object [5];
- the magnetic particle inspection method allows for the detection of both surface and subsurface heterogeneities [6]. First, the sample is exposed to an external magnetic field, whereby magnetic powder particles can be placed on the outer surface of the sample in two ways: during the magnetization or after switching the magnetic field source off. The magnetic flux dispersing on the inhomogeneities appears on the

material's surface and changes the distribution of the particles [7]. The resulting image contains foci of particles that indicate the material heterogeneities [8]. Instead of magnetic powder, a suspension liquid can also be used to enhance the inspection sensitivity. Nowadays, apart from traditional indicators such as magnetic powders and suspension liquids, GMR or Hall sensors are also applicable to the inspection. The magnetic particle method is a quick, inexpensive, and relatively uncomplicated inspection method that gives immediate indications of surface and near-surface defects.

- the eddy current testing method is based on observing the flow path of the induced currents in the examined material [9]. The excitation magnetic field source induces eddy currents in the material. Disturbances in eddy current flow caused by inhomogeneities become apparent in the resultant field [10]. The advantages of testing with eddy currents are the high efficiency of detecting even the most minor defects, no need for direct access, and the penetration of many layers of material [11,12]. The main disadvantage of this method is that it only detects defects located not too deep under the surface due to the skin effect, which is especially strong in the case of ferromagnetic materials [13];
- the Barkhausen noise method relies on observing the magnetization process, which causes the dipoles to rotate [14–16]. If the material contains inhomogeneities, the process of domain wall shifting will be disrupted. As a result, sudden magnetization changes induce voltage pulses, which become apparent and can be observed as Barkhausen noise [17,18]. The Barkhausen noise testing method may be beneficial in some cases because of its low cost, high reliability, and simplicity [19]. The method is especially useful for stress monitoring. However, it has also some drawbacks, such as limited sensitivity resulting from thermal effects [20];
- the 3MA is an approach that combines features of four NDT methods: the Barkhausen noise, eddy current, incremental permeability, and harmonic analysis of magnetic field strength methods. Using several methods simultaneously reduces the likelihood of inconclusive inspection results [21];
- the magnetic memory method is based on the measurement of the residual magnetization, which appears in the material under the influence of a stress load or external geomagnetic fields [22]. The residual magnetization is recorded using sensors and then analyzed to assess defects [23]. The advantages of this method are the possibility to detect failures at an early stage, the lack of a need to provide an external magnetic field, and its simplicity [24]. A significant drawback is that it generally can be used only as an auxiliary method because of its low accuracy [24];
- the hysteresis loop observation method is another method aimed at localizing stress and heterogeneities. The microstructure of the ferromagnetic material strongly affects the hysteresis loop shape [25]. If the tested object is subjected to stress, the coercivity field and remanence induction values change due to the displacement of the dipoles separated by Bloch walls [26,27].

This work presents an examination of steel-made samples subjected to a static tensile stress load. The raw data acquired during the observation of the local hysteresis loops in two perpendicular directions were normalized, visualized, and analyzed using an approach based on statistics.

2. Materials and Methods

The subject of the examination consisted of specimens made of SS400 (JIS3101 standard) steel with an applied tensile stress. SS400 is one of the most commonly used hot-rolled general structural carbon and low-alloy steels designated for such structures as bridges, ships, and rolling stocks.

The mechanical properties of SS400 are as follows [28]: yield strength, 350 MPa; elastic modulus, 209 GPa; Poisson's ratio, 0.29; and chemical composition (wt), C—0.148%, Mn—0.458%, Si—0.213%, S—0.018%, P—0.012%, and Fe—bal.

The shape and dimensions of the SS400 specimens used in the experiments are shown in Figure 1. They are different from the standard tensile test specimens because a sufficient area for measurements has to be provided for the relatively large transducer. The samples were manufactured using a water jet cutting machine to avoid sharp edges, eliminate the heat-affected zone, and minimize introduced stress.

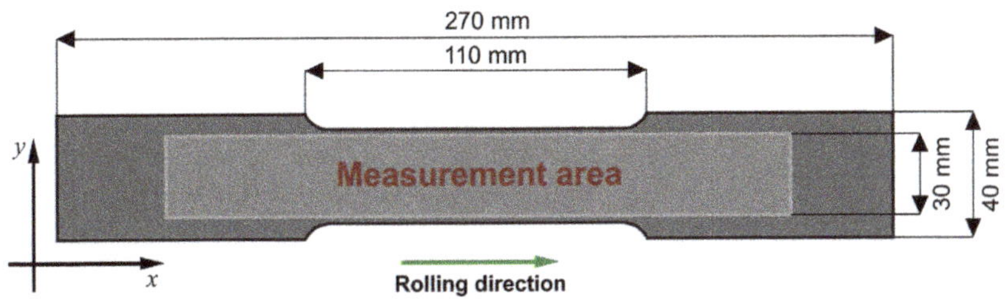

Figure 1. Shape and dimensions of the samples used in the experiments.

Before the experiment, each specimen (except the reference one, named S00) was subjected to a static tensile stress load at ambient temperature. The applied stress in the longitudinal direction (x-axis) was coincident with the rolling direction of the specimen.

Before the inspection, five samples (named S03, S04, S05, S06, and S07) were prepared and loaded using a standard Instron system. For each sample, the tensile stress test was stopped for different strain values (Figure 2). The residual strain ε was between 0.7% and 14%, while the maximum tensile stress was 389 MPa. Samples S04 and S05 were stress-loaded, respectively, up to the yield point (strain ε = 2%) and over it (strain ε = 2.4%). For illustrative purposes, all samples are marked on a single stress–strain curve, which is shown in Figure 2.

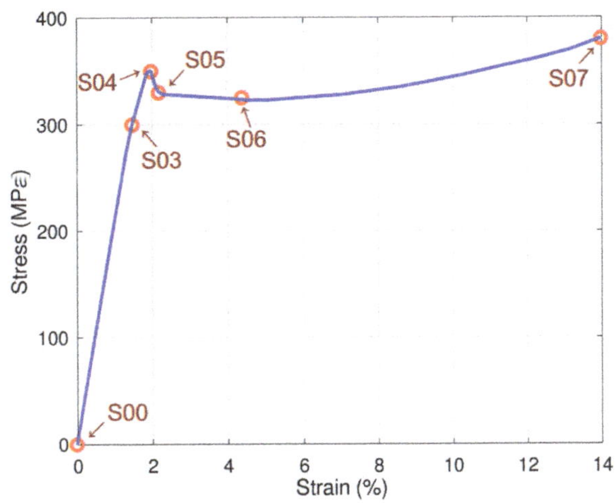

Figure 2. The stress–strain curve for the examined SS400 specimens.

Protection of the top specimen's surface from mechanical damage and the transducer from rupture of the measurement windings was provided by a polytetrafluoroethylene (PTFE) tape with a thickness of 0.2 mm. The samples were placed below the transducer one by one, attached to the scanning system's head, and the transducer was aligned with the x-axis direction (Figure 1). During the measurements, the scanning system's head was moved to the upper left corner of the area to be examined. The transducer moved along the x-axis at a 200 mm distance and stopped for 400 ms every 0.5 mm to read and record signal values. The procedure was repeated by analogy 25 times for other transducer positions (every 1 mm) varying along the y-axis. In the subsequent part of the inspection, the transducer was aligned with the y-axis direction (Figure 1) by rotating it 90 degrees clockwise. The measurement procedure with the rotated transducer was analogous to the measurements performed previously. The signal values were stored on a computer for further processing.

The main objective was to observe changes in the hysteresis loop caused by the stress applied to the samples before the inspection. The magnetization during the measurements was carried out twice on two perpendicular axes (one axis was parallel to the rolling direction).

Examination of the samples was carried out using a transducer. The transducer consists of a support plate, an auxiliary support, and three coils. The stillage carries a u-shaped ferrite core with two pick-up coils, while the excitation coil is attached to the auxiliary support (Figures 3 and 4). Selected transducer parameters are shown in Table 1.

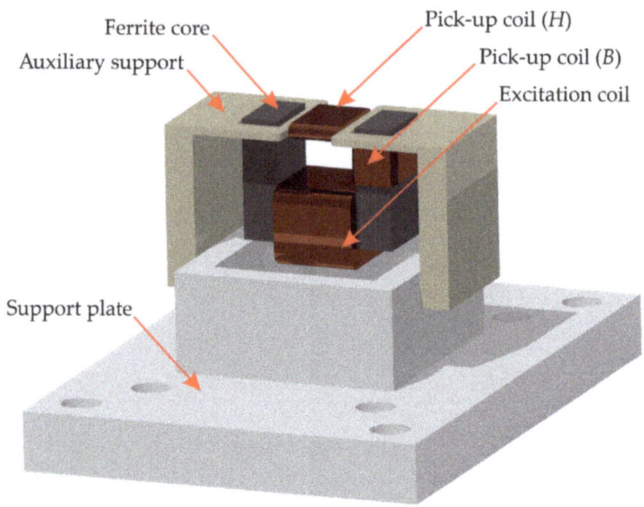

Figure 3. Transducer model.

Figure 4. Photo of the transducer.

Table 1. Transducer parameters.

Parameter Definition	Parameter	Value
Excitation current RMS value	I_e	400 mA
Frequency of the excitation	f_e	4.4 kHz
Low-pass filter cutoff frequency	f_l	50 kHz
Amplifier gain	G	30 dB
Ferrite core length	l_{c1}	10.5 mm
Ferrite core width	l_{c2}	5 mm
Ferrite core height	l_{c3}	8.2 mm
Pick-up coil (*H*) length	l_{h1}	3.2 mm
Pick-up coil (*H*) width	l_{h2}	2 mm
Pick-up coil (*H*) height	l_{h3}	4.75 mm
Pick-up coil (*H*) number of turns	n_H	90
Pick-up coil (*B*) length	l_{b1}	5 mm
Pick-up coil (*B*) width	l_{b2}	2.5 mm
Pick-up coil (*B*) height	l_{b3}	3 mm
Pick-up coil (*B*) number of turns	n_B.	100

The transducer works in the system shown in Figure 5. The system consists of a scanning device, a signal generator, A/D signal converters, amplifiers, the transducer, and a control computer. Parameters of the requested excitation signal (amplitude and frequency f_E = 4.4 kHz) are sent to the signal generator, which provides a relevant voltage signal to the power amplifier. The excitation frequency was selected considering the influence of noise, the level of induced voltages in the pick-up coils, and the results of preliminary experiments carried out with the SS400 sample. Subsequently, the boosted signal supplies an excitaon coil of the transducer. This coil induces a primary magnetic field, which flows through the transducer's ferromagnetic core and two pick-up coils (*B* and *H*) and penetrates the tested material. Then, the voltage signal from a field-sensing pick-up coil (*H*) reaches the instrumentation amplifier and, after being modified by a second-order Butterworth low-pass filter with a cutoff frequency of 50 kHz, enters an analog-to-digital converter. A signal from a flux-sensing pick-up coil (*B*) is passed directly to the second-order Butterworth low-pass filter with a cutoff frequency of 50 kHz and digitalized. Eventually, both voltage signal values are saved in the computer's memory for further analysis. Detailed information about the measurement system is provided in Table 1.

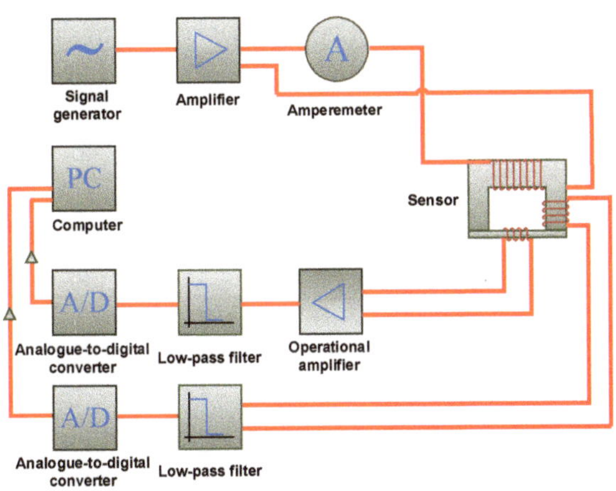

Figure 5. A block scheme of the measuring system.

The raw data collected from the two-dimensional scanning of the samples were processed using dedicated software. First, the data were normalized and corrected by removing distortions. Subsequently, the values of the hysteresis loop parameters were calculated for both axes, and graphs were plotted showing the relative change in a given parameter as a function of the coordinates of the point from the measurement area (Figure 1).

The following calculations were done to assess the condition of the examined samples. First, every single magnetic loop parameter in two perpendicular magnetization directions (1) was calculated.

$$\Delta k_{Amax}(x,y) = k_{Amax}(x,y) - k_{Amax}(x_0,y_0), \tag{1}$$

where: k—the chosen magnetic loop parameter, A—the selected magnetization direction, x,y—the measurement area point's coordinates, $k_{Amax}(x,y)$—the maximum values of the chosen magnetic loop parameter, and $k_{Amax}(x_0,y_0)$—a mean value calculated from measurements achieved for points located within a distance of 2 mm from a starting point having the coordinates (x_0,y_0). The starting point was located in the broader part of the sample, where the stress level was significantly lower. A similar point of construction will have to be selected as the reference point of the real tested structures.

Afterwards, normalized relative changes in the maximum (2) and minimum (3) magnetic induction, maximum (4) and minimum (5) magnetic field strength, as well as the area of the hysteresis loop (6) were calculated.

$$K_{A1}(x,y) = \frac{\Delta B_{Amax}(x,y)}{\max(|\Delta B_{Amax}(x,y)|)} \tag{2}$$

$$K_{A2}(x,y) = \frac{\Delta B_{Amin}(x,y)}{\max(|\Delta B_{Amin}(x,y)|)} \tag{3}$$

$$K_{A3}(x,y) = \frac{\Delta H_{Amax}(x,y)}{\max(|\Delta H_{Amax}(x,y)|)} \tag{4}$$

$$K_{A4}(x,y) = \frac{\Delta H_{Amin}(x,y)}{\max(|\Delta H_{Amin}(x,y)|)} \tag{5}$$

$$K_{A5}(x,y) = \frac{\Delta P_{A}(x,y)}{\max(|\Delta P_{A}(x,y)|)} \tag{6}$$

where: A—the magnetization direction, x,y—the measurement area point's coordinates, $\Delta B_{Amax}(x,y)$—the relative change in the maximum magnetic induction, $\Delta B_{Amin}(x,y)$—the relative change in the minimum magnetic induction, $\Delta H_{Amax}(x,y)$—the relative change in the maximum magnetic field strength, $\Delta H_{Amin}(x,y)$—the relative change in the minimum magnetic field strength, and $\Delta P_{A}(x,y)$—the relative change in the local hysteresis loop's area.

The next step concerned performing statistical analysis to assess the conditions of the material (after the stress loading). Initially, frequency histograms were plotted for the hysteresis loop parameters measured for all samples. Then, the focus was on performing statistical analysis of the data.

3. Results and Discussion

The measurements were done according to the procedure described in Section 2. As a result of these measurements, a set of signals was obtained for each sample, necessary to plot the hysteresis loop measured for magnetization in the x- and y-axis directions. The signals were achieved for all positions (x_i, y_i) over which the transducer was moved. Due to the design of the transducer and the shape of the samples, in the case of magnetization in the y-axis, the shift range in the y-axis was slightly smaller than the magnetization in the x-axis direction.

Next, the parameters defined by Equations (2)–(6) were calculated using the signals acquired for each of the measuring points. Examples of two-dimensional plots of these parameters are presented in Figures 6 and 7. The plots show changes in the parameter's value over the surface sample. It is possible to generally observe a good correlation between the quantities measured on both orthogonal axes, but also significant differences are visible. It allows for a hypothesis that these signals are complementary to each other. By analyzing the signals for samples S00, S03, and S04, it can be observed that the signal value increases significantly and then decreases in the case of S06 and S07. Similar trends occur both for the parameters K_{A4} (Figure 6) and K_{A5} (Figure 7).

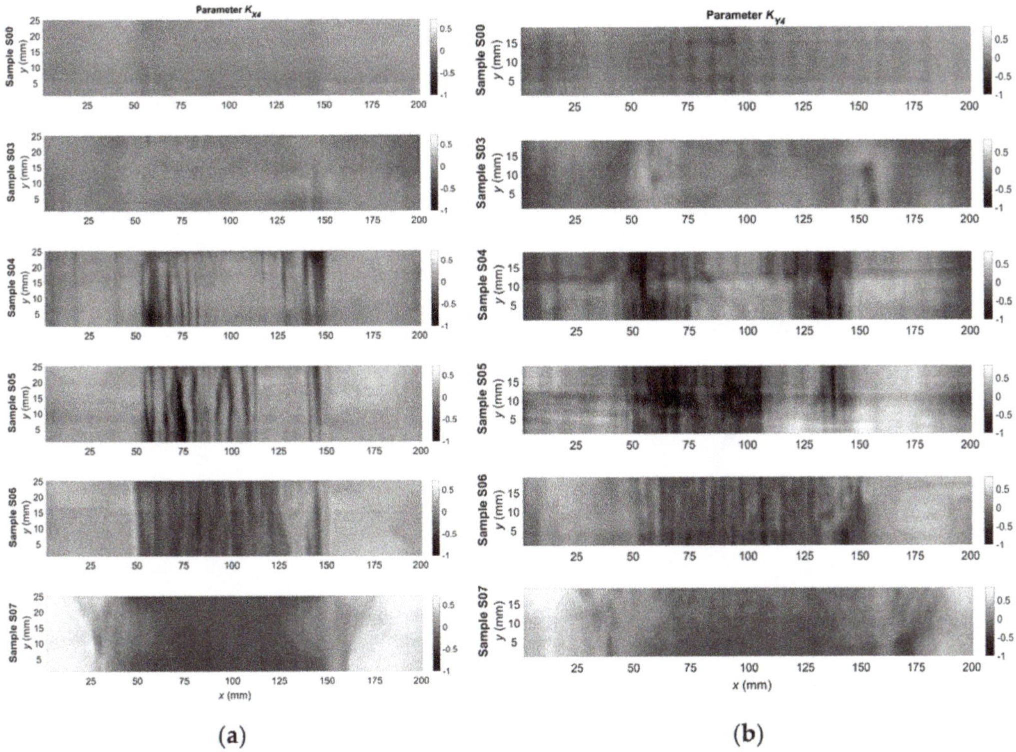

Figure 6. Relative change in the K_{A4} parameter as a function of the coordinates of the point from the measurement area for the: (**a**) *x*-axis direction; and (**b**) *y*-axis direction.

Despite the visible trends, assessing the load condition of the sample directly based on the measured signals is complex and may lead to ambiguity, for example, when using only the maximum value of the parameter. For this reason, attempts were made to statistically analyze the determined parameters to become independent from random changes in the signal value and take into account only the general trends.

In the first phase of the analysis, a frequency histogram was prepared for each of the parameters determined from measurements carried out for a single sample and one direction of magnetization (Figures 8 and 9). In the beginning, frequency histograms were subjected to normalization, relying on reducing the influence of interference signals. Then, the focus was on performing statistical analysis of the data by calculating and visualizing the following values: the maximum and minimum value, expected value, median, mode, variance, standard deviation, kurtosis, and histogram intervals. The frequency histograms achieved for samples with various levels of introduced stress show significant differences. For example, in the case of the K_{X4} parameter (Figure 8a), interesting changes in skewness and distribution modality can be observed. The distribution is almost symmetrical and unimodal for the undeformed sample, an evident skewness appears starting from sample S03, and from sample S06 the distribution becomes bimodal. The reason for the transition from a unimodal to a bimodal distribution is the increasing level of stress and dislocations in the internal structure of successive samples. In the second part of the analysis, statistical values for the K_{A4} and K_{A5} parameters on both axes were computed as already mentioned. However, some statistical values, such as variance, expected value, standard deviation, and kurtosis, could not eventually be taken into account due to non-monotonic changes and a problem with unequivocal identification of material conditions.

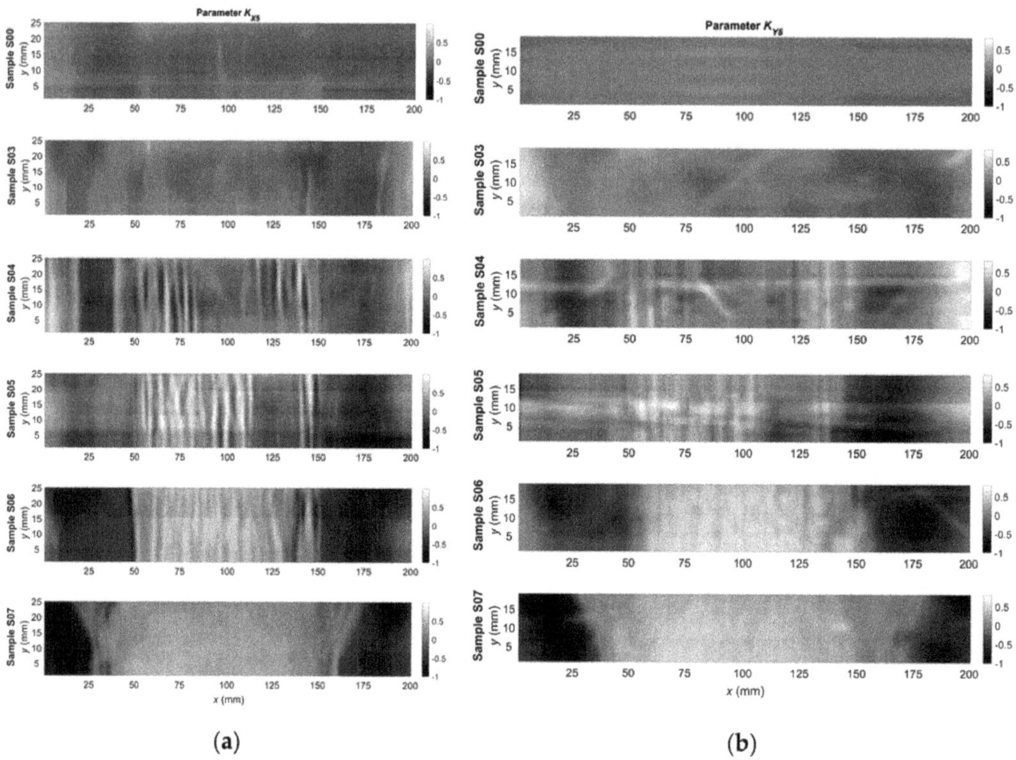

Figure 7. Relative change in the K_{A5} parameter as a function of the coordinates of the point from the measurement area for the: (**a**) x-axis direction; and (**b**) y-axis direction.

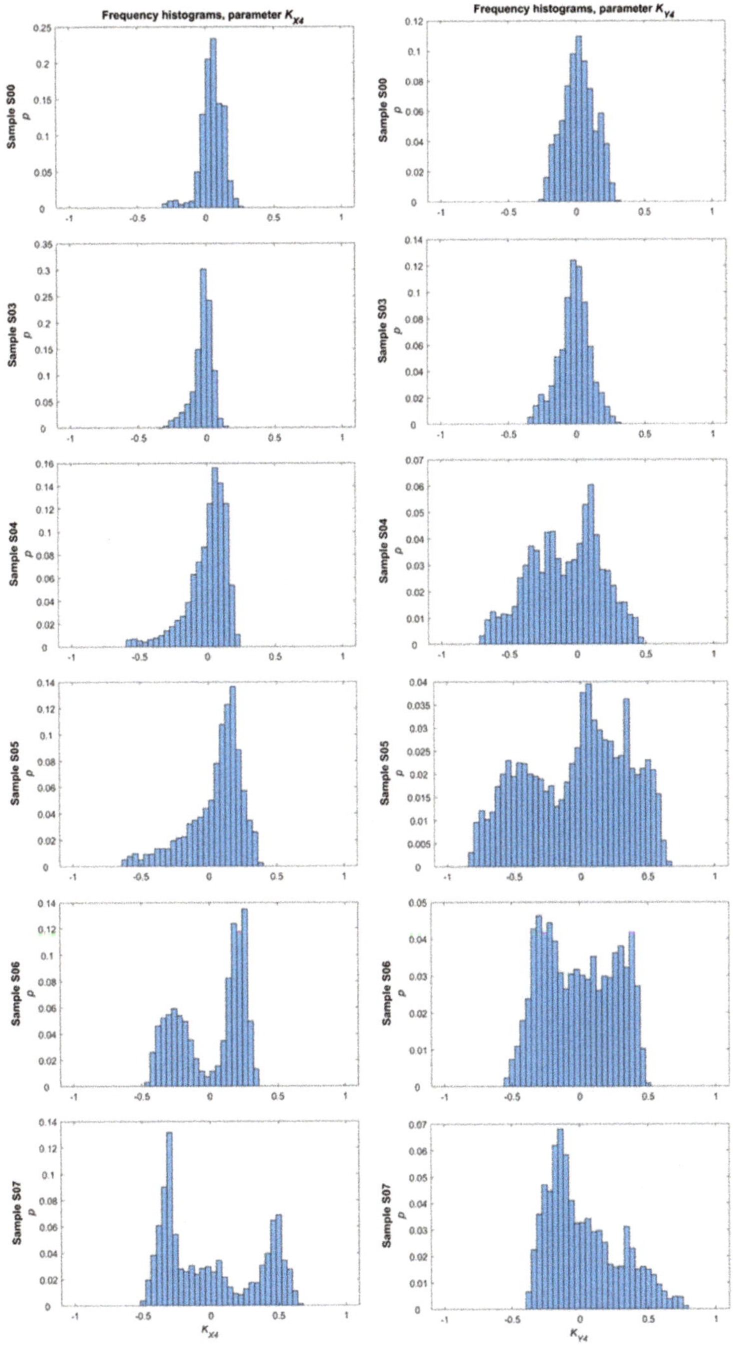

(a) (b)

Figure 8. Frequency histograms of the K_{A4} parameter for the: (**a**) *x*-axis direction; and (**b**) *y*-axis direction.

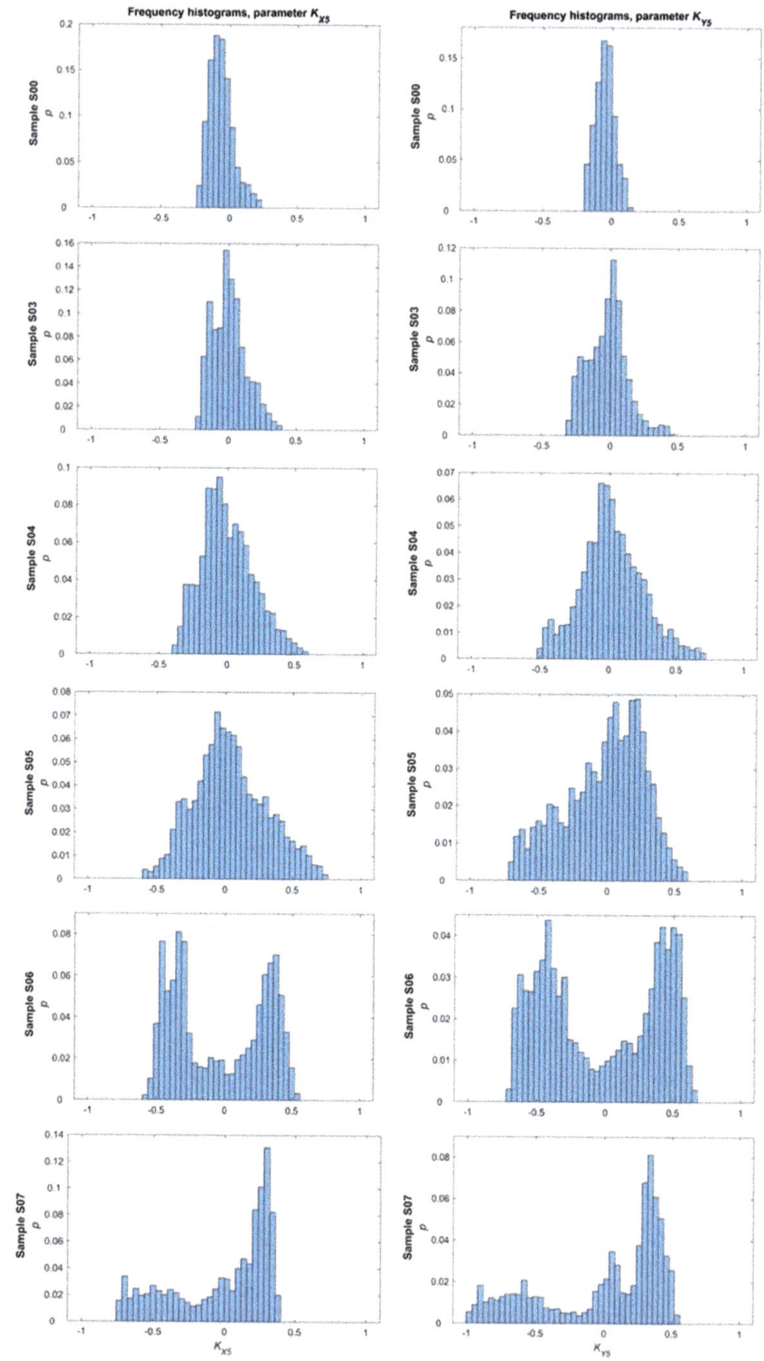

(a) (b)

Figure 9. Frequency histograms of the K_{A5} parameter for the: (**a**) x-axis direction; and (**b**) y-axis direction.

Figures 10 and 11 contain plots of selected statistical values, which create the best opportunity to evaluate the conditions of a given sample.

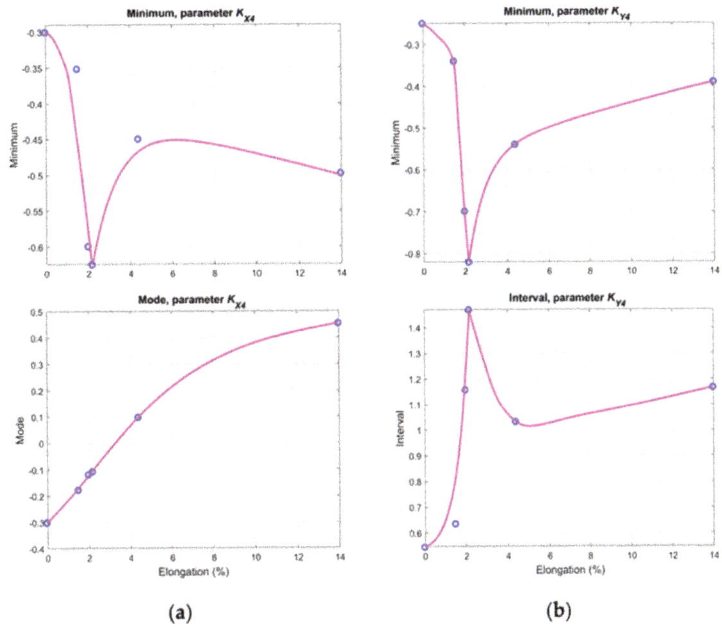

Figure 10. Chosen statistical values of the K_{A4} parameter for the: (**a**) *x*-axis direction; and (**b**) *y*-axis direction.

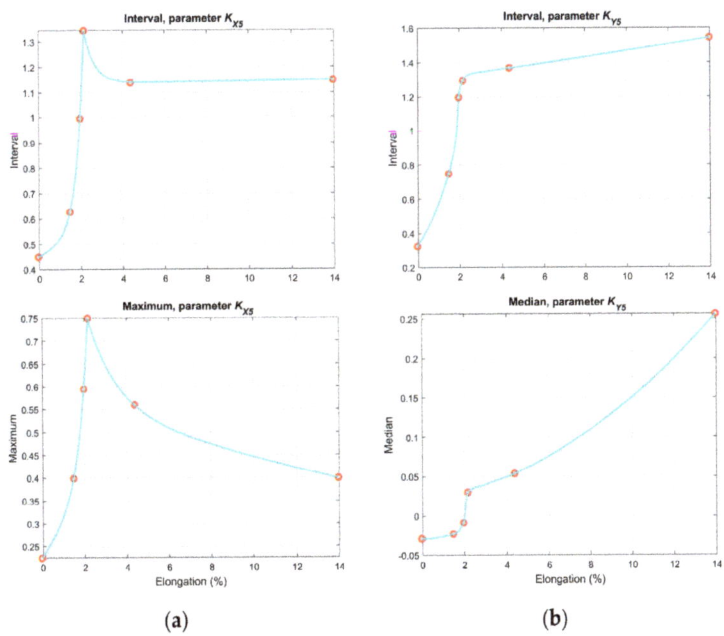

Figure 11. Chosen statistical values of the K_{A5} parameter for the: (**a**) *x*-axis direction; and (**b**) *y*-axis direction.

As shown in Figure 10, the minimum values of both the K_{X4} and K_{Y4} parameters decrease from the undamaged sample S00 to sample S05. Around specimens S05 and S06, the stress level caused the yield point to be exceeded, which can also be observed in the graphs as an inflection of the curve. After passing the yield point and increasing the stress level, the values decrease (in the case of the K_{X4} parameter) or increase (in the case of the K_{Y4} parameter). Mode and interval curves allow for the state of the samples to be determined straightforwardly as well. Graphs related to K_{X5} (Figure 11) indicate that the interval value increases up to the S05 sample and then the curves bend due to the growing number of inhomogeneities. Particularly beneficial are the interval and median values of the K_{Y5} parameter, enabling the identification of the sample's condition unequivocally (both before and after yield).

The main goal of all the tests and analyses was to find the parameters of the measured signals that would allow for the assessment of the condition of the tested structure in terms of the stresses they were subjected to. An important guess was to choose such individual parameters or groups of parameters that guarantee assessment in both the elastic and plastic regions. The use of histograms and statistical features made the results independent of randomly changing signals caused by disturbances and other external factors, such as surface unevenness. Therefore, the parameters presented in Figures 10 and 11 create a good chance that the assumed goals have been achieved. One of the important conclusions is the need to carry out measurements in at least two directions (parallel and perpendicular to the expected direction of stresses). A good example is the K_{A5} parameter measured in the y-axis direction (Figure 11b), which allows for an unambiguous identification of the state. Such parameters are crucial for building an automatic identification system in the future, which will use several of the presented parameters and the rough sets algorithm.

4. Conclusions

The achieved results allow us to conclude that the proposed approach based on non-destructive testing using observation of hysteresis loops and selected statistical analysis methods utilizing frequency histograms can be helpful to evaluate the condition of ferromagnetic materials subjected to a static stress load. Nevertheless, further research is needed to assess the usefulness of the statistical parameters and their applicability for testing other ferromagnetic materials (e.g., various construction steels). Based on the achieved results, it is recommended that not just one but several statistical parameters be used when assessing the state of a material under evaluation. In addition to the tests presented in this paper with the transducer aligned in the two perpendicular directions, additional measurements could also be performed for the transducer rotated at 45 degrees to the rolling axis. Such a methodology could provide additional information about the inhomogeneities in the material (e.g., higher sensitivity in the case of Lüders band detection). Therefore, we plan to develop an integrated transducer consisting of three directional sensors.

Author Contributions: Conceptualization and methodology, T.C. and R.Ł.; software, R.Ł.; validation, T.C. and R.Ł.; formal analysis, T.C. and R.Ł.; investigation and measurement, T.C. and R.Ł.; resources, T.C.; data curation, R.Ł.; writing—original draft preparation, R.Ł.; writing—review and editing, T.C. and R.Ł.; visualization, R.Ł.; supervision, T.C.; All authors have read and agreed to the published version of the manuscript.

Funding: This research received no external funding.

Institutional Review Board Statement: Not applicable.

Informed Consent Statement: Not applicable.

Data Availability Statement: The data presented in this study are available on request from the corresponding author. The data are not publicly available due to a complicated structure that requires additional explanations.

Conflicts of Interest: The authors declare no conflict of interest. The funders had no role in the design of the study; in the collection, analyses, or interpretation of data; in the writing of the manuscript; or in the decision to publish the results.

References

1. Pitoňák, M.; Neslušan, M.; Minárik, P.; Čapek, J.; Zgútová, K.; Jurkovič, M.; Kalina, T. Investigation of Magnetic Anisotropy and Barkhausen Noise Asymmetry Resulting from Uniaxial Plastic Deformation of Steel S235. *Appl. Sci.* **2021**, *11*, 3600. [CrossRef]
2. Zhang, J.; Peng, F.; Chen, J. Quantitative Detection of Wire Rope Based on Three-Dimensional Magnetic Flux Leakage Color Imaging Technology. *IEEE Access* **2020**, *8*, 104165–104174. [CrossRef]
3. Shimin, P.; Donglai, Z.; Enchao, Z. Analysis of the Eccentric Problem of Wire Rope Magnetic Flux Leakage Testing. In Proceedings of the 2019 IEEE 3rd Information Technology, Networking, Electronic and Automation Control Conference (ITNEC 2019), Chengdu, China, 15–17 March 2019. [CrossRef]
4. Shi, Y.; Zhang, C.; Li, R.; Cai, M.; Jia, G. Theory and Application of Magnetic Flux Leakage Pipeline Detection. *Sensors* **2015**, *15*, 31036–31055. [CrossRef] [PubMed]
5. Tsukada, K.; Yoshioka, M.; Toshihiko, K.; Joshinobu, H. A magnetic flux leakage method using a magnetoresistive sensor for nondestructive evaluation of spot welds. *NDT Int.* **2011**, *44*, 101–105. [CrossRef]
6. Wong, B.S.; Low, Y.G.; Wang, X.; Ho, J.; Tan, C.H.; Boon, O.J. 3D Finite Element Simulation of Magnetic Particle Inspection. In Proceedings of the 2010 IEEE Conference on Sustainable Utilization and Development in Engineering and Technology, Kuala Lumpur, Malaysia, 20–21 November 2010; pp. 50–55. [CrossRef]
7. Mouritz, A.P. Nondestructive Inspection and Structural Health Monitoring of Aerospace Materials. *Introd. Aerosp. Mater.* **2012**, 534–557. [CrossRef]
8. Wang, M.L.; Lynch, J.P.; Sohn, H. Sensing Solutions for Assessing and Monitoring Underwater Systems. Sens. *Technol. Civ. Infrastruct.* **2014**, 525–549. [CrossRef]
9. Xie, Y.; Li, J.; Tao, Y.; Wang, S.; Yin, W.; Xu, L. Edge Effect Analysis and Edge Defect Detection of Titanium Alloy Based on Eddy Current Testing. *Appl. Sci.* **2020**, *10*, 8796. [CrossRef]
10. Socheatra, S. Printed Circuit Board Interconnect Fault Inspection Based on Eddy Current Testing. In Proceedings of the 2014 5th International Conference on Intelligent and Advanced Systems (ICIAS), Kuala Lumpur, Malaysia, 3–5 June 2014; pp. 1–4. [CrossRef]
11. Repelianto, A.; Naoya, K. The Improvement of Flaw Detection by theConfiguration of Uniform Eddy Current Probes. *Sensors* **2019**, *19*, 397. [CrossRef] [PubMed]
12. Neugebauer, R.; Drossel, W.G.; Mainda, P.; Roscher, H.J.; Wolf, K.; Kroschk, M. Sensitivity Analysis of Eddy Current Sensors Using Computational Simulation. In Proceedings of the Progress In Electromagnetics Research Symposium, Suzhou, China, 12–16 September 2011; pp. 441–445.
13. Liu, S.; Yanhua, S.; Xiaoyuan, J.; Yihua, K. A Review of Wire Rope Detection Methods, Sensors and Signal Processing Techniques. *J. Nondestruct. Eval.* **2020**, *39*, 85. [CrossRef]
14. Krkoška, L. Investigation of Barkhausen Noise Emission in Steel Wires Subjected to Different Surface Treatments. *Coatings* **2020**, *10*, 912. [CrossRef]
15. Jurkovič, M.; Kalina, T.; Zgútová, K.; Neslušan, M.; Pitoňák, M. Analysis of Magnetic Anisotropy and Non-Homogeneity of S235 Ship Structure Steel after Plastic Straining by the Use of Barkhausen Noise. *Materials* **2020**, *13*, 4588. [CrossRef] [PubMed]
16. Čilliková, M.; Mičietová, A.; Čep, R.; Mičieta, B.; Neslušan, M.; Kejzlar, P. Asymmetrical Barkhausen Noise of a Hard Milled Surface. *Materials* **2021**, *14*, 1293. [CrossRef] [PubMed]
17. Prabhu, G.N.G.; Nlebedim, I.C.; Prabhu, G.G.V.; Jiles, D.C. Examining the Correlation Between Microstructure and Barkhausen Noise Activity for Ferromagnetic Materials. *IEEE Trans. Magn.* **2015**, *51*, 1–4. [CrossRef]
18. Bui, T.M.T. Toward a Model of Barkhausen Noise Measurement System. In Proceedings of the 2011 International Conference on Advanced Technologies for Communications (ATC 2011), Da Nang, Vietnam, 2–4 August 2011; pp. 315–318. [CrossRef]
19. Vourna, P.; Ktena, A.; Tsakiridis, P.E.; Hristoforou, E. An accurate evaluation of the residual stress of welded electrical steels with magnetic Barkhausen noise. *Measurement* **2015**, *71*, 31–45. [CrossRef]
20. Franco, G.F.A.; Padovese, L.R. Non-destructive scanning for applied stress by the continuous magnetic Barkhausen noise method. *J. Magn. Magn. Mater.* **2018**, *446*, 231–238. [CrossRef]
21. Wolter, B.; Gabi, Y.; Conrad, C. Nondestructive Testing with 3MA—An Overview of Principles and Applications. *Appl. Sci.* **2019**, *9*, 1068. [CrossRef]
22. Zhang, Z. Wavelet Energy Entropy Based Multi-Sensor Data Fusion for Residual Stress Measurement Using Innovative Intense Magnetic Memory Method. In Proceedings of the 2009 9th International Conference on Electronic Measurement Instruments, Beijing, China, 16–19 August 2009; pp. 1044–1047. [CrossRef]
23. Bao, S.; Fu, M.; Hu, S.; Gu, Y. A Review of the Metal Magnetic Memory Technique. In Proceedings of the ASME 2016 35th International Conference on Ocean, Offshore and Arctic Engineering, Busan, Korea, 19–24 June 2016.
24. Wan, B. Research on the Damage Experiment Model Design and Metal Magnetic Memory Testing of Wind Turbine Tower in Service. In Proceedings of the 2020 IEEE 4th Conference on Energy Internet and Energy System Integration (EI2), Wuhan, China, 30 October–1 November 2020; pp. 1879–1884. [CrossRef]

25. Fasheng, Q.; Wenwei, R.; Gui, Y.T.; Yunlai, G.; Bin, G. The effect of stress on the domain wall behavior of high permeability grain-oriented electrical steel. In Proceedings of the 2015 IEEE Far East NDT New Technology & Application Forum (FENDT), Zhuhai, China, 28–31 May 2015. [CrossRef]
26. Liu, J.; Yun, T.G.; Gao, B.; Zeng, K.; Zheng, Y.; Chen, Y. Micro-Macro Characteristics between Domain Wall Motion and Magnetic Barkhausen Noise under Tensile Stress. *J. Magn. Magn. Mater.* **2020**, *493*, 165719. [CrossRef]
27. Ding, S.; Tian, G.Y.; Dobmann, G.; Wang, P. Analysis of Domain Wall Dynamics Based on Skewness of Magnetic Barkhausen Noise for Applied Stress Determination. *J. Magn. Magn. Mater.* **2017**, *421*, 225–229. [CrossRef]
28. Lee, Y.H.; Ji, W.; Kwon, D. Stress measurement of SS400 steel beam using the continuous indentation technique. *Exp. Mech.* **2004**, *44*, 55–61. [CrossRef]

materials MDPI

Article

Extended Damage Detection and Identification in Aircraft Structure Based on Multifrequency Eddy Current Method and Mutual Image Similarity Assessment

Tomasz Chady [1,*], Krzysztof Okarma [1], Robert Mikołajczyk [1], Michał Dziendzikowski [2], Piotr Synaszko [2] and Krzysztof Dragan [2]

1 Faculty of Electrical Engineering, West Pomeranian University of Technology in Szczecin, 70-313 Szczecin, Poland; okarma@zut.edu.pl (K.O.); robertmikolajczyk93@gmail.com (R.M.)
2 Airworthiness Division, Air Force Institute of Technology, 01-494 Warsaw, Poland; michal.dziendzikowski@itwl.pl (M.D.); piotr.synaszko@itwl.pl (P.S.); krzysztof.dragan@itwl.pl (K.D.)
* Correspondence: tchady@zut.edu.pl

Abstract: In this paper, a novel approach to Non-Destructive Testing (NDT) of defective materials for the aircraft industry is proposed, which utilizes an approach based on multifrequency and spectrogram eddy current method combined with an image analysis method previously applied for general-purpose full-reference image quality assessment (FR IQA). The proposed defect identification method is based on the use of the modified SSIM4 image quality metric. The developed method was thoroughly tested for various locations, sizes, and configurations of defects in the examined structure. Its application makes it possible to not only determine the presence of cracks but also estimate their size.

Keywords: Non-Destructive Testing; aerospace testing; multifrequency and spectrogram eddy current testing; machine vision; image quality

Citation: Chady, T.; Okarma, K.; Mikołajczyk, R.; Dziendzikowski, M.; Synaszko, P.; Dragan, K. Extended Damage Detection and Identification in Aircraft Structure Based on Multifrequency Eddy Current Method and Mutual Image Similarity Assessment. *Materials* **2021**, *14*, 4452. https://doi.org/10.3390/ma14164452

Academic Editor: Giovanni Bruno

Received: 6 July 2021
Accepted: 5 August 2021
Published: 9 August 2021

1. Introduction

Structure diagnostics in terms of damage presence determination is a vital factor in aircraft maintenance systems. At present, most of the used diagnostic techniques include mainly visual inspections and those using Non-Destructive Testing–NDT [1]. Such methods are a part of damage tolerance or fail-safe methodology [2], which assumes the possible existence of failure mode. Each method used for the detection should be adequately selected and tailored to the specified failure mode, including estimating the possibility of detecting damage size [3].

The main phenomena that may lead to catastrophic aircraft damage are corrosion, stress corrosion cracking, and fatigue cycle cracking [4]. In most cases, aerial structures deal with the following damages: fatigue cracking under rivet head in riveted joints, multiple site damage which may lead to widespread fatigue damage, hidden corrosion, cracking, and corrosion in multilayer joints, as well as cracking phenomena due to stress and corrosion [4]. For damage detection of fatigue cracking phenomena, the eddy current method is predominant [5]. The use of eddy current is related to aluminum alloy electric properties, allowing for tests with this method. Moreover, cracking damage geometry is perpendicular to generated eddy current paths in a material that makes the detection easier due to the changes in the electric impedance of the detection unit [6].

Testing with the eddy current method has a comparative character and requires calibration. It constitutes a significant limitation, especially in diagnosing complex geometry elements and multilayer ones [7]. On the other hand, the development of numerical simulation methods allows for creating models that make it possible to determine the distribution of electromagnetic fields and to calculate eddy current characteristics, which may, in turn, make an additional reference to improve the diagnostic process [7,8].

According to the industry survey of structural health monitoring technology and NDT usage [9], crack detection is one of the most essential and current aerospace industry needs in implementing the NDT and Structural Health Monitoring (SHM) systems. The most problematic are inspections of multilayer riveted structures for detecting fatigue cracks propagating from the rivet hole, especially in lower layers [10]. For such structures, the crack size and depth determination are crucial issues from the detection accuracy point of view.

As Mottl proposed using the Dodd and Deeds model [11], detection parameters may be determined theoretically for the selected probe type and geometry. However, due to the local variations of the conductivity, surface treatment, and specific types of probes, as Mook et al. indicate [12], the current density may differ from the theoretical ones, which implies the design of probes.

Several types of probes can be used for damage position, size, and orientation determination: manual (reflection, absolute, differential), rotating and sliding [13].

Several constraints in selecting the proper inspection approach should be considered, such as time of inspection, geometrical implications, surface treatment conditions, the possibility of damage detection, and sensitivity [14,15].

Therefore, an application of the NDT methods for materials used in the aircraft industry is one of the most demanding areas of research in the era of Industry 4.0, integrating the knowledge of electromagnetic measurements, particularly the eddy current method, with data analysis methods originating from signal and image processing domain.

In many cases, the results of non-destructive measurements take the form of 2D or 3D images, and therefore image analysis methods are of particular interest. Regardless of the potential use of image quality assessment (IQA) methods for an automatic evaluation of the usefulness of individual images or measurements for further analysis, some of the IQA metrics might also be applied as tools for detecting the presence of material defects or even for their classification.

Unfortunately, the applicability of many full-reference (FR) metrics, considered the most popular and the most universal, may be limited by assuming the necessary knowledge of the reference image, which may be unavailable in many practical applications [16–19]. Another limitation is the prerequisite perfect spatial adjustment and identical size (in terms of pixel resolution) of both compared images, which may require additional image registration and cropping. Therefore, a potential application of the FR IQA metrics for NDT purposes would require their modifications, as proposed in this paper, where the idea of the mutual similarity evaluation is considered.

The rest of the paper is organized as follows: In Section 2, the inspection approach is presented (Section 2.1), followed by discussing measurements (Section 2.2) and the discussion of similarity-based IQA methods (Section 2.3). The proposed method of image analysis is described in Section 2.4, and the discussion of the obtained results is provided in Section 3, followed by concluding remarks.

2. Materials and Methods

2.1. Approach to Inspection

An example of non-destructive technologies for structural integrity monitoring of the aerial components considered in this paper is the eddy current inspection performance on the PZL-130 Orlik aircraft structure. Within the conducted Full-Scale Fatigue Test (FSFT) for that aircraft, a program of detailed NDT inspections was scheduled. A part of that program was the Eddy Current inspection delivery, especially for critical parts of the wing structure. Inspection delivered during the FSFT revealed the existence and development of more than 100 failure modes, which were detected and described.

According to the best practice standards, the exemplary failure modes that can be found during the aircraft inspection are as follows [20]: hole peripheral cracks, through-thickness cracks in radius, hole through-thickness cracks, and hole radial countersink cracks.

Several eddy current techniques were implemented for damage detection, such as high-frequency reflection probes, high- and low-frequency automated scanning, and the magneto-optic imaging MOI™ system for the general and detailed structure visualization. Figure 1 presents the example of the wing structure automated inspection result using the MAUS® mobile system and C-scan data presentation. Such presentation mode shows the signal value (impedance module) as a function of the sensor location.

The inspection was delivered with the 20 kHz for surface-breaking cracks and 1 kHz for the subsurface, consecutively. Scan resolution (pixel-to-pixel size) was equal to 1 mm. Based on collected data, the NDT operator provides data analysis using visual recognition of the anomaly, which may exist on the provided image. That analysis is being supported by signal classification and description tools embedded in the system. Moreover, the data export function enables more advanced analysis based on signal processing methods [21].

Such an inspection method has also benefited from data comparison possibility, a valuable function for the FSFT to monitor damage nucleation and damage progression. A more detailed analysis is possible for specific regions of interest (ROI) with the complex impedance plane analysis, signal evaluation, and size quantification tool.

Figure 2 presents the result of the failure presented as the hole radial countersink crack. The combined presentation of the real and imaginary part of the impedance allows evaluating damage severity and creates an opportunity to evaluate crack orientation and depth of location. However, with the lower frequency inspection, damage presence is not as clear as with higher ones and may cause some difficulties in recognizing deeper located flaws.

Once the FSFT inspection was completed, the need for an increase in the damage detection sensitivity, especially in the second or third layer, has arisen.

The approach to the detection and classification of damage considered in this article is based on using a probe adapted to detect damage in multilayer aluminum aircraft structures, an advanced multifrequency inspection method, and the mutual image similarity assessment.

From the observations and practical inspection, the use of a dedicated probe with the proper signal-to-noise ratio and efficient penetration depth is a required tool for detailed aircraft inspection. In the case of structures with multiple layers, it is also crucial to utilize many frequencies.

Figure 1. Eddy current scan of the wing structure.

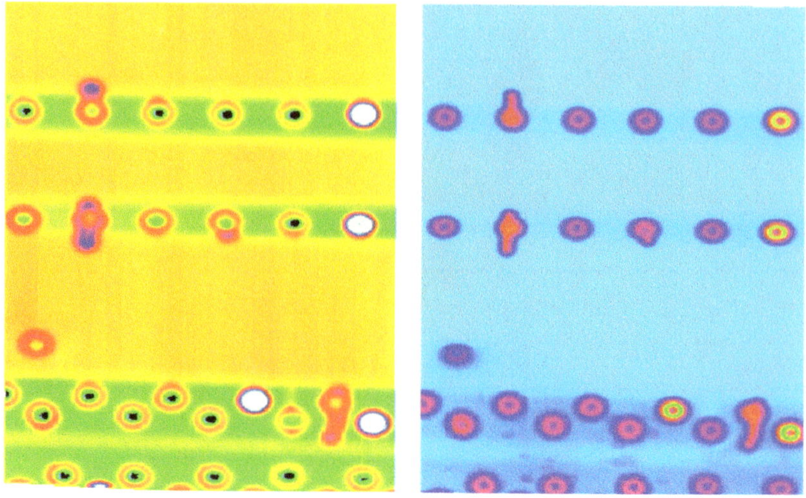

(a) (b)

Figure 2. Eddy current scan of the region of interest obtained for the 20 kHz excitation signal: (**a**) imaginary part; (**b**) real part.

Analysis of the various signals achieved from all the frequencies is a complicated and time-consuming procedure. For this reason, the automation of the defect detection and identification process is essential for the proper and effective functioning of the inspection system.

In the classical ECT systems, the use of several frequencies requires the application of several probes with valid central frequencies to obtain the estimated depth of penetration. Additional limitations also come from a decrease in the current density with an increase in depth of penetration. Such phenomena reduce inspection reliability and affect the probability of damage detection. Thus, the inspection of multilayer structures and geometry constraints requires specialized probes and signal processing to keep linearity of damage detection and thickness change. In the proposed multifrequency ECT system, many of these limitations were overcome.

2.2. Data Acquisition

In the eddy current method, the measurement sensitivity depends on several factors, such as the transducer dimensions, an excitation frequency, and properties of a tested structure (permeability and conductivity). Unfortunately, high sensitivity and a high resolution usually cannot be achieved at the same time. Enlargement of the probe dimensions will result in higher sensitivity (for deeply loaded defects) and a lower spatial resolution. Decreasing the excitation frequency will cause a similar effect. In testing multilayer aircraft structures, high and possibly similar sensitivity for detecting defects in different layers and at different depths is required. At the same time, it is necessary to maintain the appropriate spatial resolution. It is crucial to detect even the most minor cracks in the vicinity of holes after riveted connections. In this case, it is particularly advantageous to use a measuring system based on the Massive Multifrequency Excitation and Spectrogram Method (MMFES), proposed in [22]. This method uses a complex signal containing many sinusoidal components as an excitation signal and a spectrogram to detect and identify defects. A large variety of frequency components creates an opportunity for all defects to be detected using the most appropriate testing frequency. It is for the first time to apply the MMFES method for the inspection of riveted structures.

2.2.1. Measuring System Description

The measurements were carried out using the computer system implementing the Massive Multi-Frequency Excitation and Spectrogram (MMFES) method. In the MMFES system, a complex waveform is used as the excitation, created by adding sinusoidal components with precisely defined frequencies and amplitudes. The formula which describes the function defining the excitation voltage is as follows:

$$s_E(t) = \sum_{i=1}^{n} U_i \cdot \sin(2\pi f_i t + \phi_i). \quad (1)$$

where s_E—waveform of the excitation signal, n—number of harmonics, U_i—amplitude of the i-th component, f_i—amplitude of the i-th component, ϕ_i—phase of the i-th component.

The frequency range is selected in such a way as to include the values for which the maximum interaction of eddy currents with defects occurs at possible depths of the tested structure. The number of harmonic components n is selected in such a way as to achieve the highest possible signal to noise ratio at the eddy current transducer output. The relation of individual harmonics amplitude to the expected noise level and the possibility of filtering signals in the frequency domain are considered. The initial phases of individual harmonics do not significantly affect the ability to detect defects, but it is possible to limit the maximum value achieved by the excitation signal by appropriate selection. One of the critical issues is the selection of the excitation harmonic amplitudes. They are experimentally determined so that the harmonic components of the signal measured at the transducer output have the same value when it is located over the tested material that does not contain heterogeneity. The amplitude equalization process usually requires two to three iterations until the differences are satisfactorily reduced (i.e., below 1%). The experimental process allows eliminating the influence of the transducer frequency characteristic, lead wires, and other factors such as a lift-off and the inclination of the transducer concerning the tested material surface. The plot of excitation voltage measured on the output of the D/A converter as a function of time is shown in Figure 3. The amplitude spectrum of this voltage is shown in Figure 4.

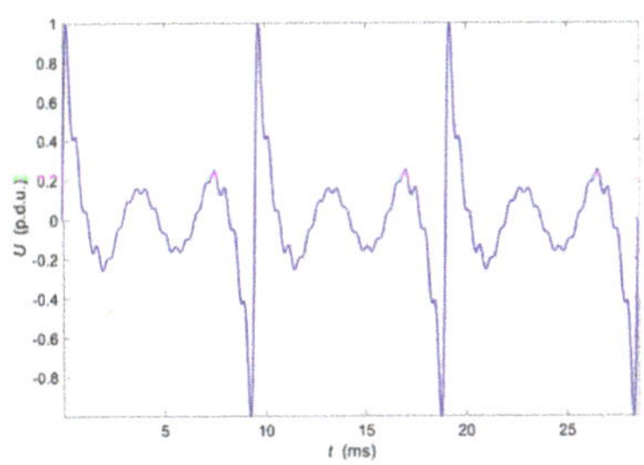

Figure 3. The plot of excitation voltage measured on the output of the D/A converter.

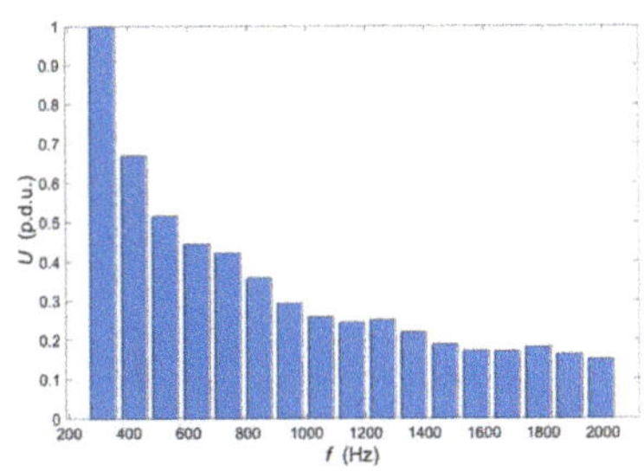

Figure 4. The amplitudes of the harmonics used to synthesize excitation voltage waveform shown in Figure 3.

The excitation driving voltage is generated in the D/A converter (NI PXI 5422-sample rate 200 MS/s, 80 MHz bandwidth, 16-Bit waveform generator) and then fed through the high-frequency power amplifier (NF HAS 4101, frequency range DC to 10 MHz, slew rate 5000 V/µs, max. current 1.4 A, amplification gain 20) to the excitation coils of the eddy current transducer.

The voltage generated in the measuring coil of the eddy current transducer is amplified and filtered with the Kronhite 3988 and then digitized in the NI PXI 5922 A/D converter (maximum sampling rate 15 MS/s, maximum resolution 24 bits). Due to the high resolution and accuracy of the A/D converter, which is achieved while maintaining a relatively high conversion rate, it is possible to obtain precise information about minor changes of amplitudes of the harmonic components in the processed signals. In these measurements, the sampling rate was set to 315 kS/s, which allows a resolution of 24 bits to be achieved. One of the critical issues in the MMFES system is to minimize spectral leakage. Therefore, the sampling frequency is precisely selected, considering the harmonic frequencies present in the excitation signal. Additionally, a clock that controls the A/D converter is used simultaneously to control the D/A converter generating the excitation. The block scheme of the MMFES system is shown in Figure 5. The digitized signal from the sensing coil is then processed using the Fast Fourier Transform (FFT) to determine the individual harmonic amplitudes. The values of harmonic amplitudes acquired for a single transducer position constitute the frequency characteristic of the transducer and the material within the effective range of the generated electromagnetic field.

Signals obtained in the Ultrasonic Testing (UT) are commonly displayed in three different formats. The formats are known in the NDT terminology as A-scan, B-scan, and C-scan presentations. Each presentation format provides different information about the region of material being inspected. The A-scan presents the amount of received ultrasonic energy as a function of time for a single position of the transducer. Peaks visible in the A-scan presentation allow to estimate the location and size of the discontinuity in the inspected material. Similar information can be achieved from the frequency characteristic obtained in the MMFES ECT system. Therefore, we will use the term A-scan for such kind of data presentation despite some significant differences. An example of the MMFES ECT A-scan in the 300 Hz–2 kHz range is shown in Figure 6.

In the UT method, the image composed of A-scans captured during the transducer's linear scanning is called a B-scan. From the B-scan, it is possible to identify the depth of the discontinuity and its linear dimension in the scanning direction. A C-scan is an image obtained during two-dimensional scanning (x, y) of the transducer and showing values for a selected time of acquisition. Finally, D-scan is a three-dimensional representation of the signals measured during (x, y) scanning.

Moving the eddy current probe in a line over the tested material and registering the frequency characteristics for the following positions enables creating a spectrogram. The spectrogram is a two-dimensional plot of the relative amplitude of the signal frequency components from the pick-up coil versus the transducer x position. The spectrogram amplitude can be calculated as a difference between the current amplitude and the amplitude measured over the uniform material to emphasize material structure changes. The spectrogram corresponds to a B-scan.

Successive scanning of the tested element, line by line, allows creating a two-dimensional plot of the signal value against the orthogonal scanning directions (x and y). The plot of the signal value for a single frequency as a function of the transducer x and y positions (Figure 7) is the equivalent of a C-scan.

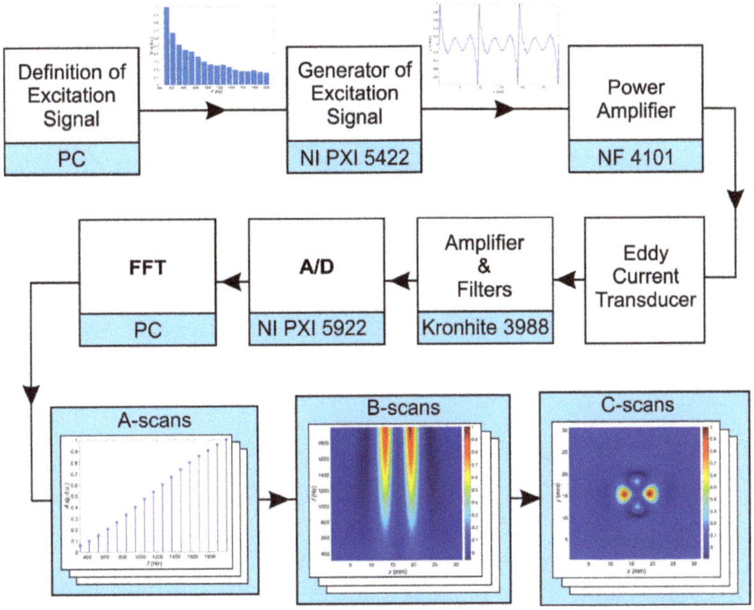

Figure 5. The block scheme of the MMFES system.

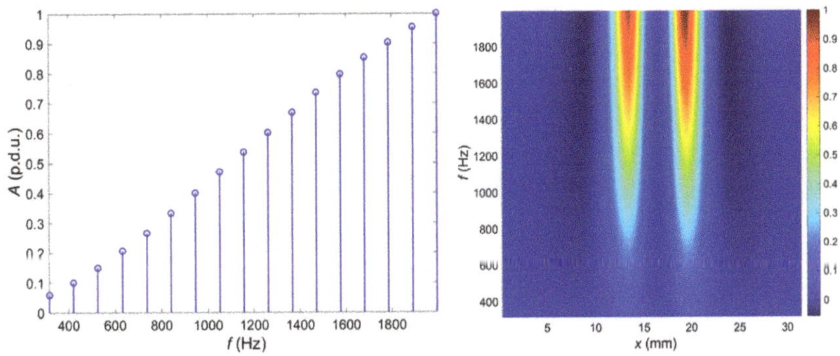

(**a**) (**b**)

Figure 6. An example of: (**a**) the A-scan–frequency characteristics, position of the sensor x = 19.5 mm, y = 15.5 mm; (**b**) B-scan–spectrogram, position of the sensor y = 15.5 mm.

Finally, the signal value three-dimensional plot against frequency f and the transducer position coordinates (x, y) corresponds to a D-scan.

The C-scans taken for the selected frequencies will be the input images to the automatic defect recognition system using image quality assessment based on similarity.

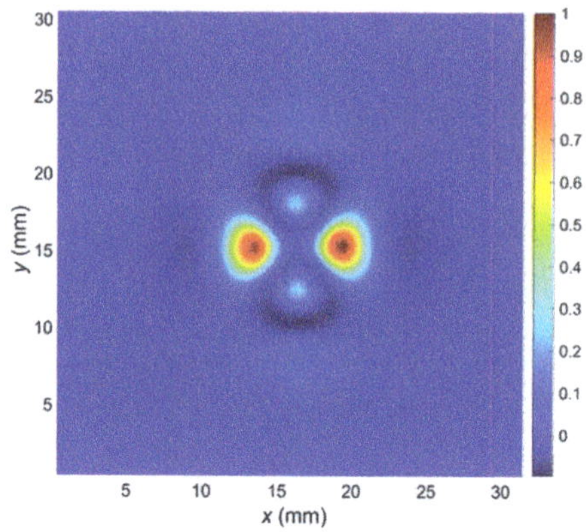

Figure 7. An example of the C-scan (2D plot of the values vs. sensor coordinates, excitation frequency f = 1995 Hz).

2.2.2. The Eddy Current Transducer

Inspection of the aluminum multilayer structures required a transducer characterized by high sensitivity, deep penetration, and the ability to detect defects in all directions with similar sensitivity. The eddy current differential transducer (Figure 8a) consists of a ferrite core with five symmetrically placed columns.

A receiving coil containing 100 turns is wound on the central column. Four excitation coils (E_A, E_B, E_C, E_D) with 25 turns each are wound on the remaining columns. The excitation coils (E_A and E_B) are connected in series, as are the E_C and E_D coils (Figure 8b). Both pairs are generating in the pick-up coil opposite directed magnetic fluxes. The resulting flux in the pick-up coil is close to zero in the equilibrium state. The output signal depends on a difference of fluxes ϕ_x and ϕ_y. The differential configuration of the transducer enables easy detection of the lack of symmetry in the tested specimen. Figure 8c shows the distribution of eddy currents induced in the specimen obtained by numerical analysis of the transducer.

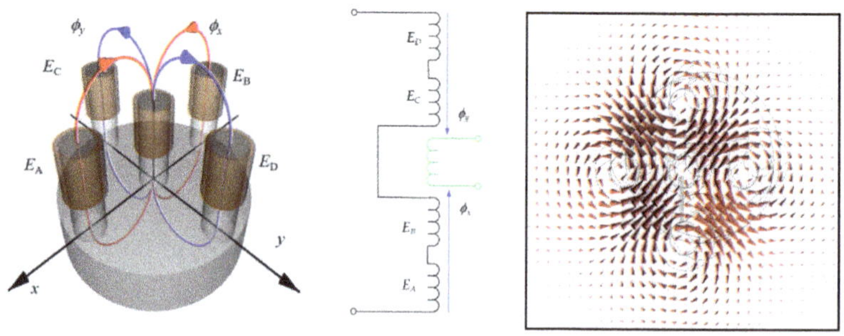

(a) (b) (c)

Figure 8. The eddy current transducer: (**a**) the schematic view; (**b**) the electric circuit; (**c**) the distribution of calculated eddy currents induced in the specimen.

Although dedicated for tube inspection, the transducer with a quite similar configuration was presented for the first time [23].

The transducer size was selected, considering the distribution of eddy currents induced in the tested specimen and the necessity to detect defects located in the deepest layers of the structure. The core dimensions are provided in Figure 9.

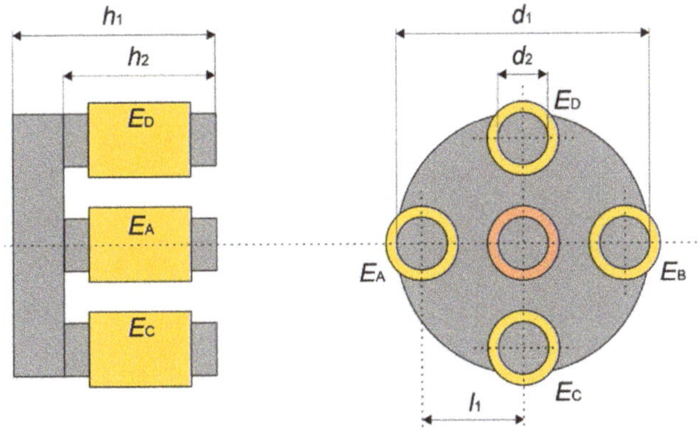

Figure 9. Dimensions of the eddy current transducer: d_1 = 12 mm, d_2 = 2 mm, l_1 = 5 mm, h_1 = 10 mm, and h_2 = 6 mm.

2.2.3. Test Samples and Measurements

Preliminary experiments were carried out on a specimen (Figure 10) taken from an aircraft operated under various load spectra. High loads initiated crack propagation at the edges of the rivet holes.

The cracks are very well visible in the X-ray image (Figure 11) of the sample. Most of the observed cracks occurred in the first layer of the examined structure. Only some of the cracks appeared in the second layer (deeper under the surface). All the cracks were propagated from the edges of the rivet holes. The vast majority of the cracks were directed in one of two perpendicular directions due to the load's directivity.

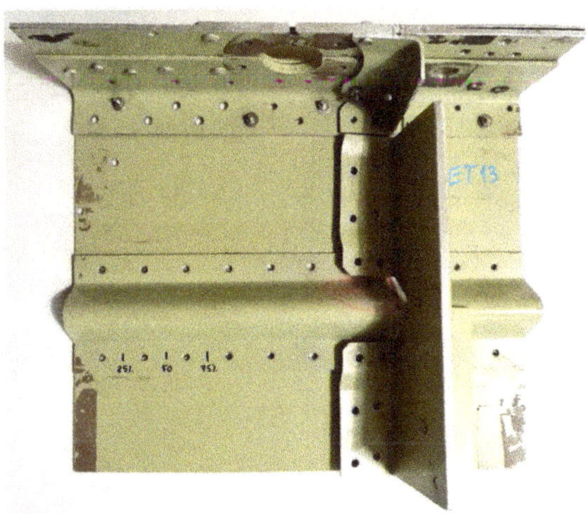

Figure 10. Photo of the airplane sample used in the initial experiments.

Figure 11. X-ray image of the airplane sample used in the initial experiments.

Most of these cracks were of considerable length (longer than 5 mm). There were also long cracks in the second layer, but it was observed that some of the cracks were less than 2 mm long.

This research aimed to develop a universal identification algorithm that works appropriately with defects of various sizes and orientations. However, as mentioned earlier, aircraft specimens contained a relatively limited set of naturally formed defects. For this reason, it was decided to produce aluminum specimens with a similar structure but with defects of a much more different nature, both in configuration and length.

Each sample consists of three 1 mm thick aluminum alloy plates joined together (Figure 12). The test plates were sufficiently large (160 mm × 160 mm) that measurements could be made without significant influence of the edges. In all the plates, the holes were made in the center using a CNC machine. The diameter of the holes was 4 mm. In several plates, cuts simulating cracks of various lengths (2 mm, 5 mm, and combined two cuts 2 and 5 mm) were manufactured, as shown in Figure 12.

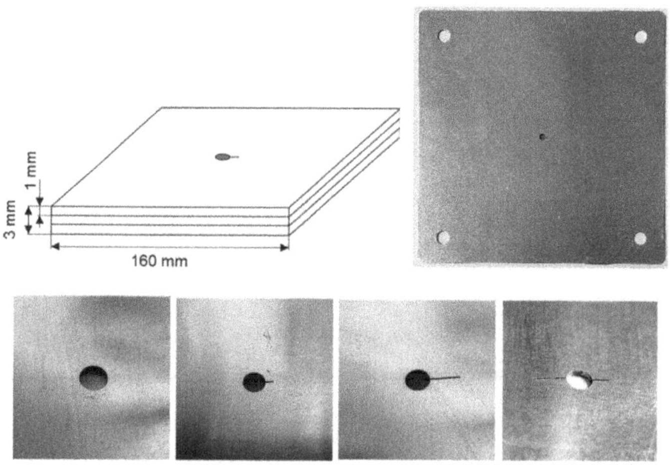

Figure 12. The aluminum test specimens used in the experiments.

In the research, we used 40 different samples, which were created from the assembly of the presented four plates with artificial defects in various configurations (defects in one plate and several plates simultaneously, on different layers, rotated by different angles, etc.). The use of various variants of the arrangement of basic defects creates a chance to test the system in a fairly wide range. However, it is planned to further expand the database of samples with artificial defects to verify responses in less frequent situations. At the same time, work is underway to collect a set of samples taken from real objects, but unfortunately, the number of experimental samples is limited by their availability. The defects found there are not very diverse and most often have larger dimensions than those that we would like to detect during routine inspections. For this reason, it is crucial to operating in two ways, collecting actual samples and using samples with artificially generated defects.

The measurements were performed by scanning the probe over the area surrounding the hole in steps of 0.5 mm in x and y-direction. The measurements were made using the multifrequency excitation signal consisting of sinusoidal components with frequencies from 315 Hz up to 1995 Hz (17 frequency components). The lift-off (the gap between the specimen and the sensor) was measured to be 0.4 mm.

The resulting C-scans for the lowest excitation frequency are presented in Figures 13–15. In Figure 13, the results of measurements made for a hole with defects (5 mm long crack) on different layers are presented. It can be seen that the C-scan measured for the rivet hole without defects (Figure 13a) is very symmetrical, and the presence of a defect on any layer causes a significant disturbance of the symmetry of the response (Figure 13b–d).

A similar phenomenon can be observed that it occurs even more intensively in the case of cuts manufactured at an angle of 90 degrees along the y axis (Figure 14).

However, as shown in Figure 15, in the case of minor cuts or when they are deeper below the surface (Figure 13d), the symmetry disturbances are not very large, and for effective detection of defects, it is necessary to use appropriate identification algorithms.

It is worth mentioning that the signals (C-scans) achieved for the other frequencies were not significantly different. Therefore, only one frequency component was selected for the presentation.

2.3. Similarity-Based Image Quality Assessment

Examples of the eddy current C-scans for a specified ROI using a single frequency are shown in the previous subsection as color plots. Nevertheless, the type of measurement data obtained as the C-scans is single-channel, similarly as in the case of greyscale images, although the data range is different. Hence, their color representation is unsuitable from the image processing point of view, where various three-channel color spaces are applied to analyze color images. Additionally, most state-of-the-art image quality assessment (IQA) methods, described further, were developed for monochrome images. Therefore, the obtained results were converted into grayscale 2D images, using 8-bit values in the range 0–255, for which the considered IQA methods may be directly applied.

Since the expected features of such obtained images are their symmetry and self-similarity, as it may be observed in Figures 13a, 14a and 15a, the similarity-based IQA metrics are of primary interest in further investigations.

General-purpose image quality assessment methods, including those used in our experiments, may be considered as useful approaches for many applications where the image or video quality plays an important role, e.g., making it possible to skip the degraded frames during the analysis of video sequences or to optimize some newly developed image filtering or lossy compression algorithms.

On the other hand, the universal full-reference IQA methods, based on the comparison of an undistorted reference ("pristine") image with the degraded one, may also be used for the detection of some types of contamination or image distortions as well as the estimation of their amount or level.

Although in many applications, where the reference image might be unavailable, the most desired approach would be the no-reference (NR) IQA, also referred to as "blind", such methods are usually quite complex, and their performance is lower than state-of-the-art FR IQA metrics. Usually, such NR metrics are sensitive only to a limited number of distortion types, hence their universality is relatively low.

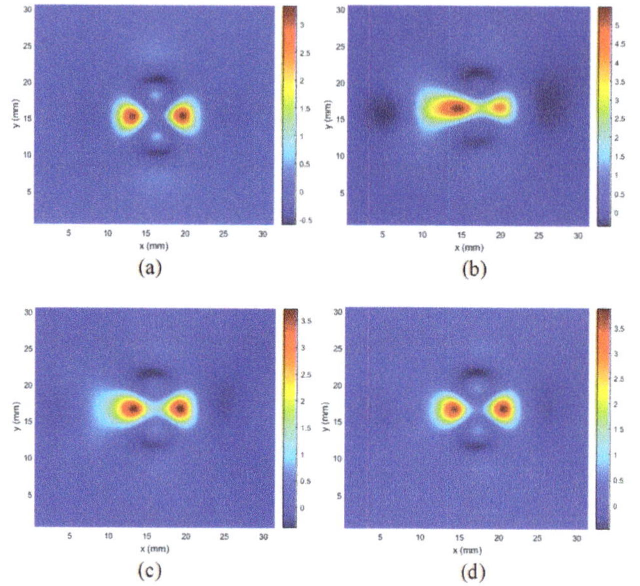

Figure 13. Comparison of C-scans: (**a**) achieved in case of the specimen without defect; (**b**) the specimens with the 5 mm long crack along the *x*-axis in the 1st layer; (**c**) in the 2nd layer; (**d**) in the 3rd layer.

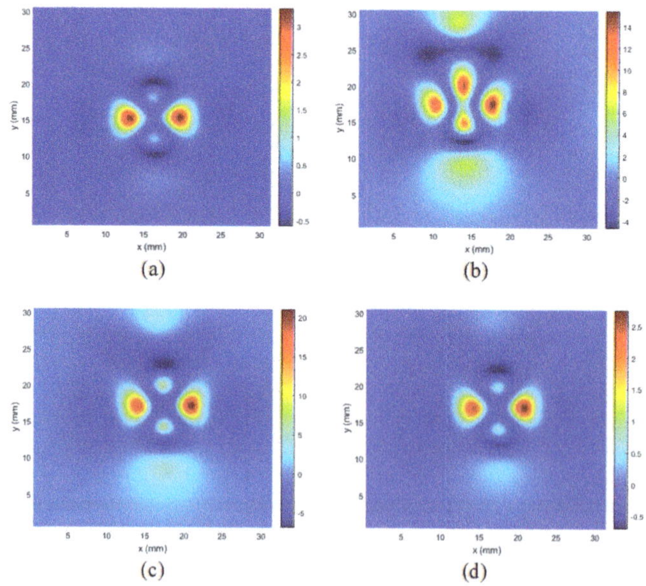

Figure 14. Comparison of C-scans: (**a**) achieved in case of the specimen without defect; (**b**) the specimens with the 5 mm long crack along the *y*-axis in the 1st layer; (**c**) in the 2nd layer; (**d**) in the 3rd layer.

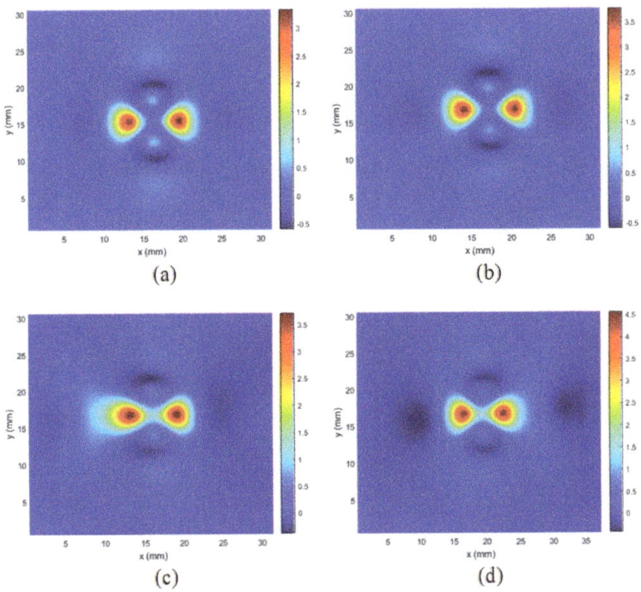

Figure 15. Comparison of C-scans: (a) achieved in case of the specimen without defect; (b) the specimens with the crack in the 1st layer, along the x-axis, and with a length of 2 mm; (c) length 5 mm; (d) combined two cracks with a length of 2 mm and 5 mm.

Nevertheless, the development of new IQA methods during recent years [17–19,24] has been concentrated on the possibly best correlation of the newly proposed metrics with subjective quality evaluations for the publicly available image datasets containing numerous degraded images, together with Mean Opinion Score (MOS) values gathered from human observers.

The most popular FR metric, known as Structural Similarity (SSIM), was proposed in [25], being the combination of three factors representing the luminance distortions, loss of contrast, and structural distortions. Since the formula of the SSIM utilizes the averaging of the local similarity indexes calculated for each position of the sliding window (by default, 11 × 11 pixels Gaussian window is used), it is possible to build the quality map as well. Nevertheless, in most practical applications, a single scalar index is preferred, possibly well correlated with human perception of various distortions.

The simplified idea of the calculation of the local SSIM value between the blocks x and y, representing two fragments of the compared images for the specified location of the sliding window, may be expressed as the product of three components:

$$SSIM = \frac{2\cdot\mu_x\cdot\mu_y + C_1}{\mu_x^2 + \mu_y^2 + C_1}\cdot\frac{2\cdot\sigma_x\cdot\sigma_y + C_2}{\sigma_x^2 + \sigma_y^2 + C_2}\cdot\frac{\sigma_{xy} + C_3}{\sigma_x\cdot\sigma_y + C_3}, \tag{2}$$

where μ_x and μ_y are the mean values whereas σ_x and σ_y denote the variance values in blocks x and y, respectively, and σ_{xy} stands for the covariance. Three stability constants: $C_1 = (0.01\cdot L)^2$, $C_2 = (0.03\cdot L)^2$ and $C_3 = 0.5\cdot C_2$, prevent the possible division by zero for very dark or flat areas of the image, assuming that L is the dynamic range of pixels values. The final SSIM result is obtained as the average of local values calculated for all window positions.

Following the first success of the SSIM, many similar approaches and extensions of this metric were proposed by various researchers during recent years, starting from its multi-scale version, known as MS-SSIM [26], making its results independent of the sliding window size. Some other modifications of the idea of Structural Similarity utilize the calculations of gradients, leading to Gradient Similarity [27] as well as Gradient Magnitude Similarity Deviation (GMSD) [28]. Another direction of development of SSIM-inspired metrics is related to the use of Riesz transform [29,30] and phase congruency, leading to well-performing Feature Similarity (FSIM) [31], where the similarity of gradient magnitude is combined with the similarity of the phase congruency, calculated similarly as in the SSIM formula.

The same general approach is also used in the Edge Strength Similarity metric [32], as well as in some other metrics utilizing local variance [33] or spectral residual [34]. Another interesting approach, originating from the idea of the SSIM, is based on the perceptual combination of multi-scale gradient magnitude maps that concern the micro- and macrostructural similarity that are further combined with color information similarity and high-distortion-based pooling, forming the final Perceptual Similarity (PSIM) metric [35].

Nevertheless, although the performance of most of these improved metrics is significantly better in comparison to the "original" SSIM for many IQA benchmarking datasets, their application for some other types of images, e.g., containing multiple distortions [36] or for the surface quality assessment of 3D prints [37] not always leads to satisfactory results. Some other examples might be the development of quality metrics for audio-visual signals, stitched images [38], or light field reconstruction, compression, and display [39], where some specific types of distortions may take place. Therefore, the application of the general-purpose IQA methods in some other areas is not always straightforward and may require their extensions, significant modifications, and an appropriate choice of potentially useful metrics.

A remarkable extension of the SSIM formula was proposed by Ponomarenko et al. [40], where the fourth multiplicative component was introduced, which is related to the similarity of predictability of image blocks. This component may be determined by calculating the mean square error (MSE) between the specified block and the neighboring blocks. Hence, the minimum MSE value is chosen as the local result and then multiplied by the three "original" SSIM components, namely similarity of the average luminance of image blocks, a correlation factor, and contrast similarity. Nevertheless, the additional necessity of searching for the most similar block increases the computational complexity. The formula of the method denoted as SSIM4 may be expressed as:

$$SSIM4 = \frac{2 \cdot \mu_x \cdot \mu_y + C_1}{\mu_x^2 + \mu_y^2 + C_1} \cdot \frac{2 \cdot \sigma_x \cdot \sigma_y + C_2}{\sigma_x^2 + \sigma_y^2 + C_2} \cdot \frac{\sigma_{xy} + C_3}{\sigma_x \cdot \sigma_y + C_3} \cdot \frac{2 \cdot \varepsilon_x \cdot \varepsilon_y + C_4}{\varepsilon_x^2 + \varepsilon_y^2 + C_4}, \quad (3)$$

where $C_4 = C_2$ is assumed, ε_x^2 and ε_y^2 are the measures of predictability of the respective image blocks expressed as mean MSE values between the given block A and its neighboring blocks B according to

$$\varepsilon_A^2 = \min_B \left(\frac{1}{M \cdot N} \cdot \sum_{i=1}^{M} \sum_{j=1}^{N} (A_{ij} - B_{ij})^2 \right), \quad (4)$$

assuming that neighboring blocks B are limited by a specified distance (the default parameters are search area 19 × 19 pixels and the block size $M = N = 9$). Nevertheless, the blocks that contain the whole central part consisting of 5 × 5 pixels block are excluded from the calculations of the predictability measure ε_A^2 [33] as illustrated in Figure 16, where three excluded red blocks and three green blocks are shown.

Although all SSIM-based metrics utilize the estimation of image similarity, they cannot be applied directly for the analysis of images obtained as the results of measurements considered in this paper. Hence, their modification related to the calculation of the mutual (internal) similarity between the image fragments is postulated.

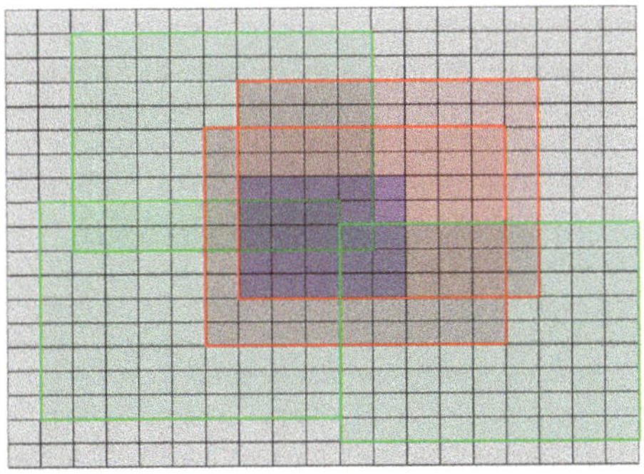

Figure 16. Illustration of the sample blocks included (green) and excluded (red) from the calculations of blocks predictability.

The concept of similarity calculation between two images without using reference images was also used in blind image quality assessment, although in this case, the use of the pseudo-reference image (PRI) is necessary [41,42] is based on the use of multiple PRIs, assuming various degrees of distortion aggravation.

Nevertheless, the approach proposed in this paper is different since it does not require any comparisons with any other images, and the only required data for the mutual similarity calculation is the image that is the result of eddy current measurements.

2.4. *The Proposed Image Analysis Method*

2.4.1. Symmetry Detection

As the samples without any defects should provide symmetrical images obtained as measurements, the first necessary task is to detect the origin as the measurement results are not necessarily centered on the image plane. Unfortunately, considering the influence of defects and the presence of the measurement noise, the application of moment-based methods, e.g., the Centre of Gravity, has not led to satisfactory results. Nevertheless, the proper results for all considered samples, i.e., the detection of an individual pixel representing the center of symmetry of measurement data, were obtained employing the correlation between the original image and its horizontally or vertically flipped equivalents. After determining the locations of the maximum correlation values, the images are shifted and cropped, forming four quadrant images, which may be used as inputs for mutual similarity calculations.

2.4.2. Thresholding and Normalization of Data Range

Before the symmetry detection, the grayscale images were initially normalized to the range 0–255, where 0 corresponds to the lowest measured value, and 255 represents the highest peak value. To decrease the influence of the highest peaks on the obtained results and make it possible to detect the presence of smaller defects, which influence the lower values the additional thresholding and normalization were conducted, equivalent to image saturation, assuming the preservation of only the lower 20% of the original range for simplicity the threshold value equal to 50 was chosen for 8-bit grayscale images). Such obtained images, divided into four quadrants expected to be symmetrical, form the input data for the mutual similarity calculations.

2.4.3. Mutual Image Similarity Calculation

Since most of the FR IQA metrics were proposed assuming that the compared images represent the same scene and only one of these images is corrupted, they might also be applied for comparison purposes, assuming that both images are highly similar. Another essential requirement is the proper registration of both images; hence the precise determination of the center of symmetry is a crucial issue, influencing the obtained results.

Studying the obtained experimentally images, both horizontal and vertical symmetry may be easily observed as expected for the images without defects. Hence, the division of images into four fragments was proposed making it possible to calculate four similarity factors between each part and the flipped neighboring fragment, as shown in Figure 17. For the images where the center of symmetry is shifted from the center of the image, an appropriate cutting of boundary rows or columns should be made to ensure the same size of both compared image fragments.

A similar approach to mutual similarity calculation was also used for the aesthetic quality evaluation of the 3D printed surfaces [37]. However, in this case, a precise alignment has not been necessary, and no symmetry assumptions were utilized. Nonetheless, in the problem investigated in this paper, image flipping should be applied. Nevertheless, the diagonal similarities, i.e., these between the top left and bottom right (or top right and bottom left) fragments, marked by gray arrows in Figure 17, are not necessarily calculated to accelerate the computations.

During the conducted experiments, several similarity-based IQA metrics discussed in Section 2.3 were verified, starting from the "classical" SSIM formula [25], through FSIM [31], PSIM [35], and other modifications, to the SSIM4 metric [40], leading to the most accurate classification of samples, presented further.

Nevertheless, it should be noted that the direct application of the FR IQA would be impossible without the knowledge of the reference images of metal structures without cracks which should be precisely registered with the analyzed images. Therefore, we proposed a novel approach assuming the calculation of the mutual similarities reflecting the expected symmetry of the images obtained using the multifrequency eddy current method. As the typical methods used for this purpose failed in the initial experiments, we adapted some metrics that are well-known and commonly accepted for general-purpose IQA, although the application of the Structural Similarity and the other IQA metrics for the classification of samples required the use of the average and maximum values of the mutual similarities obtained for 17 frequencies, not just the simple calculation of their values.

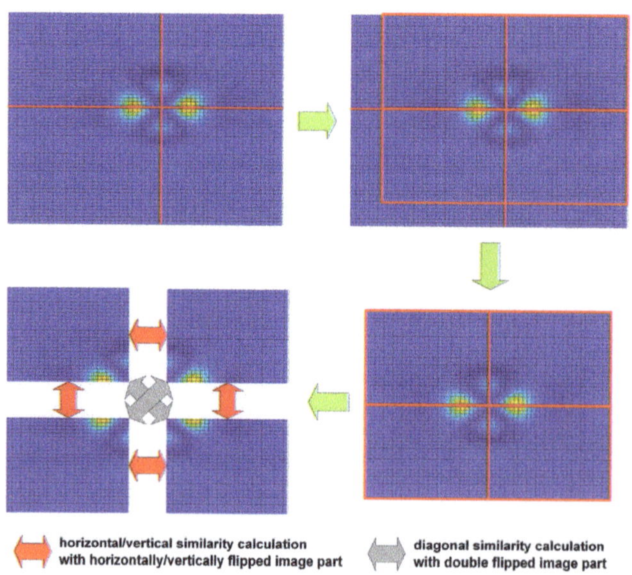

Figure 17. Illustration of the idea of mutual similarity calculation for the symmetrical fragments of the sample image.

3. Results and Discussion

During the conducted experiments, several similarity-based IQA methods were analyzed and verified towards their potential application as the similarity measures in the investigated problem. To illustrate the non-applicability of most of the considered metrics the results achieved for the PSIM, assuming the use of six mutual similarities, are presented in Figure 18. The results achieved using the SSIM and FSIM metrics applied for the mutual similarity calculations are illustrated in Figures 19 and 20, respectively. As it may be easily observed, the discrimination between the samples without defects and these containing them is not possible in such cases.

Similarly, the application of some other metrics considered in the paper, except the proposed application of the SSIM4, has not led to satisfactory results, regardless of the analysis of the maximum, minimum, average, or median values of the mutual similarities.

Much better results, presented in Figure 21, were obtained using four mutual similarities (horizontal and vertical) calculated according to the SSIM4 formula. Analyzing the plots, one may observe that reliable discriminations between the samples without defects and these containing some cracks are possible. It was achieved using a combination of two features, namely the highest values of the average and maximum of four mutual similarities, determined for each of 17 images obtained from measurements conducted using various frequencies (from 315 Hz to 1995 Hz).

Even better discrimination may be observed using six mutual similarities, as shown in Figure 22, where an improved separation between the green points representing the samples without cracks and the others may be easily observed. Both plots contain the results of measurements and further analysis conducted for 36 samples, where various colors indicate the depth of cracks and their size corresponds to the size of individual points.

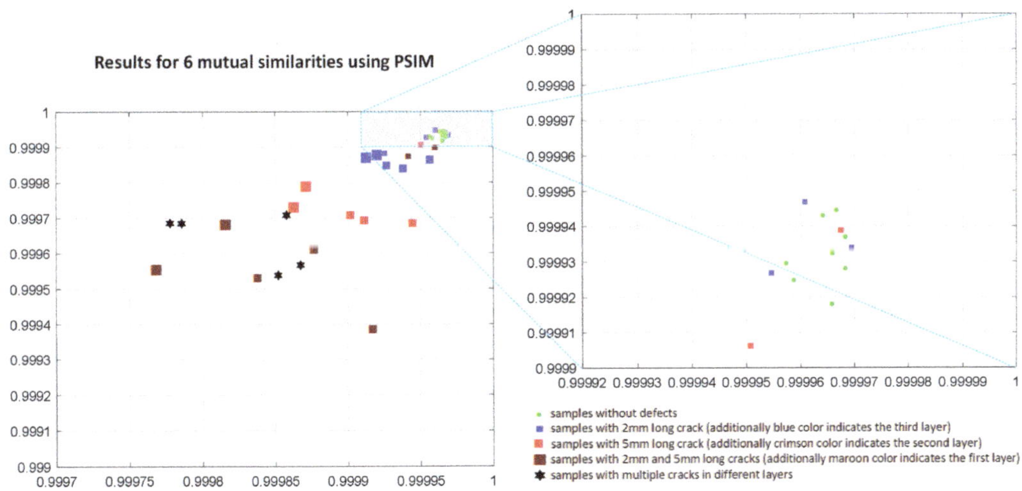

Figure 18. Results obtained applying six mutual comparisons employing the PSIM metric using the maximum and average of its values calculated for 17 images corresponding to various frequencies used in experiments.

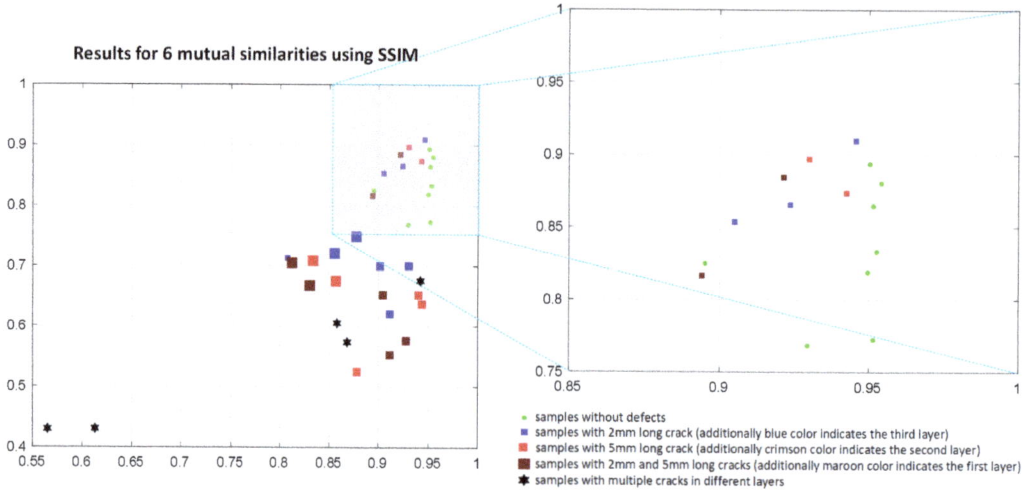

Figure 19. Results obtained applying six mutual comparisons employing the SSIM metric using the maximum and average of its values calculated for 17 images corresponding to various frequencies used in experiments.

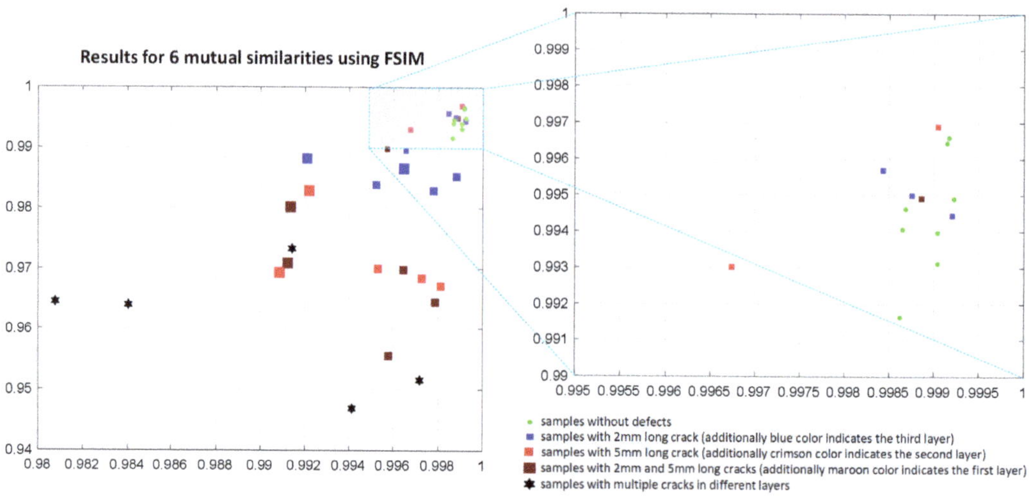

Figure 20. Results obtained applying six mutual comparisons employing the FSIM metric using the maximum and average of its values calculated for 17 images corresponding to various frequencies used in experiments.

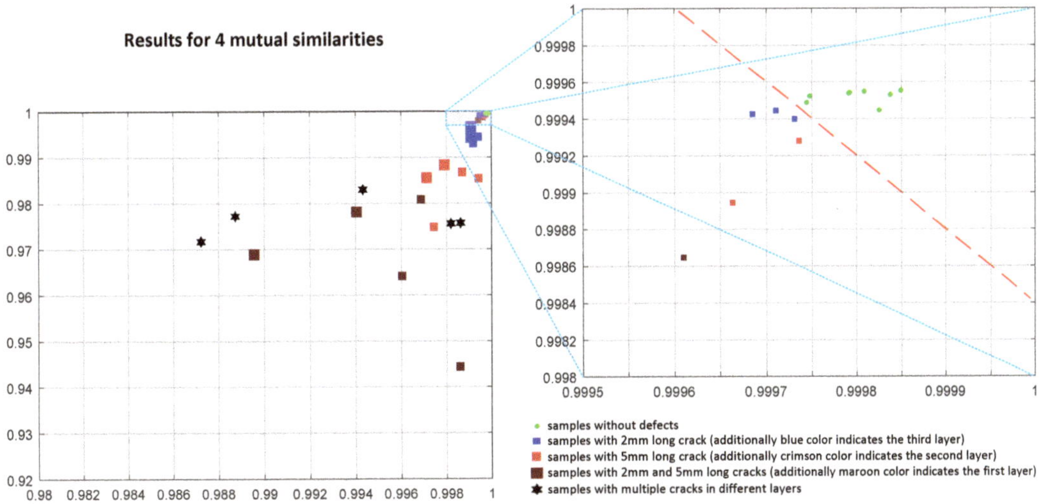

Figure 21. Results obtained applying four mutual comparisons employing the SSIM4 metric using the maximum and average of its values calculated for 17 images corresponding to various frequencies used in experiments.

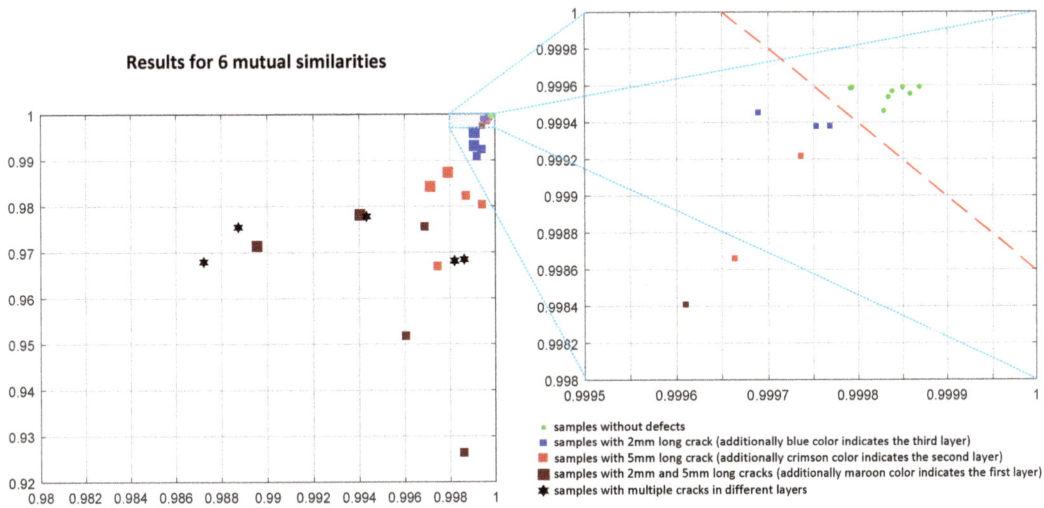

Figure 22. Results obtained applying six mutual comparisons employing the SSIM4 metric using the maximum and average of its values calculated for 17 images corresponding to various frequencies used in experiments.

Since some samples used in experiments contain multiple cracks, they were marked with black hexagons. As it may be observed, there is a relation between the size of the crack (as well as its depth) and the values of both considered features, as well.

Nevertheless, correct classification of samples into high-quality ones and those containing some cracks may be conducted using the product of two considered features (as well as their sum), particularly assuming the calculation of six mutual comparisons.

The applied multifrequency method allows for even more detailed identification of defects. Thanks to the access to information obtained at many frequencies, it is possible to determine the layer of the structure containing the damage more precisely. Figure 23 shows an example of the frequency characteristics of the obtained similarity. It can be observed that the minimum similarity for the defect located in the third layer is achieved at a significantly lower frequency (f = 630 Hz) than in the case of the defect located in the second layer (f = 1050 Hz). This effect can be further emphasized by using a wider range of excitation frequencies. It is crucial in the case of surface defects where the applied frequencies should be much higher. Nevertheless, due to the limited maximum excitation frequency, the plot of the frequency characteristics for the defect in the first layer did not contain the extreme point. This curve is not shown in Figure 23 due to significantly lower similarity values, which would make the remaining plots unreadable.

The broader range of excitation frequencies (and higher maximum excitation frequency) may cause lower sensitivity to defects occurring in the third layer or below. The provision of the ability to detect deeply located defects has resulted from specialists' requirements from the aviation industry.

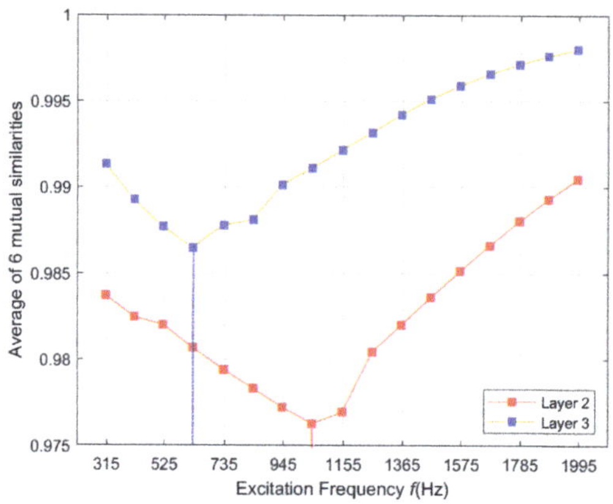

Figure 23. Frequency dependence of the average of mutual similarities calculated using the proposed method.

4. Concluding Remarks

Currently, the eddy current techniques used for damage detection in the aerospace industry require impedance plane analysis or C-scan data visualization to evaluate damage existence. Each data collected must be evaluated by the qualified operator. Such an approach requires a relatively large amount of time for analysis and data comparison. Moreover, the sensitivity of inspection is decreased with the use of lower frequencies. Therefore, the application of the proposed method is interesting for the further development of an efficient damage detection methodology in the aerospace industry.

The solution proposed in the paper makes it possible to determine the presence of cracks in the aircraft structures using the mutual image similarity analysis of images obtained as the result of the multifrequency eddy current measurements. After validating the usefulness of several image quality assessment metrics, the results achieved by the application of the modified SSIM4 method were encouraging, allowing not only for the detection of structural defects but also for the estimation of their size.

A relatively large number of experiments with defects of various sizes, configurations, and locations in the studied structure were carried out. The possibility of both single cracks and complicated, complex damages with cracks propagating in many directions was considered. Such a complementary set of data made it possible to thoroughly test the proposed solution.

The experimental verification of the proposed approach confirms its potential industrial application, mainly due to a fast calculation of the quality metric compared to the total measurement time. The computation of the final combination of two features based on six mutual similarities using the SSIM4 metric together with visualization takes a few seconds. The calculations of the mutual similarities for different layers and areas may be easily parallelized as well. Assuming that the number of rivets in 1 m^2 of the specimen is around 150, the analysis using non-optimized Matlab code will take about 15 min. Analysis can partially overlap with the measurement process. The time can be considerably reduced using GPU and parallel computing for data acquisition (FFT analysis) and defect detection. Another aspect is the time needed to perform the measurements. A significant limitation, in this case, is the lowest excitation frequency, as it is necessary to measure at least half of the period. In the case of the excitation frequency as low as 300 Hz, the measurement of 1 m^2 with a resolution of 0.5 mm will take at least 50 min.

According to our best knowledge, the proposed method is the first attempt to use image quality assessment methods to analyze eddy current measurements, particularly concerning the application of the multifrequency eddy current method to the inspection of riveted structures. In contrast to many popular deep learning approaches, our method is explainable as its main assumption is based on detecting the expected symmetry and its numerical estimation related to the size of the cracks.

Author Contributions: Conceptualization, T.C. and K.O.; methodology, T.C., K.O. and K.D.; software, T.C. and K.O.; validation, T.C. and K.O.; formal analysis, T.C. and K.D.; investigation, T.C. and K.O.; resources, T.C. and K.D.; data curation, T.C., K.O., R.M. and K.D.; writing—original draft preparation, T.C, K.O. and K.D.; writing—review and editing, T.C., K.O. and K.D.; visualization, T.C. and K.O.; project administration, T.C., K.O. and K.D.; scientific consultation, T.C., K.O., M.D., P.S. and K.D.; Contribution of the Authors in %: K.O.—40%, T.C.—33%, K.D.—10%, R.M.—7%, M.D.—5%, P.S.—5%. All authors have read and agreed to the published version of the manuscript.

Funding: This research received no external funding.

Institutional Review Board Statement: Not applicable.

Informed Consent Statement: Not applicable.

Data Availability Statement: The data presented in this study are available on request from the corresponding author. The data are not publicly available due to a complicated structure that requires additional explanations.

Conflicts of Interest: The authors declare no conflict of interest.

References

1. Grandt, A.F. *Fundamentals of Structural Integrity: Damage Tolerant Design and Nondestructive Evaluation*; John Wiley & Sons: Hoboken, NJ, USA, 2003; ISBN 978-0-471-21459-5.
2. Beattie, A.; Dahlke, L.; Gieske, J. *Emerging Nondestructive Inspection Methods for Aging Aircraft*; Reort Number SAND-92-2732; FAA William J. Hughes Technical Center: Egg Harbor Township, NJ, USA, 1994; p. 10136510. [CrossRef]
3. Uemura, C.; Underhill, P.; Krause, T. Estimating POD of a screening technique for cracks about ferrous fasteners without fastener removal. *NDT E Int.* **2019**, *107*, 102124. [CrossRef]
4. National Research Council Aging of U.S. *Air Force Aircraft: Final Report*; National Academies Press: Washington, DC, USA, 1997; p. 5917, ISBN 978-0-309-05935-0.
5. García-Martín, J.; Gómez-Gil, J.; Vázquez-Sánchez, E. Non-Destructive Techniques Based on Eddy Current Testing. *Sensors* **2011**, *11*, 2525–2565. [CrossRef] [PubMed]
6. Mohseni, E.; Boukani, H.H.; França, D.R.; Viens, M. A Study of the Automated Eddy Current Detection of Cracks in Steel Plates. *J. Nondestruct. Eval.* **2019**, *39*, 1–12. [CrossRef] [PubMed]
7. Szlagowska-Spychalska, J.; Dragan, K.; Kukla, D.; Spychalski, W.; Kurzydłowski, K.J. A Novel Approach for The Eddy Current Inspection of The Aerospace Structures Based on The Signal Modeling and Signal Processing. In Proceedings of the 18th World Conference on Nondestructive Testing, Durban, South Africa, 16–20 April 2012.
8. Wang, B.; Basart, J.P.; Moulder, J.C. Wavelet Expansions in Volume Integral Method of Eddy-Current Modeling. In *Review of Progress in Quantitative Nondestructive Evaluation*; Thompson, D.O., Chimenti, D.E., Eds.; Springer: Boston, MA, USA, 1996; pp. 377–384, ISBN 978-1-4613-8027-6.
9. Roach, D.P.; Neidigk, S. *Industry Survey of Structural Health Monitoring Technology and Usage*; Sandia National Laboratories: Washington, DC, USA, 2012.
10. Abdou, A.; Safer, O.A.; Bouchala, T.; Bendaikha, A.; Abdelhadi, B.; Guettafi, A.; Benoudjit, A. An Eddy Current Nondestructive Method for Crack Detection in Multilayer Riveted Structures. *Instrum. Mes. Métrol.* **2019**, *18*, 485–490. [CrossRef]

11. Mottl, Z. The quantitative relations between true and standard depth of penetration for air-cored probe coils in eddy current testing. *NDT Int.* **1990**, *23*, 11–18. [CrossRef]
12. Mook, G.; Hesse, O.; Uchanin, V. Deep Penetrating Eddy Currents and Probes. *Mater. Test.* **2007**, *49*, 258–264. [CrossRef]
13. Sun, Y.; Ouyang, T.; Udpa, S.S. *Recent Advances in Remote Field Eddy Current NDE Techniques and Their Applications in Detection, Characterization and Monitoring of Deeply Hidden Corrosion in Aircraft Structures*; Mal, A.K., Ed.; International Society for Optics and Photonics: Newport Beach, CA, USA, 1999; pp. 200–210. [CrossRef]
14. Sasi, B.; Rao, B.P.C.; Jayakumar, T.; Raj, B. Development of Eddy Current Test Procedure for Non-Destructive Detection of Fatigue Cracks and Corrosion in Rivets of Air-Intake Structures. *Def. Sci. J.* **2009**, *59*, 106–112. [CrossRef]
15. Mohd Ali, M.I.; Hussin, A.I.; Md Hairudin, M.K.A. Defect Analysis on Boeing 737-400 Skin Panel Rivet Row Aft Ward Fuselage. *Int. J. Innov. Technol. Explor. Eng.* **2019**, *8*. Available online: http://ir.unikl.edu.my/jspui/handle/123456789/23652 (accessed on 2 July 2021).
16. Lin, W.; Kuo, C.-C.J. Perceptual visual quality metrics: A survey. *J. Vis. Commun. Image Represent.* **2011**, *22*, 297–312. [CrossRef]
17. Bovik, A. Automatic Prediction of Perceptual Image and Video Quality. *Proc. IEEE* **2013**, *101*, 1–17. [CrossRef]
18. Zhai, G.; Min, X.; Liu, N. Free-energy principle inspired visual quality assessment: An overview. *Digit. Signal. Process.* **2019**, *91*, 11–20. [CrossRef]
19. Zhai, G.; Min, X. Perceptual image quality assessment: A survey. *Sci. China Inf. Sci.* **2020**, *63*, 1–52. [CrossRef]
20. Shoales, G. Procedures for Aircraft Structural Teardown Analysis: Development of A Best Practices Handbook. In *ICAF 2009, Bridging the Gap between Theory and Operational Practice*; Bos, M.J., Ed.; Springer: Dordrecht, The Netherlands, 2009; pp. 330–353, ISBN 978-90-481-2745-0.
21. Katunin, A.; Wronkowicz-Katunin, A.; Dragan, K. Impact Damage Evaluation in Composite Structures Based on Fusion of Results of Ultrasonic Testing and X-ray Computed Tomography. *Sensors* **2020**, *20*, 1867. [CrossRef] [PubMed]
22. Chady, T.; Enokizono, M. Multi-frequency exciting and spectrogram-based ECT method. *J. Magn. Magn. Mater.* **2000**, *215–216*, 700–703. [CrossRef]
23. Chady, T.; Enokizono, M.; Todaka, T.; Tsuchida, Y.; Sikora, R. Evaluation of the JSAEM round robin test samples using multifrequency excitation and spectrogram method. In *Electromagnetic Nondestructive Evaluation (V)*; Pávó, J., Vértesy, G., Takagi, T., Udpa, S.S., Eds.; Studies in Applied Electromagnetics and Mechanics; IOS Press: Amsterdam, The Netherlands; Washington, DC, USA, 2001; Volume 21, pp. 171–178, ISBN 978-1-58603-155-8.
24. Niu, Y.; Zhong, Y.; Guo, W.; Shi, Y.; Chen, P. 2D and 3D Image Quality Assessment: A Survey of Metrics and Challenges. *IEEE Access* **2018**, *7*, 782–801. [CrossRef]
25. Wang, Z.; Bovik, A.; Sheikh, H.; Simoncelli, E. Image Quality Assessment: From Error Visibility to Structural Similarity. *IEEE Trans. Image Process.* **2004**, *13*, 600–612. [CrossRef] [PubMed]
26. Wang, Z.; Simoncelli, E.P.; Bovik, A.C. Multiscale Structural Similarity for Image Quality Assessment. In Proceedings of the 37th Asilomar Conference on Signals, Systems and Computers, Pacific Grove, CA, USA, 9–12 November 2003; pp. 1398–1402. [CrossRef]
27. Liu, A.; Lin, W.; Narwaria, M. Image Quality Assessment Based on Gradient Similarity. *IEEE Trans. Image Process.* **2011**, *21*, 1500–1512. [CrossRef]
28. Xue, W.; Zhang, L.; Mou, X.; Bovik, A. Gradient Magnitude Similarity Deviation: A Highly Efficient Perceptual Image Quality Index. *IEEE Trans. Image Process.* **2013**, *23*, 684–695. [CrossRef] [PubMed]
29. Zhang, L.; Zhang, L.; Mou, X. RFSIM: A Feature Based Image Quality Assessment Metric Using Riesz Transforms. In Proceedings of the 2010 IEEE International Conference on Image Processing, Hong Kong, China, 26–29 September 2010; pp. 321–324. [CrossRef]
30. Yang, G.; Li, D.; Lu, F.; Liao, Y.; Yang, W. RVSIM: A feature similarity method for full-reference image quality assessment. *EURASIP J. Image Video Process.* **2018**, *2018*, 1–15. [CrossRef]
31. Zhang, L.; Zhang, L.; Mou, X.; Zhang, D. FSIM: A Feature Similarity Index for Image Quality Assessment. *IEEE Trans. Image Process.* **2011**, *20*, 2378–2386. [CrossRef]
32. Zhang, X.; Feng, X.; Wang, W.; Xue, W. Edge Strength Similarity for Image Quality Assessment. *IEEE Signal. Process. Lett.* **2013**, *20*, 319–322. [CrossRef]
33. Aja-Fernandez, S.; Estepar, R.S.J.; Alberola-Lopez, C.; Westin, C.-F. Image Quality Assessment Based on Local Variance. In Proceedings of the 2006 International Conference of the IEEE Engineering in Medicine and Biology Society, New York, NY, USA, 30 August–3 September 2006; pp. 4815–4818. [CrossRef]
34. Zhang, L.; Li, H. SR-SIM: A Fast and High Performance IQA Index Based on Spectral Residual. In Proceedings of the 2012 19th IEEE International Conference on Image Processing, Orlando, FL, USA, 30 September–3 October 2012; pp. 1473–1476. [CrossRef]
35. Gu, K.; Li, L.; Lu, H.; Min, X.; Lin, W. A Fast Reliable Image Quality Predictor by Fusing Micro-and Macro-Structures. *IEEE Trans. Ind. Electron.* **2017**, *64*, 3903–3912. [CrossRef]
36. Yue, G.; Hou, C.; Gu, K.; Ling, N.; Li, B. Analysis of Structural Characteristics for Quality Assessment of Multiply Distorted Images. *IEEE Trans. Multimed.* **2018**, *20*, 2722–2732. [CrossRef]
37. Okarma, K.; Fastowicz, J.; Lech, P.; Lukin, V. Quality Assessment of 3D Printed Surfaces Using Combined Metrics Based on Mutual Structural Similarity Approach Correlated with Subjective Aesthetic Evaluation. *Appl. Sci.* **2020**, *10*, 6248. [CrossRef]

38. Madhusudana, P.C.; Soundararajan, R. Subjective and Objective Quality Assessment of Stitched Images for Virtual Reality. *IEEE Trans. Image Process.* **2019**, *28*, 5620–5635. [CrossRef]
39. Min, X.; Zhou, J.; Zhai, G.; Le Callet, P.; Yang, X.; Guan, X. A Metric for Light Field Reconstruction, Compression, and Display Quality Evaluation. *IEEE Trans. Image Process.* **2020**, *29*, 3790–3804. [CrossRef]
40. Ponomarenko, M.; Egiazarian, K.; Lukin, V.; Abramova, V. Structural Similarity Index with Predictability of Image Blocks. In Proceedings of the 2018 IEEE 17th International Conference on Mathematical Methods in Electromagnetic Theory (MMET), Kiev, Ukraine, 2–5 July 2018; pp. 115–118. [CrossRef]
41. Min, X.; Gu, K.; Zhai, G.; Liu, J.; Yang, X.; Chen, C.W. Blind Quality Assessment Based on Pseudo-Reference Image. *IEEE Trans. Multimed.* **2017**, *20*, 2049–2062. [CrossRef]
42. Min, X.; Zhai, G.; Gu, K.; Liu, Y.; Yang, X. Blind Image Quality Estimation via Distortion Aggravation. *IEEE Trans. Broadcast.* **2018**, *64*, 508–517. [CrossRef]

materials

MDPI

Article

Magnetic Recording Method (MRM) for Nondestructive Evaluation of Ferromagnetic Materials

Tomasz Chady [1,*], Ryszard D. Łukaszuk [2], Krzysztof Gorący [3] and Marek J. Żwir [3]

1 Faculty of Electrical Engineering, West Pomeranian University of Technology, 70-313 Szczecin, Poland
2 Doctoral School, West Pomeranian University of Technology, 70-313 Szczecin, Poland; ryszard.lukaszuk@zut.edu.pl
3 Department of Polymer and Biomaterials Science, Faculty of Chemical Technology and Engineering, West Pomeranian University of Technology, 70-311 Szczecin, Poland; krzysztof.goracy@zut.edu.pl (K.G.); marek.zwir@zut.edu.pl (M.J.Ż.)
* Correspondence: tchady@zut.edu.pl

Abstract: This paper proposes and experimentally investigates a novel nondestructive testing method for ferromagnetic elements monitoring, the Magnetic Recording Method (MRM). In this method, the inspected element must be magnetized in a strictly defined manner before operation. This can be achieved using an array of permanent magnets arranged to produce a quasi-sinusoidal magnetization path. The magnetic field caused by the original residual magnetization of the element is measured and stored for future reference. After the operation or loading, the magnetic field measurement is repeated. Analysis of relative changes in the magnetic field (for selected components) allows identifying applied stress. The proposed research methodology aims to provide information on the steel structure condition unambiguously and accurately. An interpretation of the results without referring to the original magnetization is also possible but could be less accurate. The method can be used as a standard technique for NDT (Non-Destructive Testing) or in structural health monitoring (SHM) systems.

Keywords: nondestructive testing (NDT); nondestructive evaluation (NDE); magnetic recording method (MRM); ferromagnetic materials; stress test; structural health monitoring (SHM)

Citation: Chady, T.; Łukaszuk, R.D.; Gorący, K.; Żwir, M.J. Magnetic Recording Method (MRM) for Nondestructive Evaluation of Ferromagnetic Materials. *Materials* **2022**, *15*, 630. https://doi.org/10.3390/ma15020630

Academic Editors: Giovanni Bruno and Christian Müller

Received: 20 November 2021
Accepted: 12 January 2022
Published: 14 January 2022

1. Introduction

The need to reduce greenhouse gas emissions and, due to the Paris Agreement, the need for countries to achieve climate neutrality in the second half of the 21st century have resulted in modifications to structural components. One such change is the production of components with a reduced thickness or cross-sectional area. However, the negative effect of such an approach is the significant impact of even small heterogeneities on the structural strength of the part, which may threaten the safe use of the structure. Therefore, it is necessary to frequently evaluate the structure with nondestructive testing.

Carbon structural steels are the primary construction materials that have a specific chemical composition defined for these varieties, and are delivered in the form of sheets and other rolled products with fixed, typical cross-sections. The chemical composition of structural carbon steels is designed for their intended use. In Europe, the requirements for such steels are specified in the European standard EN 10025. Examples of carbon structural steels are S195, S235, S355, S420, and S460. The letter S in the steel designation indicates "carbon structural steel" and the number following it specifies the minimum yield stress for this steel grade in MPa. The EN 10025 standard defines the yield stress as a value at which irreversible plastic deformation of a rod with a diameter of 16 mm will occur.

In engineering practice, the yield strength is a point on the graph of stress dependence on the strain, which means exceeding the stresses below, with material behaving according to Hook's law. That is, if the stress does not exceed the yield strength, the material behaves perfectly elastic. After exceeding the yield strength, at least part of the deformation of the material will be permanent. The yield strength is a number characteristic for a given material. In practice, it means the maximum stress that a part or structure can carry without permanent damage. For structural carbon steels, this limit is relatively easy to determine.

Carbon structural steels are ferromagnetic and retain their ferromagnetic properties up to a temperature of about 770 °C—in this respect, they have properties such as their main component, iron. This distinguishes them from alloy steels in which the Curie temperature strongly depends on other alloying elements present in their composition: Ni, Cr, Mn, Co. This dependency in some configurations of constituents may even lead to the loss of ferromagnetic properties at ambient temperature (e.g., austenitic steels).

The conditions of magnetic materials can be examined in a nondestructive way using the following methods:

The magnetic flux leakage method relies on analyzing changes in the magnetic field distribution around the tested object. Magnetizing the material with an external magnetic field excites the magnetic flux in the material. If the flux encounters any geometrical inhomogeneities with significantly lower permeance, it breaks out of the material and can be registered by the magnetic sensor [1,2]. Flux leakage allows the inspector to localize and identify surface and subsurface flaws [3]. The inevitable advantages of this technique are high efficiency and no requirement for direct contact with the tested object [4,5]. However, it also has some disadvantages, such as susceptibility to the flaw orientation, the need to demagnetize the object after inspection, a sensitivity that is dependent on the distance between sensor and material, and difficulty detecting small and stress-induced changes [6–8].

The Barkhausen noise method is based on the phenomenon occurring in ferromagnetic material. The structure of any such material is made up of magnetic domains separated by domain walls. Each domain contains dipoles oriented in one privileged magnetization direction [9,10]. The external magnetic field will cause the movement of the domain walls. If any inhomogeneities occur in the material's internal structure, the walls change their position discontinuously. This process is accompanied by a sudden change in magnetization and an induction of voltage pulses in the sensor coil [11]. This technique is suitable for detecting surface and subsurface changes, determining grain dimensions or hardness, and assessing stress levels [12,13]. Some benefits of this method include good sensitivity, a simple examination procedure, no requirement for surface preparation, and quick residual stress recognition [14,15]. This method suffers several drawbacks: the necessity of sensor calibration and a non-standardized measurement approach [16,17].

The Magnetic Memory Method is a relatively novel approach to the nondestructive inspection of ferromagnetic materials. It was proposed by Dubov in 1997 [18]. Under the influence of Earth's magnetic field or applied stress, the intrinsic magnetic domains irreversibly change their position and direction [19]. The process of stress influence on magnetic materials has been known for a long time as an inverse magnetostrictive effect or Villari effect [20]. At the core of the metal magnetic memory method is the detection of a self-magnetic leakage field, indicating the inhomogeneities of the internal structure caused by the effect mentioned above [21]. The significant advantages of this method are no requirement to prepare the surface or premagnetize or demagnetize the material, low-cost measurement equipment, simplicity, time-saving inspection procedure, and the possibility to detect and localize the stress zones, thus avoiding a sudden catastrophic accident [22–25]. The disadvantages of this technique include a weak field forcing the use of sensitive sensors and its applicability only if no external, strong magnetic fields act on the material before or during the inspection [25,26].

2. Materials and Methods

The proposed new method for nondestructive testing of magnetic materials is somehow like those discussed in Section 1, particularly the magnetic memory method.

In the case of the proposed Magnetic Recording Method (MRM), the tested object has to be magnetized in a strictly defined way, e.g., quasi-triangular or quasi-sinusoidal pattern. If external factors such as static stresses act on the material, the residual magnetization changes. By analyzing changes in the magnetic field caused by residual magnetization, it is possible to determine the intensity and direction of the structural influences.

The samples used in the experiments were made of structural S355 steel. Due to its beneficial properties and low-cost production, S355 is widely used in modern industry branches such as civil engineering, offshore, shipbuilding, and automotive [27–34]. The chemical composition of S355 is as follows: Mn—1.45, Al—0.33, P—0.23, Si—0.21, C—0.17, S—0.08 [32]. The exemplary magnetic properties of the steel S355 are as follows [35]: a relative peak permeability of 1500, a saturation point of 1.7 T at 6.9 kA/m, a coercive field of 310 A/m, and a residual flux density of 1 T (measured on the major loop).

Each sample was cut out of a hot-rolled plate using a waterjet cutter to avoid jagged metal edges. The shape and dimensions of the samples produced in this way are shown in Figure 1.

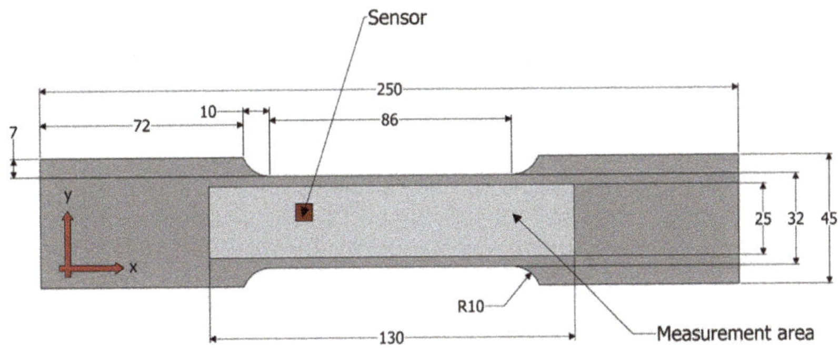

Figure 1. Sample shape and dimensions with depicted measurement area.

The measuring procedure consisted of four steps. In the first step, the sample was magnetized in a strictly defined manner. The magnetizing element consisted of the magnets configured in the array to generate a quasi-sinusoidal magnetization pattern in the sample. A simplified view of the magnetizing element is shown in Figure 2. It was constructed using 100 neodymium plate magnets, 2 mm high, 15 mm wide, and 30 mm long, made of N38 material, and magnetized in the length direction (30 mm). The material N38 ($Nd_2Fe_{14}B$) has the following magnetic parameters: remanence B_r = 1.2 T; coercivity $H_{cb} \geq 899$ kA/m; coercivity $H_{cJ} \geq 955$ kA/m; energy density (BH) max $\geq$ 287–310 kJ/m^3. The magnets were separated from each other with a tape 0.12 mm thick. On one side (facing the magnetized sample), a 0.8 mm thick PTFE (polytetrafluoroethylene) spacer was glued to the array of magnets to facilitate sliding and ensure a permanent lift-off. The magnetic field in the gap between the magnets and the magnetized sample was 0.97 T. It was measured with a GM08 Gaussmeter manufactured by Hirst Magnetic Instruments (Falmouth, United Kingdom) with a PT7810 Hall effect probe. The array was manually moved above the sample surface with a lift-off of 0.8 mm in a direction parallel to the y-axis from one edge to the other edge of the sample (Figure 2). The magnets were moved at a speed of around 5 mm/s. In this way, the plate was magnetized relatively evenly in the y-axis direction. If necessary, the uniformity of the magnetization could be improved by using a motorized mechanical scanner.

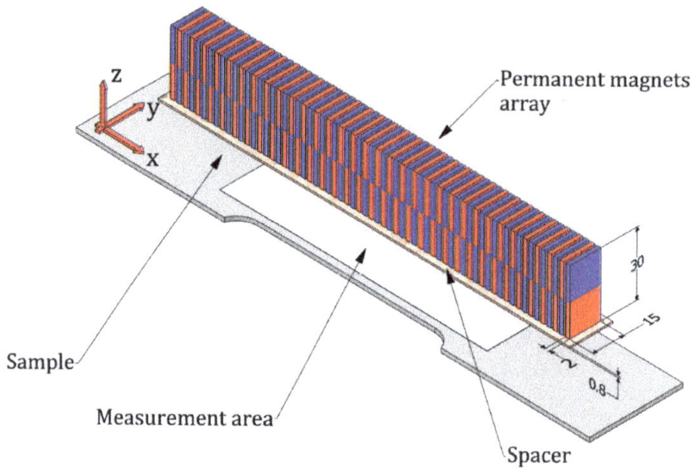

Figure 2. The array of magnets over the sample under magnetization.

In the second step, the magnetic field caused by the residual magnetization of the sample was measured with a three-axis magnetometer (HMC5883L) moved in the x- and y-directions over the sample surface (lift-off 0.3 mm) in the area depicted in Figures 1 and 2. The third step of the procedure included filtering two-dimensional signals and then averaging, which results in obtaining one-dimensional signals. In the last stage, one-dimensional signals were analyzed and their characteristic parameters, such as amplitude and frequency, were determined. A flowchart of the procedure designated for this purpose is shown in Figure 3.

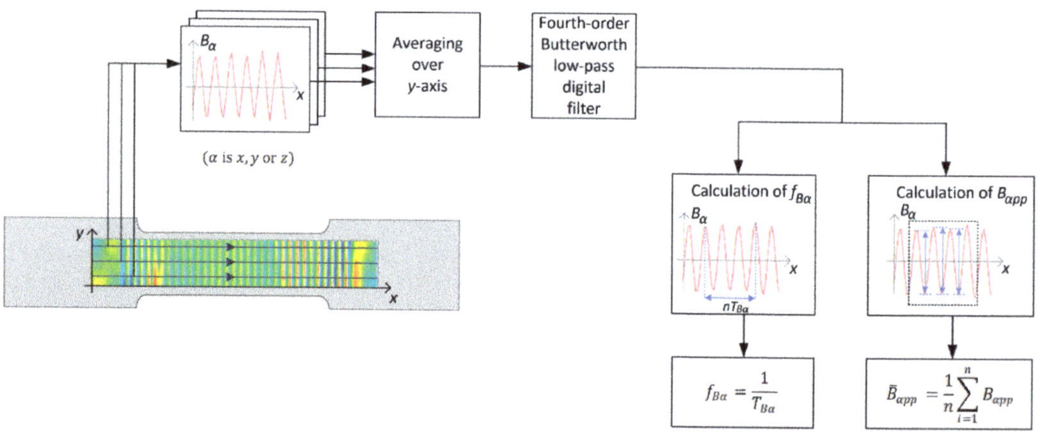

Figure 3. The measuring procedure.

In all cases, data measured for selected y-coordinates were used for the analysis. The scanning paths were chosen in such a way as to avoid the influence of the edge effect on the calculation of characteristic parameters. The selected signals were used to calculate an average signal. Next, a low-pass, fourth-order, digital Butterworth filter ($f/f_N = 0.4$, f_N—Nyquist frequency) was used to remove external interferences of the measured signals. After filtration, the characteristic parameters of the signal were calculated.

Several cycles of the signal were selected to determine the signal period, and thus its frequency (Equation (1)):

$$f_{B\alpha} = \frac{1}{T_{B\alpha}} \quad (1)$$

where: α—x, y, or z component of the magnetic field, $f_{B\alpha}$—frequency of a given magnetic field component, and $T_{B\alpha}$—magnetic field period of a given component. Then, the windowed central part of the signal (corresponding to the magnetic field measured in the middle part of the sample) was utilized to calculate the mean peak-to-peak value (Equation (2)):

$$\overline{B}_{\alpha pp} = \frac{1}{n}\sum_{i=1}^{n} B_{\alpha pp} \quad (2)$$

where: $B_{\alpha pp}$—the peak-to-peak value of magnetic field component (α could be x, y, or z), n—the number utilized in calculations of peak-to-peak values of B_α, $\overline{B}_{\alpha pp}$—mean peak-to-peak value of magnetic field component.

Furthermore, additional calculations: relative mean change in magnetic field (Equation (3)) and relative frequency change in the magnetic field (Equation (4)) were performed to assess the variations in magnetization of the samples after their stress-loading.

$$\Delta\overline{B}_\alpha = \frac{\overline{B}^{before}_{\alpha pp} - \overline{B}^{after}_{\alpha pp}}{\overline{B}^{before}_{\alpha pp}} \cdot 100\% \quad (3)$$

$$\Delta f_{B\alpha} = \frac{f^{before}_{B\alpha} - f^{after}_{B\alpha}}{f^{before}_{B\alpha}} \cdot 100\% \quad (4)$$

where: α—could be x, y, or z component of the magnetic field, $\overline{B}^{before}_{\alpha pp}$—mean peak-to-peak value of the magnetic field for the non-stressed samples, $\overline{B}^{after}_{\alpha pp}$—mean peak-to-peak value of the magnetic field for the samples after tensile loading, $f^{before}_{B\alpha}$—signal frequency for the non-stressed samples, $f^{after}_{B\alpha}$—signal frequency for the samples after tensile loading.

3. Results

This experiment was performed according to the following methodology. Eight samples (S01–S08) made of S355 were magnetized to record a quasi-sinusoidal pattern. Next, the 2D distribution of the magnetic field caused by the residual magnetization of the sample was measured using a magnetometer. Subsequently, each sample was loaded to a different degree in elastic and plastic regions' volume using an Instron Universal Testing machine (Figure 4 and Table 1). In order to investigate possible changes in the magnetization pattern, the magnetic field was measured once again. The signals measured for each sample before and after stress-loading were stored and used to prepare plots presented in this section.

Table 1. Tensile test results.

Sample	Stress (MPa)	Strain (%)
S01	100	0.18
S02	200	0.43
S03	300	0.69
S04	400	0.94
S05	473	1.20
S06	479	2.00
S07	494	3.00
S08	571	10.00

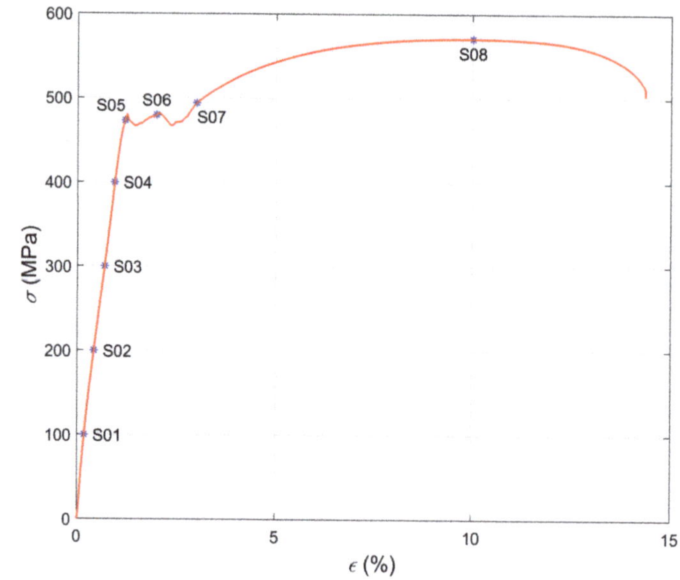

Figure 4. The stress–strain curve obtained for the sample made of S355. S01–S08—eight samples loaded to a different degree in elastic and plastic regions' volume depicted on the curve.

The measurements of the magnetic field changes were carried out following the methodology described in Section 2. As a result, two sets of two-dimensional signals for each sample (S01–S08) were obtained: the first plot for the specimen before tensile loading and the second for the specimen after tensile loading. Figure 5 shows examples of two-dimensional signals measured in both cases for the sample S05. Similar graphs obtained for other samples were omitted because they would increase the article's length without introducing important information. The plots show only two components B_x and B_z because the third component, B_y, was a small amplitude signal unused for evaluation.

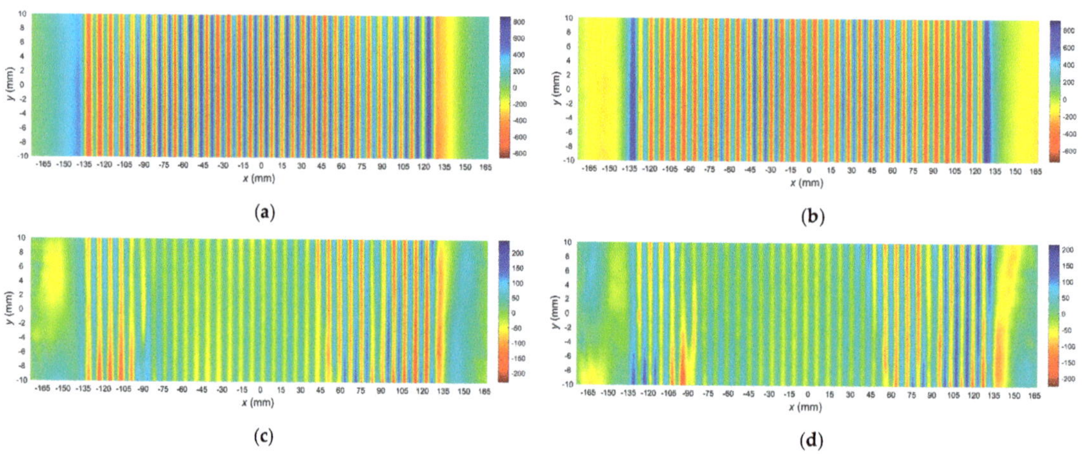

Figure 5. Results of 2D measurements of the magnetic field in the case of sample S05. (**a**) Component B_x before tensile loading; (**b**) component B_z before tensile loading; (**c**) component B_x after tensile loading; (**d**) component B_z after tensile loading.

In order to straightforwardly demonstrate the usability of the proposed method, the analysis was limited only to one-dimensional signals taken from the central part of the samples. The average signals of the x and z magnetic field components were calculated for each sample. Plots of the averaged signals for all samples before and after tensile loading are shown in Figure 6. The plots depict variations in the amplitude of the components depending on the sensor position along the x-axis. In the central part of the sample, an evident change in the signals can be observed.

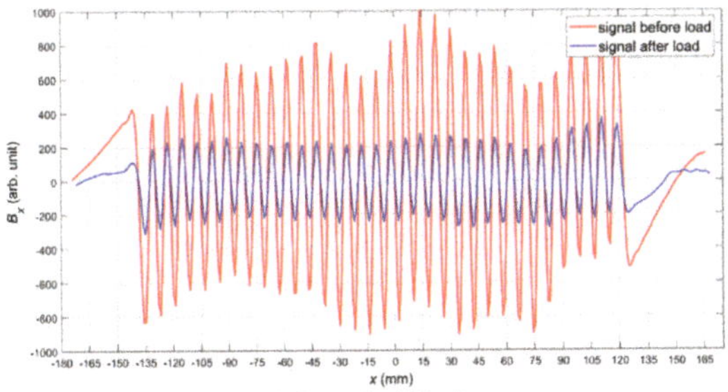

(**a**) Sample S01, B_x

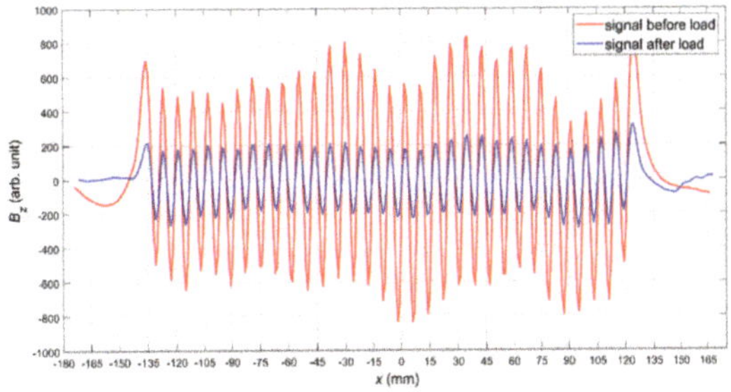

(**b**) Sample S01, B_z

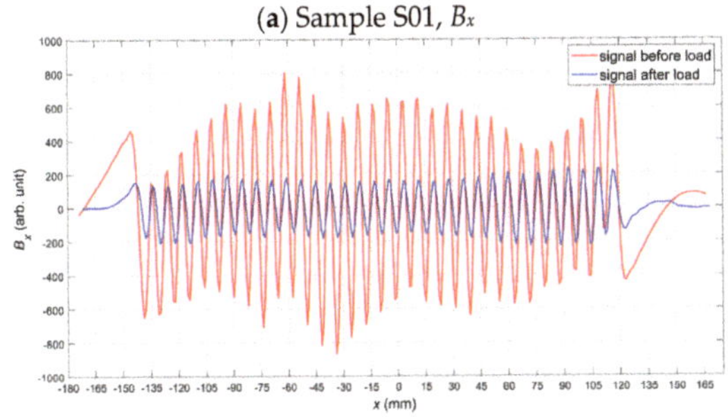

(**c**) Sample S02, B_x

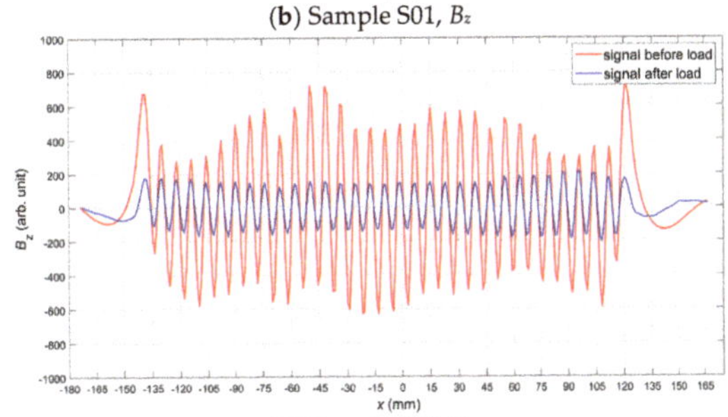

(**d**) Sample S02, B_z

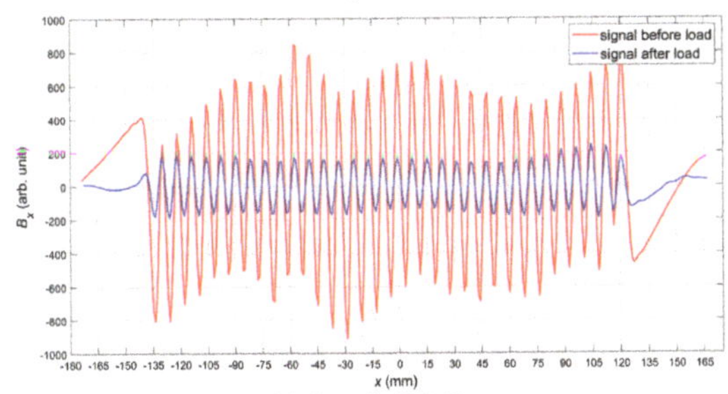

(**e**) Sample S03, B_x

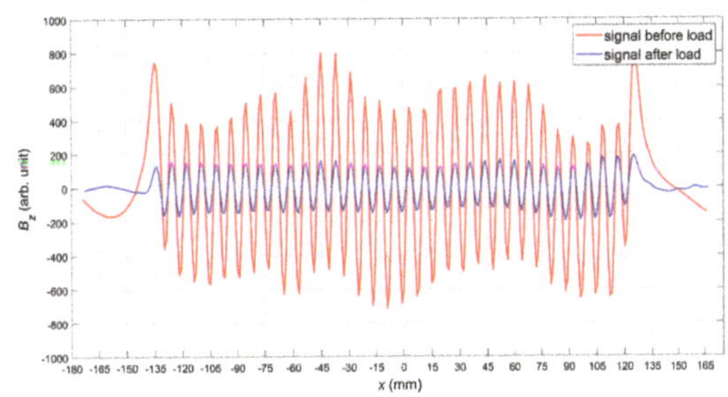

(**f**) Sample S03, B_z

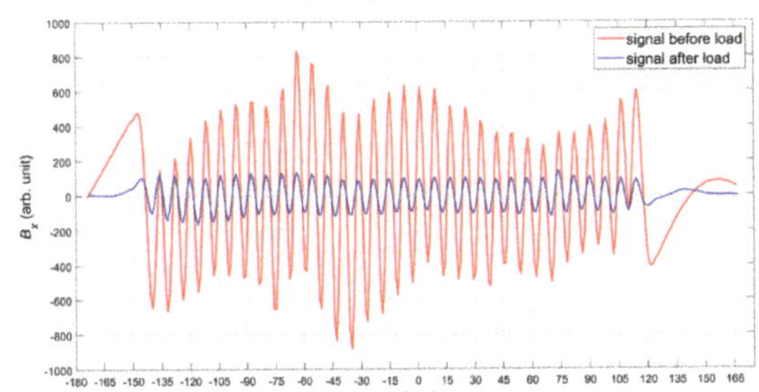

(**g**) Sample S04, B_x

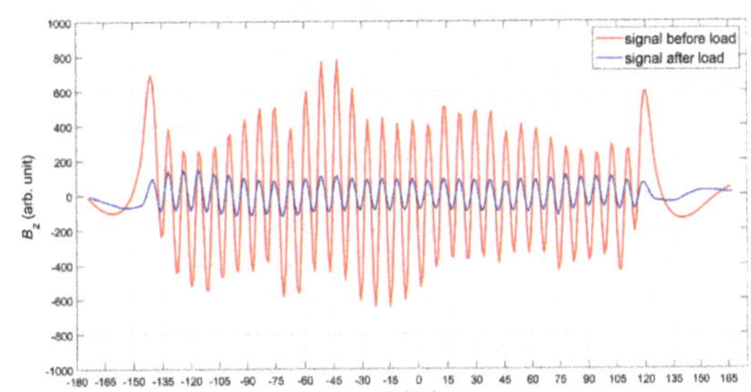

(**h**) Sample S04, B_z

Figure 6. *Cont.*

(**i**) Sample S05, B_x (**j**) Sample S05, B_z

(**k**) Sample S06, B_x (**l**) Sample S06, B_z

(**m**) Sample S07, B_x (**n**) Sample S07, B_z

(**o**) Sample S08, B_x (**p**) Sample S08, B_z

Figure 6. Components of the magnetic field measured for the magnetized samples before (red line) and after tensile loading (blue line): (**a**) B_x for sample S01, (**b**) B_z for sample S01, (**c**) B_x for sample S02, (**d**) B_z for sample S02, (**e**) B_x for sample S03, (**f**) B_z for sample S03, (**g**) B_x for sample S04, (**h**) B_z for sample S04, (**i**) B_x for sample S05, (**j**) B_z for sample S05, (**k**) B_x for sample S06, (**l**) B_z for sample S06, (**m**) B_x for sample S07, (**n**) B_z for sample S07, (**o**) B_x for sample S08, (**p**) B_z for sample S08.

Detailed analysis of the signals measured for samples S01–S05 (Figure 6) allows us to conclude that as the stress level increased, the magnetic field amplitude decreased in the central part of the measuring area, and frequencies f_{Bx}, f_{Bz} remained practically unchanged. In the case of the samples loaded over the yield point (S06–S08), the amplitudes and frequencies f_{Bx}, f_{Bz} of the signals measured after tensile loading significantly decreased compared to the parameters measured before tensile loading (Figure 6).

Evaluating the condition of samples based solely on direct observation of the signal before and after tensile loading can be problematic due to the minor differences. For this reason, characteristic parameters were determined, and additional charts were prepared to visualize the changes taking place. First, the relative change in the magnetic field amplitude as a function of strain is presented (Figure 7). As can be seen from Figure 7a,b, the curve of the above relation consists of two parts separated by the point defining the elastic limit of the samples. For samples S01 to S04, the values increased approximately linearly. Then, starting with sample S04, the curve slopes sharply down to the value corresponding to the yield point sample S05. After the yield point was exceeded, the curve increases again to a point corresponding to the sample S08, but slower than its initial part. Thus, it can be concluded that an increase in the deformation level of the samples increased the value of the relative change in the residual magnetization.

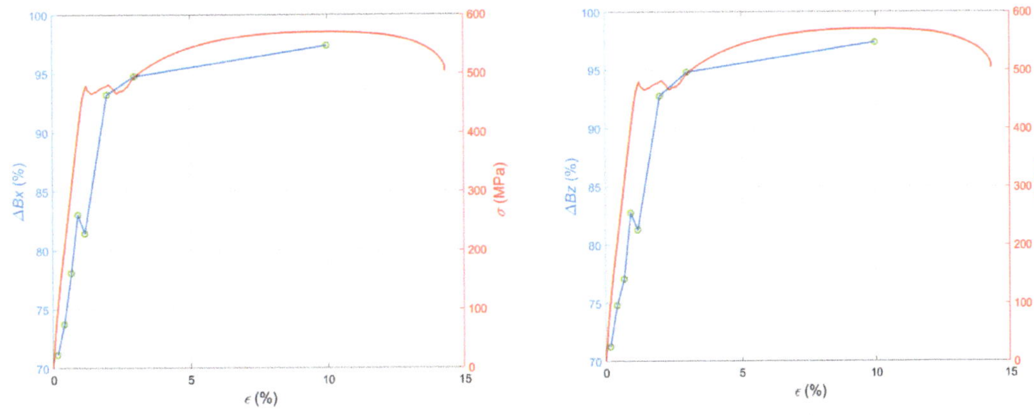

Figure 7. Relative mean changes in the magnetic field in the case of the samples S01–S08 plotted versus the strain: (**a**) component ΔB_x; (**b**) component ΔB_z.

Another two sets of plots contain the relative mean change in magnetic field ΔB as the function of applied stress σ for the samples S01–S04 (Figure 8) and strain ε for the samples S05–S08 (Figure 9), respectively. The reason for separating the parameter analysis of samples S01–S04 from samples S05–S08 is the change in mechanical properties at the point corresponding to sample S05. In the case of the first four specimens, the stresses induced an elastic deformation of the structure, and in the case of the remaining four specimens, plastic deformation was induced.

Figure 8a shows the relative mean change in the magnetic field ΔB_x, ΔB_z (Equation (3)) in the case of the samples S01–S04. Component ΔB_x increased exponentially with the rise in the stress level. On the contrary, the curve for the component ΔB_z (Figure 8b) rises slower and resembles the cubic polynomial. Due to the monotonicity of the curves, these graphs allow evaluating the sample conditions straightforwardly. Plots presented in Figure 9 show the relative mean change ΔB_x, ΔB_z in the magnetic field as the function of strain ε for the samples S05–S08.

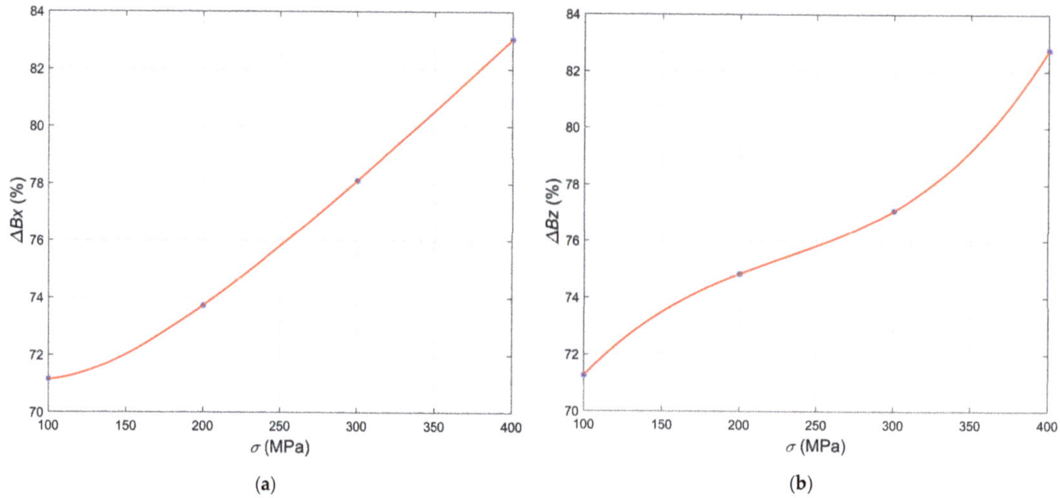

Figure 8. Relative mean changes in the magnetic field in the case of the samples S01–S04 plotted versus the stress: (**a**) component ΔB_x; (**b**) component ΔB_z.

Figure 9a,b indicates that the values of ΔB_x and ΔB_z increase exponentially with growing strain values. After passing the yield point corresponding to sample S05, the curve bends. This change is the transition from the elastic region through the yield point to the plastic region in the following samples. Plots showing the relative change in frequency Δf_{Bx}, Δf_{Bz} (Equation (4)) as a function of strain ε can also be used to evaluate the conditions of the samples S05–S08 (Figure 10). In the case of both components (B_x and B_z), the curves increase to the point of the maximum strain (sample S08). There is an inflection of the curve at the point corresponding to sample S07.

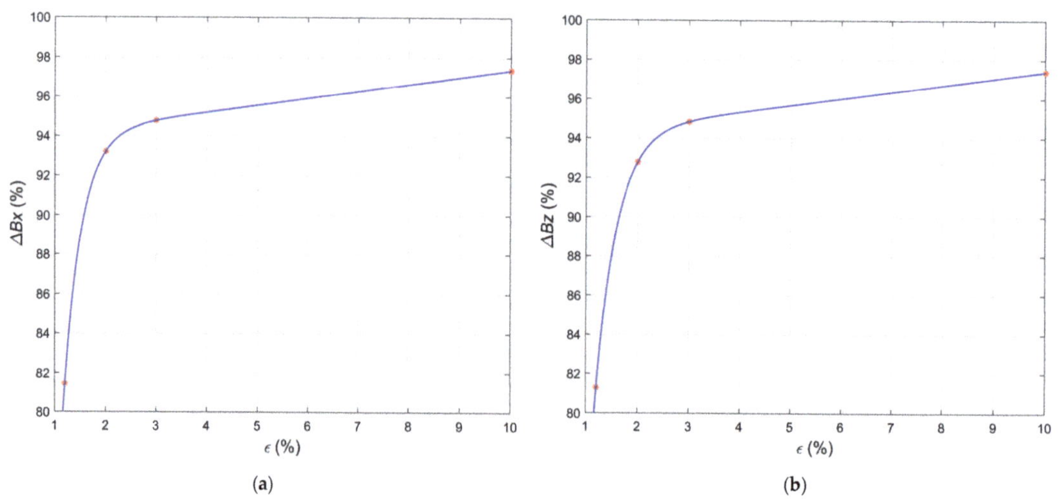

Figure 9. Relative mean changes in the magnetic field in the case of the samples S05–S08 plotted versus the strain. (**a**) component ΔB_x; (**b**) component ΔB_z.

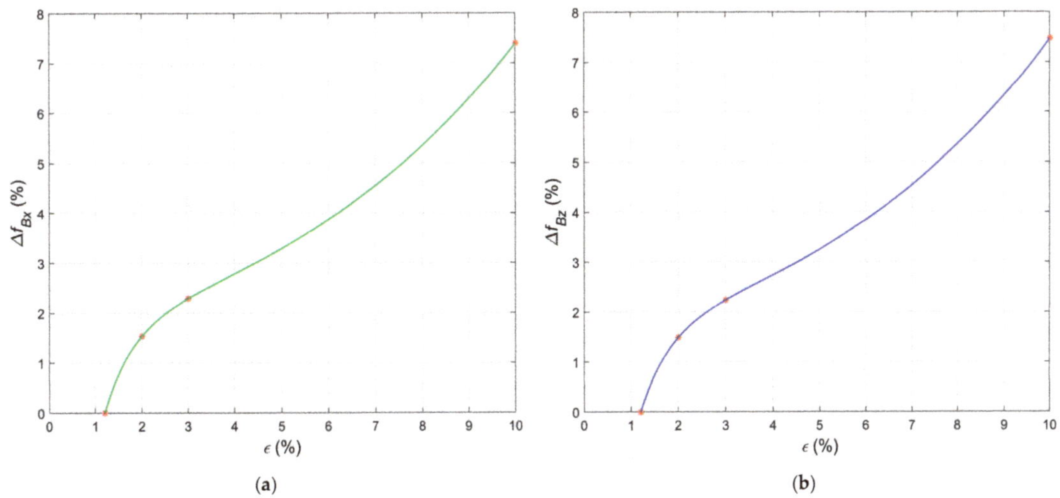

Figure 10. Relative changes in the signal frequency in the case of the samples S05–S08 plotted versus the strain. (**a**) Δf_{Bx}; (**b**) Δf_{Bz}.

4. Discussion

The tests of the proposed method of nondestructive testing, which was presented in the previous section, covered several dozen samples made of the same material (S355) and should be treated as a first attempt to verify the suitability of the method.

The strictly defined signals (e.g., a sinusoid of a specific frequency) enable the use of dedicated filtering algorithms that effectively eliminate external disturbances. For example, a simple pass-band digital filter could eliminate a DC (Direct Current)) component from the signals presented in Figure 6. The parameters of the measured signals (e.g., the amplitude and frequency of the sine wave) can be determined by proven and reliable algorithms. These parameters allow for unambiguous identification of the material condition both in the elastic (Figure 8) and plastic regions (Figure 9). It should also be noted that external sources of DC magnetic fields have a limited impact on the results obtained in the proposed method. For example, such DC fields would not affect the frequency of the measured sinusoidal signal in any way. Such frequency change (Figure 10) is a very reliable parameter, but, unfortunately, it can only be observed in the case of samples loaded over the yield point.

The achieved results of the tests can generally be assumed as promising, and the method can help identify the condition of elements made of ferromagnetic materials subjected to loads. However, as the method is new, it is necessary to conduct further detailed tests to clarify existing doubts and improve the test procedure. The following aspects of the inspection procedure should be investigated and analyzed: the magnetization process, the residual magnetization measurement process, and the algorithms for analyzing the received signals.

One of the problems that has to be addressed is the decreasing magnetization of the tested elements over time. For this purpose, samples have been retained, and measurements will be repeated during the following year. Unfortunately, it was impossible to conduct comparative tests using other NDT methods before this time elapsed.

When comparing the results obtained by the proposed method with the results from other testing methods applied to very similar samples but made of SS400 steel, some significant differences can be observed. For example, in the case of the hysteresis loop observation method [36], unambiguous identification of the sample state is possible, but this method is less sensitive in the elastic range (measurements carried out after removing the load). Moreover, in this method, the spatial resolution is lower due to the larger dimensions of the transducer, and its implementation requires the use of a more complex measurement system.

Similarly, lower sensitivity in the elastic range can be observed in the case of the results obtained from the eddy current method [37] and the residual magnetization observation method with the GMR (Giant MagnetoResistance) transducer [38]. Additionally, measured parameters of the signals did not allow for unequivocal identification of the sample state as the same value was obtained for the samples before and after the yield point. A considerable advantage of the eddy current testing is the independence of the results on the magnetic history of the tested object.

The advantage of all the compared methods over the proposed Magnetic Recording Method is that there is no need to magnetize the sample with a specific pattern beforehand. Therefore, the proposed method can be applied only in some specific cases, for example, when it is necessary to constantly monitor crucial elements of the structure.

Due to the limited number of tests of a new method, it is not easy to make a reliable comparison with other methods. A comparison should also be made using the same or very similar samples. Unfortunately, during the experiment, it was not possible. Therefore, the comparison of the proposed method with the other nondestructive electromagnetic methods presented in Table 2 should be considered only as a preliminary attempt and will be updated after the next set of experiments.

Table 2. Comparison of the Magnetic Recording Method with other magnetic methods.

	Metal Magnetic Memory	**Hysteresis Loop**	**Barkhausen Noise**	**Magnetic Flux Leakage**	**Magnetic Particle**	**Residual Magnetization**	**Eddy Current**	**Magnetic Recording**
Sensitivity to surface cracks	High	Low	Low	High	High	High	High	High
Sensitivity to subsurface cracks	High	Very Low	Very Low	Medium	Medium	Medium	High	High
Sensitivity to residual stress and plastic deformations(loading over the yield point)	High	High	High	Low	Very Low	High	High	High
Sensitivity to residual stress (loading below the yield point)	High	High	High	No	No	Medium	Low	High
Unambiguous identification of stress (loading below and over the yield point)	Medium	Medium	Low	No	No	Medium	Low	High
The necessity of preliminary preparation before operation	No	No	No	No	No	No	No	Magnetization of the pattern
The necessity of preliminary treatment before measurement	No	No	No	No	DC magnetization	DC magnetization	No	No
Influences of external DC fields during the measurement	Very High	Low	Low	Low	Medium	High	Very Low	Low/Medium
Influences of external AC fields during the measurements	Low	Low	High	Low	No	No	High	No
Influences of DC magnetization before the measurements	Very High	Low	Low	Low	Medium	Low	No	Low/Medium
Measurement speed	High	Low	Low	High	Medium	High	High	High
The complexity of the instrumentation	Low/ Medium	High	High	Low	Very Low	Low	Medium	Low
Repeatability of the results	Low	Medium	Medium	High	High	High	Very High	High
Spatial resolution	High	Low/Very Low	Low/Very Low	High	Medium	High	Medium/ High	High

5. Conclusions

The tightening of the requirements regarding the reliability of the structure creates the necessity for frequent inspections that will detect not only existing defects but also any changes that may indicate their occurrence. One such change is the residual stress distribution.

Several nondestructive testing methods can detect residual stress distribution and material changes due to stress. The authors assumed that it is also possible to analyze changes in the prerecorded magnetization of the tested element. Experiments have verified this, and the article proposes a Magnetic Recording Method that opens up new possibilities for monitoring critical structural elements.

Based on the results of the research conducted so far, it can be concluded that:

- The parameters (amplitude and frequency) of the quasi-sinusoidal pattern change significantly with the applied tensile stress, especially the amplitude in the elastic region and the frequency over the yield point.
- Regardless of the state, the load can be unequivocally determined based on formulated simple parameters.
- Additionally, the state after exceeding the yield point can be unequivocally determined based on changes in the amplitude of the signal and the frequency of the magnetization pattern.
- The obtained quasi-sinusoidal magnetization pattern is easy for later analysis.
- During the magnetization process, magnets can be placed at a relatively large distance (on the order of 1 mm) from the magnetized element.
- Further experiments are necessary to find the optimal and maximum distance between the magnets and the tested material.
- The proposed magnetization method can be used for flat surfaces.
- In the case of more complicated shapes of the tested element, it is necessary to make dedicated magnetizing systems.
- In order to obtain quasi-sinusoidal magnetization patterns on elements of larger sizes or to obtain a higher frequency of changes, it would be more effective to use a magnetizing head mounted on a motorized manipulator instead of magnets.
- The regularity of the magnetization pattern is not critical if the primary magnetization is measured and the signals are archived for normalization in later tests.
- While maintaining the signals measured after the magnetization process, the previous magnetization state of the sample is not important, but it is better to demagnetize the sample before magnetization to simplify the diagnostic process.

Despite the satisfactory initial test results, more research is needed to identify the method's strengths and weaknesses and improve the testing process. It is planned to test other magnetization methods (e.g., using recording heads) to examine objects without a flat surface. Plans are underway to analyze the measured two-dimensional signals and utilize chosen statistics features to develop more criteria for the material condition assessments. An experiment will also be carried out to assess the effect of the passage of a long time period (over one year) on the sample's magnetization state.

Author Contributions: Idea for the MRM method, T.C.; manuscript conceptualization, T.C., R.D.Ł., K.G. and M.J.Ż.; implementation of the method, T.C. and R.D.Ł.; software preparation, R.D.Ł.; resources—transducer and system elements, T.C.; designed and conducted the experiment in part concerning the incremental deformation of steel samples, K.G. and M.J.Ż.; magnetic measurements, T.C. and R.D.Ł.; data curation, R.D.Ł.; data analysis in the scope related to the mechanical properties of carbon structural steels, K.G. and M.J.Ż.; writing—original draft preparation, R.D.Ł.; writing—correction and extensive editing, T.C., R.D.Ł., K.G. and M.J.Ż.; visualization, R.D.Ł.; supervision, T.C. All authors have read and agreed to the published version of the manuscript.

Funding: This research received no external funding.

Institutional Review Board Statement: Not applicable.

Informed Consent Statement: Not applicable.

Data Availability Statement: The data presented in this study are available on request from the corresponding author. The data are not publicly available due to a complicated structure that requires additional explanation.

Conflicts of Interest: The authors declare no conflict of interest. The funders had no role in the design of the study; in the collection, analyses, or interpretation of data; in the writing of the manuscript, or in the decision to publish the results.

References

1. Sam, M.A.I.M.N.; Jin, Z.; Oogane, M.; Ando, Y. Investigation of a Magnetic Tunnel Junction Based Sensor for the Detection of Defects in Reinforced Concrete at High Lift-Off. *Sensors* **2019**, *19*, 4718. [CrossRef]
2. Shi, Y.; Zhang, C.; Li, R.; Cai, M.; Jia, G. Theory and Application of Magnetic Flux Leakage Pipeline Detection. *Sensors* **2015**, *15*, 31036–31055. [CrossRef] [PubMed]
3. Ma, Q.; Tian, G.; Zeng, Y.; Li, R.; Song, H.; Wang, Z.; Gao, B.; Zeng, K. Pipeline In-Line Inspection Method, Instrumentation and Data Management. *Sensors* **2021**, *21*, 3862. [CrossRef] [PubMed]
4. Wu, J.; Yang, Y.; Li, E.; Deng, Z.; Kang, Y.; Tang, C.; Sunny, A.I. A High-Sensitivity MFL Method for Tiny Cracks in Bearing Rings. *IEEE Trans. Magn.* **2018**, *54*, 1–8. [CrossRef]
5. Qidwai, U.; Akbar, M.A.; Maqbool, M. Robotic MFL Probe Design for Inspection in Structural Health Monitoring System. In Proceedings of the 2018 8th IEEE International Conference on Control System, Computing and Engineering (ICCSCE), Penang, Malaysia, 23–25 November 2018; IEEE: New York, NY, USA, 2018; pp. 5–9.
6. Tang, J.; Wang, R.; Liu, B.; Kang, Y. A novel magnetic flux leakage method based on the ferromagnetic lift-off layer with through groove. *Sens. Actuators A Phys.* **2021**, *332*, 113091. [CrossRef]
7. Okolo, C.K.; Meydan, T. Axial Magnetic Field Sensing for Pulsed Magnetic Flux Leakage Hairline Crack Detec-tion and Quantification. In Proceedings of the 2017 IEEE SENSORS, Glasgow, UK, 29 October–1 November 2017; IEEE: New York, NY, USA, 2017; pp. 1–3.
8. Wu, D.; Liu, Z.; Wang, X.; Su, L. Composite magnetic flux leakage detection method for pipelines using alternating magnetic field excitation. *NDT E Int.* **2017**, *91*, 148–155. [CrossRef]
9. Zgútová, K.; Pitoňák, M. Attenuation of Barkhausen Noise Emission due to Variable Coating Thickness. *Coatings* **2021**, *11*, 263. [CrossRef]
10. Rößler, M.; Putz, M.; Hochmuth, C.; Gentzen, J. In-process evaluation of the grinding process using a new Barkhausen noise method. *Procedia CIRP* **2021**, *99*, 202–207. [CrossRef]
11. Hwang, Y.-I.; Kim, Y.-I.; Seo, D.-C.; Seo, M.-K.; Lee, W.-S.; Kwon, S.; Kim, K.-B. Experimental Consideration of Conditions for Measuring Residual Stresses of Rails Using Magnetic Barkhausen Noise Method. *Materials* **2021**, *14*, 5374. [CrossRef] [PubMed]
12. Gaunkar, N.G.P.; Nlebedim, I.C.; Jiles, D.C.; Gaunkar, G.V.P. Examining the Correlation Between Microstructure and Barkhausen Noise Activity for Ferromagnetic Materials. *IEEE Trans. Magn.* **2015**, *51*, 1–4. [CrossRef]
13. Novák, M.; Eichler, J. Magnetic Barkhausen Noise Spectral Emission of Grain Oriented Steel Under Ultra Low Frequency Magnetization. In Proceedings of the 2019 12th International Conference on Measurement, Smolenice, Slovakia, 27–29 May 2019; IEEE: New York, NY, USA; 2019; pp. 158–161.
14. Sánchez, J.C.; De Campos, M.F.; Padovese, L. Comparison Between Different Experimental Set-Ups for Measuring the Magnetic Barkhausen Noise in a Deformed 1050 Steel. *J. Nondestruct. Eval.* **2017**, *36*, 66. [CrossRef]
15. Vourna, P.; Ktena, A.; Tsakiridis, P.; Hristoforou, E. An accurate evaluation of the residual stress of welded electrical steels with magnetic Barkhausen noise. *Measurement* **2015**, *71*, 31–45. [CrossRef]
16. Neslušan, M.; Trojan, K.; Haušild, P.; Minárik, P.; Mičietová, A.; Čapek, J. Monitoring of components made of duplex steel after turning as a function of flank wear by the use of Barkhausen noise emission. *Mater. Charact.* **2020**, *169*, 110587. [CrossRef]
17. Santa-Aho, S.; Laitinen, A.; Sorsa, A.; Vippola, M. Barkhausen Noise Probes and Modelling: A Review. *J. Nondestruct. Eval.* **2019**, *38*, 94. [CrossRef]
18. Dubov, A. A study of metal properties using the method of magnetic memory. *Met. Sci. Heat Treat.* **1997**, *39*, 401–405. [CrossRef]
19. Shi, P.; Su, S.; Chen, Z. Overview of Researches on the Nondestructive Testing Method of Metal Magnetic Memory: Status and Challenges. *J. Nondestruct. Eval.* **2020**, *39*, 1–37. [CrossRef]
20. Wang, H.; Dong, L.; Wang, H.; Ma, G.; Xu, B.; Zhao, Y. Effect of tensile stress on metal magnetic memory signals during on-line measurement in ferromagnetic steel. *NDT E Int.* **2021**, *117*, 102378. [CrossRef]
21. Kolokolnikov, S.; Dubov, A.; Steklov, O. Assessment of welded joints stress–strain state inhomogeneity before and after post weld heat treatment based on the metal magnetic memory method. *Weld. World* **2016**, *60*, 665–672. [CrossRef]
22. Shi, P.; Jin, K.; Zhang, P.; Xie, S.; Chen, Z.; Zheng, X. Quantitative Inversion of Stress and Crack in Ferromagnetic Materials Based on Metal Magnetic Memory Method. *IEEE Trans. Magn.* **2018**, *54*, 1–11. [CrossRef]
23. Pang, C.; Zhou, J.; Zhao, R.; Ma, H.; Zhou, Y. Research on Internal Force Detection Method of Steel Bar in Elastic and Yielding Stage Based on Metal Magnetic Memory. *Materials* **2019**, *12*, 1167. [CrossRef]

24. Chongchong, L.; Lihong, D.; Haidou, W.; Guolu, L.; Binshi, X. Metal magnetic memory technique used to predict the fatigue crack propagation behavior of 0.45%C steel. *J. Magn. Magn. Mater.* **2016**, *405*, 150–157. [CrossRef]
25. Ren, S.; Ren, X.; Duan, Z.; Fu, Y. Studies on influences of initial magnetization state on metal magnetic memory signal. *NDT E Int.* **2019**, *103*, 77–83. [CrossRef]
26. Li, Z.; Dixon, S.; Cawley, P.; Jarvis, R.; Nagy, P.B.; Cabeza, S. Experimental studies of the magneto-mechanical memory (MMM) technique using permanently installed magnetic sensor arrays. *NDT E Int.* **2017**, *92*, 136–148. [CrossRef]
27. Macek, W. Fracture Areas Quantitative Investigating of Bending-Torsion Fatigued Low-Alloy High-Strength Steel. *Metals* **2021**, *11*, 1620. [CrossRef]
28. Gao, W.; Wang, D.; Cheng, F.; Di, X.; Deng, C.; Xu, W. Microstructural and mechanical performance of underwater wet welded S355 steel. *J. Mater. Process. Technol.* **2016**, *238*, 333–340. [CrossRef]
29. Corigliano, P.; Cucinotta, F.; Guglielmino, E.; Risitano, G.; Santonocito, D. Thermographic analysis during tensile tests and fatigue assessment of S355 steel. *Procedia Struct. Integr.* **2019**, *18*, 280–286. [CrossRef]
30. Jacob, A.; Mehmanparast, A.; D'Urzo, R.; Kelleher, J. Experimental and numerical investigation of residual stress effects on fatigue crack growth behaviour of S355 steel weldments. *Int. J. Fatigue* **2019**, *128*, 105196. [CrossRef]
31. Cadoni, E.; Forni, D.; Gieleta, R.; Kruszka, L. Tensile and compressive behaviour of S355 mild steel in a wide range of strain rates. *Eur. Phys. J. Spéc. Top.* **2018**, *227*, 29–43. [CrossRef]
32. Rodrigues, D.; Leitão, C.; Balakrishnan, M.; Craveiro, H.; Santiago, A. Tensile properties of S355 butt welds after exposure to high temperatures. *Constr. Build. Mater.* **2021**, *302*, 124374. [CrossRef]
33. Forni, D.; Chiaia, B.; Cadoni, E. High strain rate response of S355 at high temperatures. *Mater. Des.* **2016**, *94*, 467–478. [CrossRef]
34. Xin, H.; Correia, J.A.; Veljkovic, M. Three-dimensional fatigue crack propagation simulation using extended finite element methods for steel grades S355 and S690 considering mean stress effects. *Eng. Struct.* **2021**, *227*, 111414. [CrossRef]
35. Anglada, J.R.; Arpaia, P.; Buzio, M.; Pentella, M.; Petrone, C. Characterization of Magnetic Steels for the FCC-ee Magnet Prototypes. In Proceedings of the 2020 IEEE International Instrumentation and Measurement Technology Conference (I2MTC), Glasgow, UK, 25 May–28 June 2020; IEEE: New York, NY, USA, 2020; pp. 1–6.
36. Chady, T.; Łukaszuk, R. Examining Ferromagnetic Materials Subjected to a Static Stress Load Using the Magnetic Method. *Mater.* **2021**, *14*, 3455. [CrossRef] [PubMed]
37. Chady, T. Evaluation of Stress Loaded Steel Samples Using Selected Electromagnetic Methods. In Proceedings of the AIP Conference Proceedings, Green Bay, WI, USA, 27 July–1 August 2003; AIP Publishing: College Park, MD, USA, 2004; Volume 700, pp. 1296–1303.
38. Chady, T. Evaluation of stress loaded steel samples using GMR magnetic field sensor. *IEEE Sens. J.* **2002**, *2*, 488–493. [CrossRef]

materials

MDPI

Article

Analysis of Magnetic Nondestructive Measurement Methods for Determination of the Degradation of Reactor Pressure Vessel Steel

Gábor Vértesy [1,*], Antal Gasparics [1], Ildikó Szenthe [1], Madalina Rabung [2], Melanie Kopp [2] and James M. Griffin [3]

1 Centre for Energy Research, 1121 Budapest, Hungary; gasparics.antal@energia.mta.hu (A.G.); szenthe.ildiko@energia.mta.hu (I.S.)
2 Fraunhofer Institute for Nondestructive Testing (IZFP), 66123 Saarbrücken, Germany; Madalina.Rabung@izfp.fraunhofer.de (M.R.); Melanie.Kopp@izfp.fraunhofer.de (M.K.)
3 Future Transport and Cities Research Centre, Coventry University, Coventry CV1 5FB, UK; ac0393@coventry.ac.uk
* Correspondence: gabor.vertesy@energia.mta.hu

Abstract: Nondestructive magnetic measurement methods can be successfully applied to determine the embrittlement of nuclear pressure vessel steel caused by neutron irradiation. It was found in previous works that reasonable correlation could be obtained between the nondestructively measured magnetic parameters and destructively determined ductile-to-brittle transition temperature. However, a large scatter of the measurement points was detected even in the cases of the non-irradiated reference samples. The reason for their scattering was attributed to the local inhomogeneity of material. This conclusion is verified in the present work by applying three different magnetic methods on two sets of Charpy samples made of two different reactor steel materials. It was found that by an optimal magnetic pre-selection of samples, a good, linear correlation can be found between magnetic parameters as well as the ductile-to-brittle transition temperature with low scattering of points. This result shows that neutron irradiation embrittlement depends very much on the local material properties.

Keywords: magnetic nondestructive evaluation; reactor pressure vessel; neutron irradiation embrittlement; magnetic adaptive testing; micromagnetic multiparameter microstructure and stress analysis 3MA; Barkhausen noise measurement; steel degradation; ductile to brittle transition temperature

Citation: Vértesy, G.; Gasparics, A.; Szenthe, I.; Rabung, M.; Kopp, M.; Griffin, J.M. Analysis of Magnetic Nondestructive Measurement Methods for Determination of the Degradation of Reactor Pressure Vessel Steel. *Materials* **2021**, *14*, 5256. https://doi.org/10.3390/ma14185256

Academic Editor: Giovanni Bruno

Received: 30 June 2021
Accepted: 3 September 2021
Published: 13 September 2021

1. Introduction

Nuclear power plants (NPPs) have a key role within the energy production landscape. An extremely important aspect is their safety, so inspection of a power plant's integrity is crucial, especially for the long-term operation. The most important part of the pressurized and boiling water reactors is the reactor pressure vessel (RPV). Their primary aging process is the irradiation generated material embrittlement and it is one of the most important lifetime limiting factors. This process, caused by the influence of the long-term and high-energy neutrons, generates changes in the mechanical properties [1], which are inspected periodically. However, the inspection of radiation embrittlement is not an easy task at all. So-called surveillance samples are put inside the vessel and after a certain period they are tested. Mechanical Charpy impact testing is the standard way of evaluation of the embrittlement [2]. The ductile-to-brittle transition temperature (DBTT) determined by Charpy impact testing is the authorized parameter that refers to embrittlement in the nuclear industry. However, this destructive measurement technique requires many samples, and the error of measurement is high. Concerted efforts have been made to continuously develop effective nondestructive methods for inspection of RPVs. Magnetic methods seem to be useful for this purpose since the reactor pressure vessel is made of ferromagnetic steels. A general overview can be found in Reference [3] about the application of nondestructive magnetic methods.

In several recent works, different nondestructive magnetic methods have been applied for detection of neutron irradiation generated embrittlement of nuclear reactor pressure vessel material. One of them is the so called magnetic adaptive testing, MAT based on the measurement of minor magnetic hysteresis loops [4,5]. Another one, the magnetic Barkhausen noise technique, is also suitable to detect the irradiation effects on RPV steel [6]. Finally, there is the 3MA method (micromagnetic multiparameter microstructure and stress analysis), which combines several different magnetic methods [7].

The general conclusion of these efforts was that a reasonable correlation had been found between the nondestructively measured magnetic parameters and the destructively measured DBTT if the above mentioned methods are applied [8–10]. It seems that magnetically measured parameters have a better potential to characterize the material embrittlement than the conventional destructive methods. However, the scatter of measurements points has been found to be rather large in all of these experiments.

The possible reason of this big scatter has been interpreted in a recent paper [11]. An important finding of this work is that the scatter of measurements points very probably can be explained by local material inhomogeneity. The embrittlement depends also very much on the initial material conditions. This fact is surprising, because the measured samples were prepared from the same RPV block, and from a predefined depth. The initial material conditions probably are connected with the microstructure of the samples, but the microstructure itself was not investigated; instead, we concentrated our attention to the interpretation of the magnetic measurements.

Considering the importance of these results for the future potential application of magnetic measurements and even for the whole nuclear industry, the results should be verified carefully. This is the purpose of the present work: Two series of standard Charpy samples made of two different types of RPV steel were measured both before and after neutron irradiation, and the three different magnetic measurement methods were applied systematically on the same specimens. The outcomes of these non-destructive methods have been evaluated jointly.

2. Materials and Methods

2.1. Materials and Mechanical Tests

For our investigations, two types of RPV materials were chosen, an Eastern RPV material (15Kh2NMFA) and a Western RPV material (A508 Cl.2). ISO-V Charpy samples were manufactured at SCK CEN [12,13] by cutting them out from $\frac{3}{4}$ depth in the case of A508 Cl.2 specimens, and from the $\frac{1}{4}$ depth in the case of 15Kh2NMFA specimens. According to the ASTM E23-16b standard the orientation of samples was selected as T-L.

The chemical composition of the samples are given in Tables 1 and 2 for both materials. It was measured by a "Spectromax LMX06" Spark Atomic Emission Spectrometer (Ametek/Spectro [14]) according to the standard ASTM E415. Heat treatment of steel forgings means quenching and tempering including post-weld heat treatment.

Table 1. 15Kh2NMFA base metal chemical composition (wt %) of the.

C	Si	Mn	Cr	S	P	Ni	Mo	Cu	V
0.16	0.29	0.42	1.97	0.008	0.012	1.29	0.52	0.12	0.12

Table 2. A508 Cl.2 base metal chemical composition (wt %) of the.

C	Si	Mn	Cr	S	P	Ni	Mo	Cu
0.201	0.27	0.578	0.372	0.0085	0.0091	0.668	0.599	0.0472

The as-received samples' microstructure of is a mixed tempered ferrite–bainite structure. As an illustration, typical microstructures of the two investigated materials, performed on non-irradiated samples are shown in Figure 1. After preparation of Charpy samples from Western and Eastern RPV material, one portion was mechanically tested and the other portion was nondestructively investigated. Following the magnetic measurements, these samples were divided into three sets for the neutron irradiation. E > 1 MeV neutron irradiation was performed in the primary water pool of the BR2 reactor at different irradiation levels with a fluence at a temperature between ~100–120 °C. Applied fluence levels were between 1.55×10^{19} n/cm^2 and 7.90×10^{19} n/cm^2.

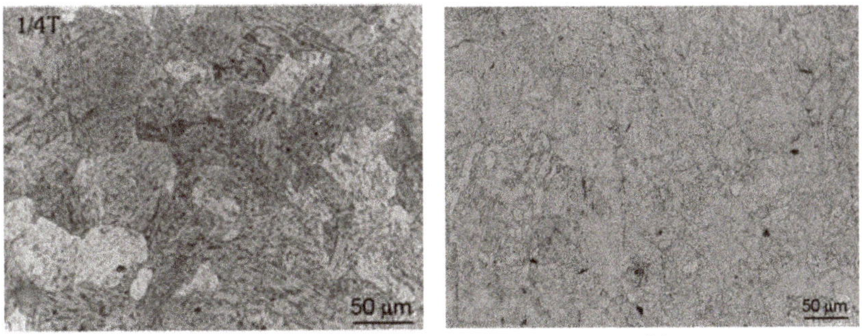

(**a**) (**b**)

Figure 1. Optical microscopy performed in etched condition to observe the grain boundaries on an A508 CL.2 (**a**) and on an 15Kh2NMFA (**b**) sample.

Four Charpy samples for each irradiation condition were investigated. The irradiated samples were nondestructively tested. After that, destructive mechanical tests were performed. They were investigated by an instrumented pendulum (ISO 148-1 and ASTM E23) for the as-received non-irradiated and neutron irradiated materials.

The DBTT (i.e., its curve and the temperature where this curve bypasses the 41J criteria) can be determined by a series of Charpy impact tests carried out at different temperatures of the test set specimens. The transition temperature curve itself is determined by mathematical regression analysis, since a-priory unknown temperature value is to be derived which value becomes available just following the physical experiments. However, this statistical approach fades out the differences between the single samples of the test set and instead, provides a single DBTT value for the whole test specimen set. In addition, note the scattering of all measurements along the whole transition function to be fitted is relevant from a DBTT perspective, but only the measurement uncertainties of this transition curve around the point it by passes the 41J criteria or, where it has a slope. For instance, the scattering of the upper shelf energy (USE) is irrelevant in this case.

The scattering of the impact tests and the results of the regression analysis can be seen in Figure 2 in the case of A508 Cl.2 and 15Kh2NMFA type material. This figure also illustrates the preliminary assumption of this paper: comparing the outcomes of the individual non-destructive measurements to a statistical mean value obtained on non-ideal specimens will lead to scattering which cannot be attributed to the uncertainty of the non-destructive approach solely. This scattering can be seen in Figure 2, demonstrating the differences between the tested specimens. Therefore, these differences are related to the material inhomogeneities, and these are reflected in the mentioned NDE results as scattering. Correlation between transition temperature change and neutron fluence was found for both steels. Results are given in Tables 3 and 4 for the A508 Cl.2 and for the 15kHNMFA steels, respectively.

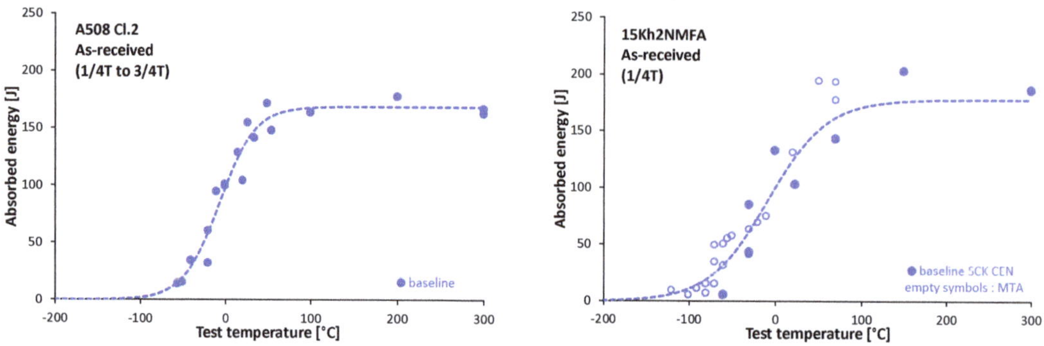

Figure 2. The scattering of the Charpy impact test measurements and the fitted ductile to brittle transition temperature curve that fades the specimens' inhomogeneities in the case of A508 Cl.2, and of 15Kh2NMFA type material.

Table 3. Fast fluence (E > 1 MeV) and DBTT for A508 Cl.2 material.

Fast Fluence (E > 1 MeV) ($\times 10^{19}$ n/cm^2)	DBTT T_{41J} (°C)
0	-33 ± 9
1.55	76 ± 15
4.38	125 ± 15
7.04	126 ± 15

Table 4. Fast fluence (E > 1MeV) and DBTT for 15Kh2NMFA material.

Fast Fluence (E > 1 MeV) ($\times 10^{19}$ n/cm^2)	DBTT T_{41J} (°C)
0	-51 ± 12
2.78	88 ± 15
6.83	136 ± 15
7.9	124 ± 15

Altogether 13 samples from 15Kh2NMFA material were measured before and after neutron irradiation, samples Nos. 166, 167, 168, 169, 171, 172, 173, 175, 176, 178, 181, 183, 185, and 11 samples from A508 Cl.2 material Nos. 572, 573, 575, 578, 579, 581, 583, 586, 587, 588, 591.

2.2. *Magnetic Adaptive Testing*

Magnetic adaptive testing (MAT) is a recently developed method of magnetic hysteresis measurement. The main point of this technique is that series of minor hysteresis loops are measured systematically, in contrast to the conventional hysteresis measurements, where major (saturation to saturation) hysteresis loops are recorded. The details of the measurement can be found in Reference [5]. As it was proven in many experiments, investigating several types of degradation of ferromagnetic materials, led to good correlation between the optimally chosen MAT descriptors and those parameters (usually determined destructively), characterize the actual material degradation. Sensitivity of MAT descriptors supersedes the sensitivity of conventional hysteresis measurements.

Samples are measured by a magnetizing yoke, attached directly to the sample surface. The size of the yoke fits the size of samples. Measurement starts with a careful demagnetization of samples by decreasing amplitude alternating magnetizing field. Samples then magnetized by a magnetizing current with a triangular waveform, starting from zero and increasing the amplitude step-by-step. Permeability loops are detected by a pick-up coil, wounded around a yoke leg. In the case of linearly increasing the magnetizing current, the pick-up coil's output signal changes proportionally with the differential magnetic permeability of the whole magnetic circuit.

From points of the obtained minor permeability loops a permeability matrix is calculated and matrix elements are compared with the corresponding elements of the reference (in our case non irradiated) sample. From this, a big data pool is generated, and relevant parameters are chosen that characterize (with large sensitivity and simultaneously with good reproducibility) the modification of material properties due to different material degradation.

As mentioned above, the first and most probable reason for the scatter of magnetic parameters vs. DBTT could be the error of magnetic measurement itself. In considering this, a careful analysis of MAT measurements was conducted. The result of this analysis is given in the Appendix of Reference [11]: The error of the total MAT evaluation has been found lower than 1% by taking into account of all possible uncertainties. This means that the error of MAT measurement and evaluation cannot be responsible for the big scatter of points, which can exceed in certain cases: 20%, as shown in Figure 3, Figure 4 and Figure 5. Similar conclusions can be made for the experimental error of 3MA and MNB measurements also.

2.3. Micromagnetic Multiparameter Microstructure and Stress Analysis

The Fraunhofer Institute for Nondestructive Testing developed the 3MA approach (3MA = micromagnetic multiparameter microstructure and stress analysis) which allows materials characterization to determine industry-relevant characteristics (hardening depth (CHD, SHD or NHD), hardness, yield and ultimate strength and DBTT. This method is suitable for measurements on active materials in hot cells. The measuring principle is rested on the correlation between the mechanical properties of ferromagnetic materials and their magnetic properties. This correlation is connected with the microstructure interaction with both the magnetic structure (consisting of magnetic domain separated by Bloch walls) as well as the dislocations [15,16].

The 3MA approach uses several parameters derived from three micromagnetic methods listed below [17]:

- Eddy currents (EC) are generated in the material under the influence of AC magnetic field. They depend on the σ electrical conductivity and on the μ magnetic permeability of the material, and they result a magnetic field with opposite direction to the originally applied magnetic field. It means, that σ and μ of the material has an influence on the excitation coil's impedance. This impedance is measured.
- Analysis of incremental permeability (IP) is a method of separating the magnetic permeability information from the electrical conductivity information. For application of this method, the material should be magnetized with a low-frequency AC magnetic field and a continuous EC impedance analysis should be performed at a higher frequency. Considering the change of the coil impedance as a function of the magnetic field strength leads to an incremental permeability plot. In such a way, a qualitative correlation of the impedance change throughout the magnetic hysteresis and the magnetic field strength at maximum permeability (usually correlated with coercivity measured by means of magnetic hysteresis analysis) is obtained.
- Harmonics analysis (in time domain signal) of magnetizing current is used to describe the magnetic hysteresis behavior of the materials by applying one-sided access sensor. For this purpose, a magnetization electromagnet should be applied, which is driven by a sinusoidal voltage. A receiver coil measures the magnetizing current.

The impedance of the electromagnet coil changes as a consequence of the hysteretic correlation between the B magnetic flux density and H magnetizing field. In such a way, the current in the electromagnet contains harmonics, but it is not sinusoidal. The measured magnetizing current exhibits distortion due to the hysteresis in magnetic circuit. Fundamental and harmonic components can be numerically determined by a fast Fourier analysis, and thus distortions of the magnetizing current can be quantified. The harmonic components calculated by this procedure make possible the determination of the material properties.

These methods differ in terms of the analysis depth and mechanisms and deliver more than 20 parameters, which correlate qualitatively with material properties. Generally, 3MA systems are consisting of a probe, which contains a magnetization unit with a coil to capture the magnetic response of the material, a 3MA device for the excitation of the magnetization and preprocessing the measuring signals via a PC for measurement control and data processing. Different material depths and areas can be investigated depending on the properties of the magnetization unit as well on the parameters of the measurement. Micromagnetic methods can therefore analyze a controllable fraction of the sample volume.

The 3MA process should be calibrated on a calibration set of samples (with well-known properties, such as DBTT or hardness) [10,18]. For mechanical-technological materials characterization, the measuring parameters are registered by the PC and are further processed having performed all measurements and analyses, the software delivers the magnetic fingerprint (MFP) of the material properties, which can be used for quantitative and qualitative materials characterization. More than one measurement parameter is used proper for materials characterization. It is necessary to ensure increased robustness contra disturbing influences such as material variations and surface condition. For the calibration regression analyses, pattern recognition or other machine-learning algorithms can be used.

2.4. Barkhausen Noise Measurement

MBN, the magnetic Barkhausen noise (MBN) method is a mature non-destructive examination technique for microstructural modifications, observation of surface defects caused by abusive manufacturing processes and residual stress [19–21]. MBN has its origins from the B–H hysteresis loop, which is not a smooth curve as the magnetic flux density versus the intensity of the magnetic field results in a curve that is instead described as a non-linear step function. These steps correlate with the irregular fluctuations in the magnetization when energized from cyclic excitation provided by ferrous yokes to excite the material area under interest. These steps or jumps of domains form Barkhausen noise and are provided from magnetic domain motion which is the basis of the Barkhausen signal. Moreover, until the applied field is increased sufficiently, pining sites restricts the moving domain wall. When the magnetizing field is reached, the sudden and discontinuous movement of domain walls result sudden changes in magnetization. In the case of microstructural characteristics "defects" such as dislocations, precipitations and segregations cause pinning of the moving domain walls and promote Barkhausen signal changes [22]. MBN is measured via a pick-up coil (independent to the energizing yoke) in the form of a voltage signal significant of surface eddy currents experienced near the surface of the material.

Magnetic Barkhausen measurements were performed by using a Rollscan 350 MBN analyzer, equipped with a Stresstech general-purpose sensor [23]. The magneto elastic parameter (mp) signifying the root mean square (RMS) value is a function of the magnetizing current, voltage and frequency. Each measurement consisted of periodic bursts of MBN signals for a set duration of ten seconds. MBN RMS can be calculated from such signal bursts. The RMS of the MBN signals is expressed as:

$$\mathrm{RMS} = \sqrt{\frac{\sum_{i=1}^{n} yi^{2}}{n}}$$

Here n is the total number of MBN signals obtained in the particular frequency range, and y_i is the amplitude of the individual burst.

The main instrumentation input parameters are voltage and frequency, and these are determined from voltage and frequency sweeps giving an optimum value for a specific material under test. In addition, the sinusoidal excitation field can be changed to a triangular one however sinusoidal was considered the optimum waveform for the tests carried out during this work. It should also be noted that the applied field frequency has an influence on the depth where the MBN reading is obtained. The lower the frequency the larger the measured depth. Between 0.01 mm and 1 mm penetration depth was achieved where a band pass filter of between 70 and 200 kHz was selected for channeling the pick-up signals of interest.

Scatter of magnetic output responses vs. DBTT was also found with MBN. It was considered such scatter is due to the material microstructure differences as measurement uncertainty was minimized as much as possible, this is in terms of sensor pick-off, surface quality and applied force. The measurement testing regime used a three times sensor pick-off (physical movement of the sensor but same position test point maintained) followed by 5 measurements each time the sensor touched the surface of the material.

3. Results

3.1. Results of MAT, 3MA and MBN Measurements Made on All Samples

3.1.1. Evaluation of Data without Normalization

In Figure 3 it is shown how the optimally chosen MAT descriptor depends on the transition temperature for the two investigated materials, 15kH2NMFA and A508 Cl.2. We use the terminology "Optimally chosen MAT descriptor" for those parameters, picked up from the generated big data pool, which characterize the best the correlation with the given independent parameter. In the present case this is the material embrittlement generated by neutron irradiation [8,9]. This parameter ensures the largest sensitivity together with good reproducibility. In the case plotted in Figure 3, this descriptor is characterized by $h_a = -30$ mA and $h_b = 1080$ mA magnetic field values for 15kH2NMFA material and by $h_a = -780$ mA and $h_b = 1200$ mA values for A508 Cl.2. material. (h_a: magnetizing field, h_b: minor loop amplitude).

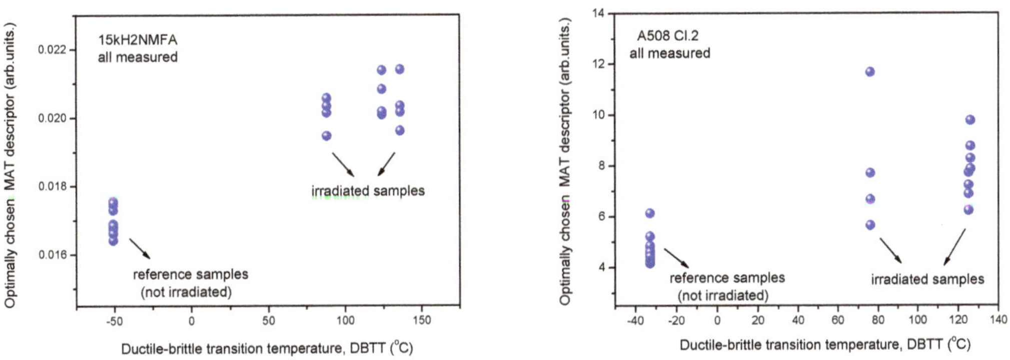

Figure 3. Optimally chosen MAT descriptor vs. transition temperature for all measured 15kH2NMFA and A508 Cl.2 samples.

Similarly, the results of 3MA are shown in Figure 4. By applying 3MA, clear trend between several magnetic parameters and DBBT was found. In Figure 4 the P3 parameter is given. This is the amplitude of the third harmonics obtained from upper harmonics analysis in the time domain signal of the magnetizing current. Results of MBN measurements, the RMS parameter as a function of transition temperature for all measured 15kH2NMFA and A508 Cl.2 samples can be seen in Figure 5.

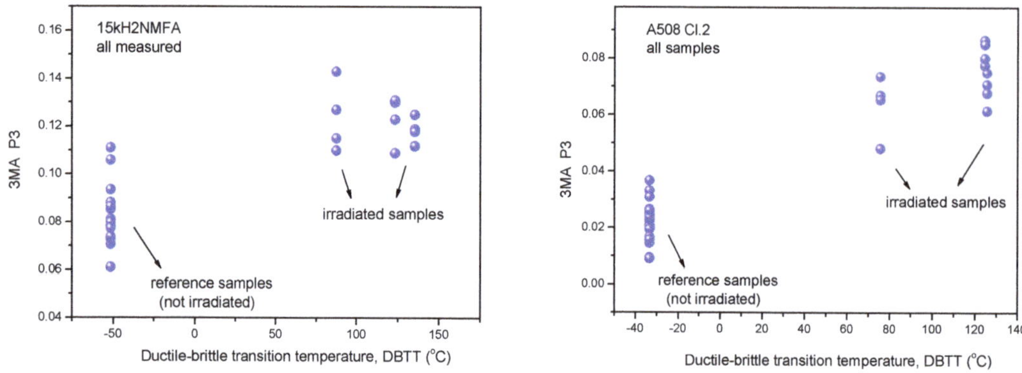

Figure 4. Dependence of the amplitude of third harmonics P3 parameter of 3MA as a function of transition temperature for all measured 15kH2NMFA and A508 Cl.2 samples.

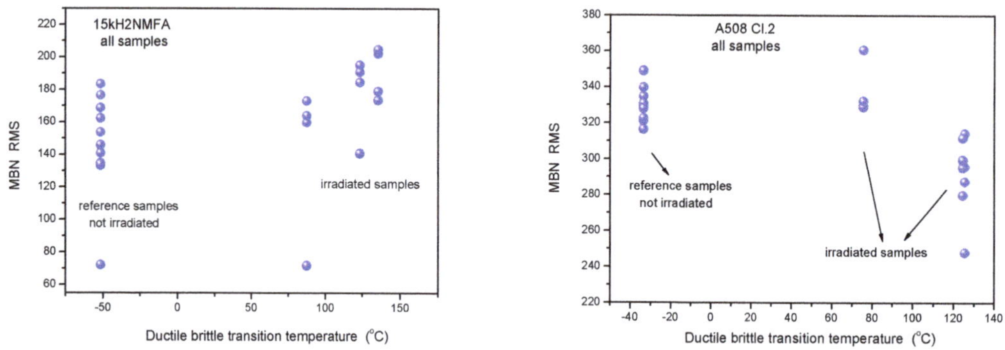

Figure 5. MBN RMS parameter as a function of transition temperature for all measured 15kH2NMFA and A508 Cl.2 samples.

It is seen that irradiation caused salient measurable modification of magnetic parameters. Magnetic parameters are significantly affected by the material degradation that changes the DBTT and there is a more or less linear correlation between magnetic parameters and DBTT (except the MBN measurements performed on A508 Cl.2 samples). However, the most visible conclusion, drawn from all measurements is the big scatter of points, regardless of the actual measurement method.

It can also seen very well in Figures 3–5 that even the magnetic parameters of not irradiated (reference) samples scatter a lot. This fact gives a possible reason of scatter of measurements points: the samples behave rather differently, despite the fact that the Charpy specimens were cut from the same block. Magnetic measurements do not make anything else but reflect this material inhomogeneity. It is not a surprise that the points will scatter also after irradiation. To have an impression about the behavior of individual samples, the next section investigates how the magnetic properties of individual samples are modified due to neutron irradiation.

3.1.2. Evaluation of Normalized Data

In Figures 3–5, all measurement results are given and samples are not marked. Another—and perhaps more useful—way is to consider the change of magnetic parameters for each individual sample. For this purpose, other graphs are shown below (Figures 6–8). In these graphs, the modifications of magnetic characteristics are given, this is with respect to the same magnetic parameter that is obtained on the same sample before irradiation giving a baseline condition. This means that the first point (Ratio = 1) is the same for all samples, while each of the other points are connected with specific numbered samples. These points represent how the magnetic behavior of a given sample was modified due to neutron irradiation. (The labeling of points is avoided in order to preserve the clarity of the graphs.)

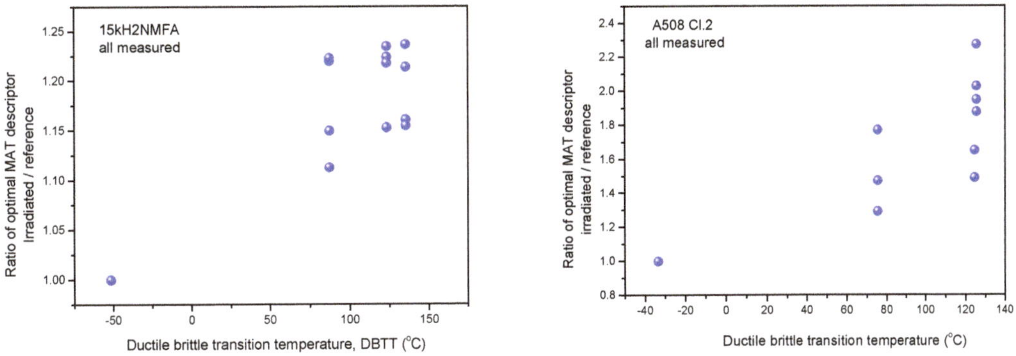

Figure 6. Normalized MAT descriptor vs. transition temperature for all 15kH2NMFA and A508Cl.2 samples.

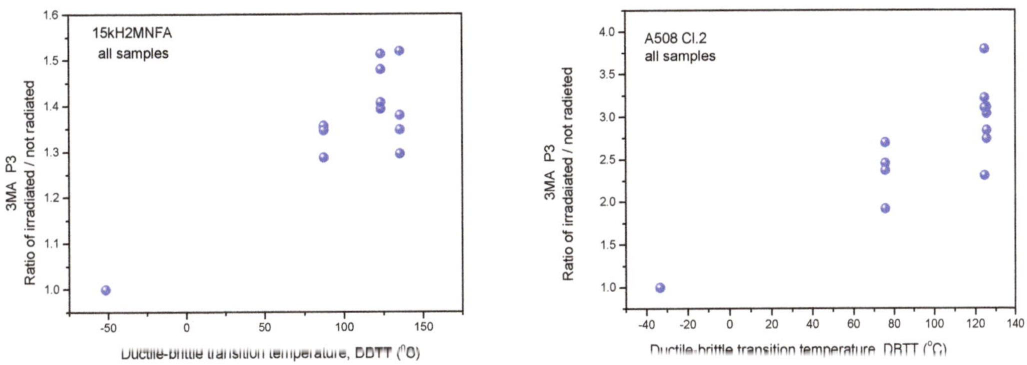

Figure 7. Normalized 3MA P3 parameter as function of transition temperature for all 15kH2NMFA and A508Cl.2 samples.

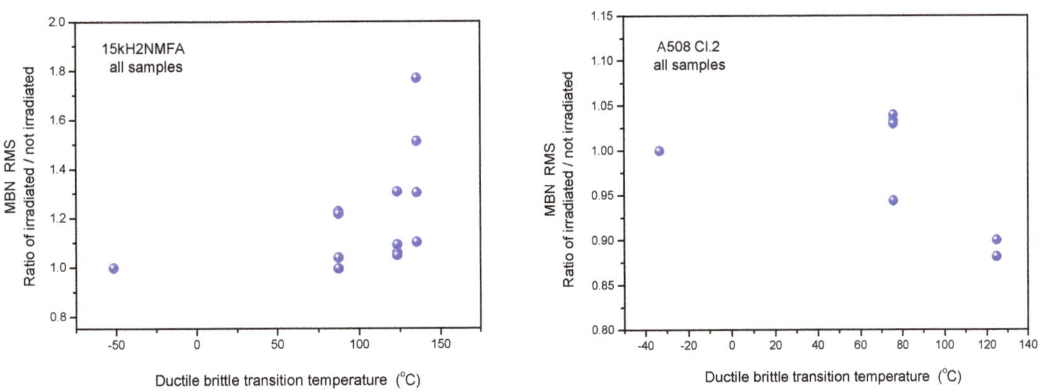

Figure 8. Normalized MBN RMS parameter as function of transition temperature for all 15kH2NMFA and A508Cl.2 samples.

As can be seen very well in the above graphs, the scatter of points is rather large in the normalized cases, too. This is proof that the scatter of points in Figures 3–5 is not the result of the originally different behavior, but also of the fact that neutron irradiation generates different material embrittlement, depending on the individual samples' behavior.

3.2. Selection of Samples

In the above sections, the influence of neutron irradiation has been investigated as if all measured samples are taken into account. As already mentioned above, it has been found that even reference samples are different from the point of view of magnetic properties, so it is not surprising that they behave differently also after irradiation. In this section, the method of the selection of samples is presented, based on permeability measurements of the samples. In the following section, it will be shown how the correlation of magnetic parameters with DBTT looks if only the selected samples are taken into consideration.

The selection of samples is based on measured permeability loops. Evidently this selection was made before any further evaluation of irradiated samples. These permeability loops were measured on reference samples (before irradiation). The criteria in this case was the similarity of the magnetic behavior. These samples were selected which were similar to each other from a magnetic point of view. A good characteristic is the maximal permeability, which can be determined easily from directly measured permeability loops. This means that this selection does not take into account the neutron irradiation generated material embrittlement, it reflects solely on the behavior of samples with initial conditions.

It is emphasized that we did not use backward reasoning to decide which data points fit the best to our hypothesis. Clarifying this statement, the selection process is shown: (1) A large scatter of all magnetic parameters measured on irradiated and reference samples was observed. (2) Independently of the result of magnetic measurements, the magnetic behaviors of the reference samples were compared to each other. Several samples were found with very similar permeability curves. (3) The MAT, 3MA and MBN evaluations were made again, but only the selected, magnetically similar samples were taken into account. No information about the behavior of the irradiated samples was available, since selection was performed prior to irradiation.

Selection reduces only the number of samples, which are taken into account. A serious argument for this selection is, that in the case of the 3MA and MBN method, this selection resulted in a very similar result as in the case of MAT method.

The series of permeability loops measured on 15kH2NMFA samples are shown in the left side of Figure 9. Magnified parts of the loops can be seen in the right side of the figure, but here only the envelope of the large amplitude minor loops are presented, to make visible the difference between loops, and to provide easy selection from a visual perspective. Four samples have been found that are similar from magnetic point of view. These samples are numbers 172, 173, 178, and 183.

Series of permeability loops measured on A508 Cl.2 samples are shown in Figure 10. Again, four samples have been found, which are similar from magnetic point of view. These samples are numbers 579, 583, 586, and 588.

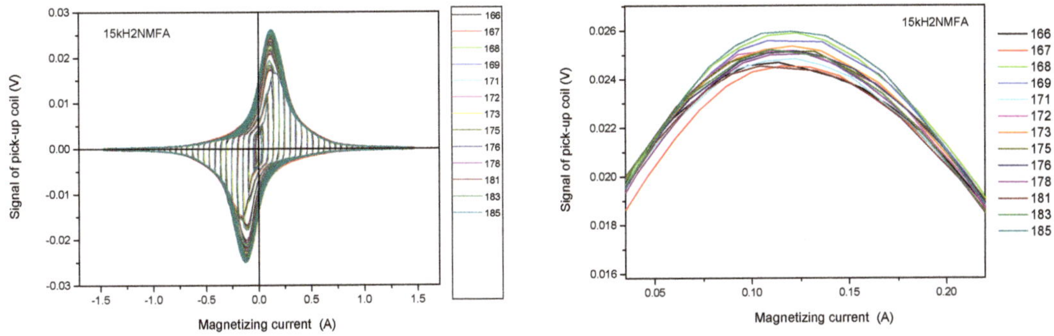

Figure 9. Measured permeability loops of 15kH2NMFA samples before irradiation. The right panel shows the magnified part of the left graph [11].

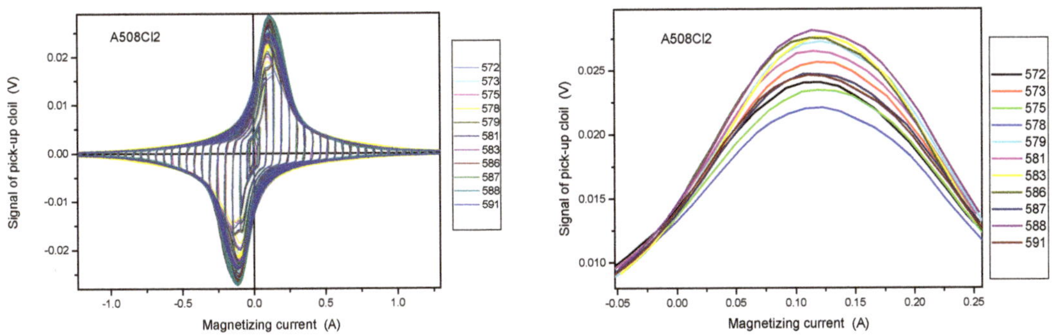

Figure 10. Measured permeability loops of A508 Cl.2 samples before irradiation. The right panel shows the magnified part of the left graph.

3.3. Results of 3MA, MAT and MBN Measurements Considering Selected Samples Only

In this section it is shown how the scatter of points is modified if the evaluation of magnetic parameters has been repeated taking into account only the magnetically pre-selected samples. Results are shown in Figures 11–13, respectively.

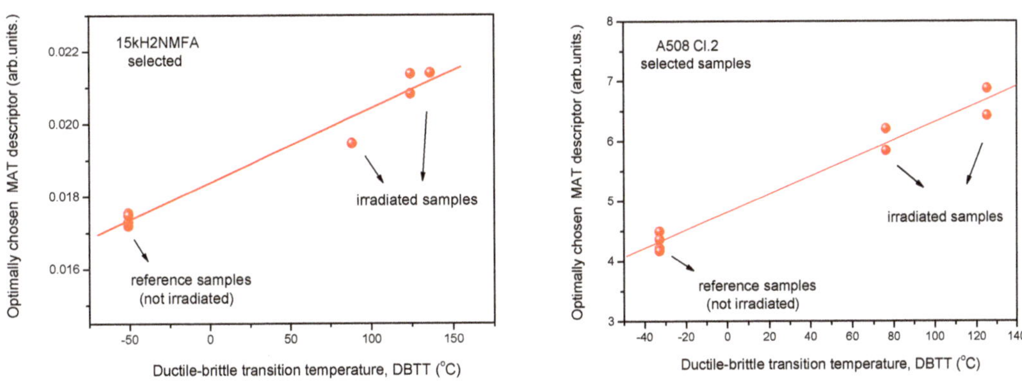

Figure 11. Optimally chosen MAT descriptor vs. transition temperature for selected 15kH2NMFA and A508 Cl.2 samples.

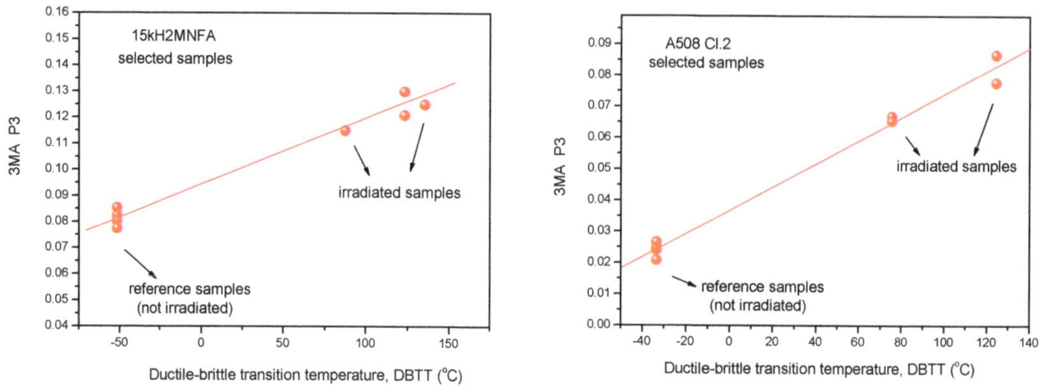

Figure 12. 3MA P3 parameter vs. transition temperature for selected 15kH2NMFA and A508 Cl.2 samples.

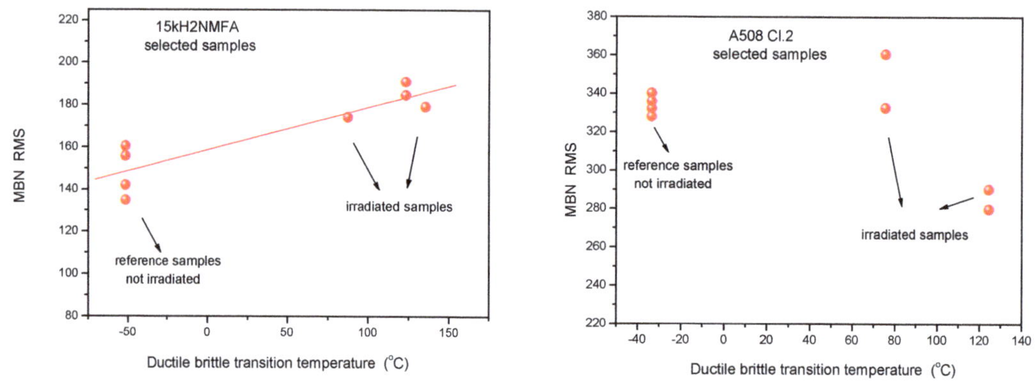

Figure 13. MBN RMS parameter as a function of transition temperature for selected 15kH2NMFA and A508 Cl.2 samples.

4. Discussion

By this analysis it has been proven that the experienced big scatter is connected with the different behavior of the samples, and the reason is not really the measurement errors of the applied magnetic methods.

It should be emphasized that MAT descriptors were determined for all the samples independently, before any selection and later any time (see Figure 11) the same parameters ($h_a = -30$ mA, $h_b = 1080$ mA for 15kH2NMFA and $h_a = 780$ mA, $h_b = 1200$ mA for A508 Cl.2) were used. Selection reduces only the number of samples, which are taken into account. A serious argument for this selection is, that in the case of 3MA and MBN method, this selection resulted in a very similar result as that in the case of the MAT method.

If we compare Figure 11 with Figure 3, Figure 12 with Figure 4 and Figure 13 with Figure 5, it can be seen that the scatter of points dramatically decreased if evaluation was performed only on the selected samples with similar magnetic behavior. An obvious linear correlation with low scatter of points has been found between magnetic parameters and DBTT for both investigated materials and for the three considered magnetic methods. One exception is the MBN RMS parameter for the A508 Cl.2 material. In this latter case, the scatter has been also decreased, similarly to all other cases, but we cannot speak about neither linear nor even monotonous correlation. This observation needs some more discussion. However, the correlation between MAT and 3MA measurements are more than satisfactory. Neither the correlation between magnetic parameters and DBTT, nor the behavior of scatter does not depend on the actual method of measurement. This fact is very promising for the future practical application of magnetic methods. The results of the different methods verify one another.

We have found relevant differences in magnetic behavior, which resulted in big scatter in MAT, 3MA and also in MBN vs. DBTT plots. These differences are rather surprising and unexpected, because the samples were cut from the same block. As three different NDT methods indicated the differences, these can not be assigned to the uncertainity of any one of them, although the structure and chemical composition of the different samples should be the same. Describing this effect we cannot use any other word than "inhomogeneity of the material", without knowing anything about the character of inhomogeneity. This result is considered one of the most important messages of our work.

In this paper we have presented figures about the scattering of the destructive mechanical tests and of the non-destructive magnetic measurements. Both types of experiments indicate that the source of the observed scattering is related to the differences between the tested specimens either from a mechanical or magnetic point of view. We cannot provide evidence that the scattering of the mechanical properties and of the magnetic features have identical causes. However, the quality of the linear relationship between the determined DBTT and the MAT, 3MA, and MBN values can be considered as a telling argument in this direction.

We know that further analysis to verify the effect of the local inhomogeneity of the material is extremely important and perhaps this result would be crucial for the whole nuclear industry. We believe that if we call the attention of the scientific community to this fact, it is important by itself. Evidently, the work should be continued, and we want to do this.

5. Conclusions

Neutron irradiation-generated embrittlement was investigated by three different types of nondestructive magnetic methods on two different types of reactor pressure vessel steel materials and the results were compared with the destructively measured transition temperature. A reasonable correlation was found between magnetic parameters and DBTT, which can be used in future potential applications to estimate DBTT from the results of magnetic measurement. A good correlation was found, as well, between the results obtained by the different methods.

The present work is considered as a direct continuation of Reference [11]. In this, recently published paper, a possible explanation was given for the big scatter of nondestructively measured magnetic characteristics as functions of transition temperature. By applying the so-called MAT method, a possible reason has been found for the scatter. Here, two other principally rather than different magnetic methods have been applied on the same series of samples, and also, on another nuclear pressure vessel steel material having different chemical composition and different properties in order to establish a much larger context of the source of the observed scatter, which was not entirely explained in our previous paper. The results from the other methods were surprisingly similar as in the case of MAT. This means:

- Verification of MAT measurements;
- Proving that the former MAT result is not methodology dependent;
- Proving that the former MAT result is not material dependent.

A common feature of different techniques—large scatter of points—was also analyzed. As an explanation, this scatter was attributed to the local material inhomogeneity. It was shown that the measurement error is not responsible for the scatter. It was clearly demonstrated in these experiments that, if the behavior of the reference (non-irradiated) samples are similar to each other, the irradiation-induced embrittlement can be determined very well, and in this case the scatter of the magnetic parameters is very low.

One of the most important conclusions of this work is, that the parameters determined by magnetic measurements seem to characterize better the neutron generated material embrittlement than the conventionally used destructive methods. The scatter of the magnetic results is lower than the scatter of the Charpy tests. In addition, the magnetic method characterized the actual, individual samples, in contrary to the transition temperature values determined by the Charpy impact testing methods, which can provide only statistical values on the set of samples.

Another important conclusion is that local material inhomogeneities have a great influence on the neutron irradiation-induced material embrittlement. Different parts of the reactor pressure vessel, even if they are cut from the same larger block, are hardened differently. Taking into account the measurement conditions' analysis, local material conditions can be responsible for the different neutron irradiation generated embrittlement of the pressure vessel steel material caused by the same dosage of neutron irradiation.

These facts mean a telling argument for the application of non-destructive magnetic measurements in the reactor industry for future operations.

Author Contributions: Conceptualization and original draft preparation, G.V.; investigation and methodology, G.V., I.S., M.R., M.K. and J.M.G.; project administration, A.G. and I.S. All authors have read and agreed to the published version of the manuscript.

Funding: This research was carried out in frame of the "NOMAD" project. This project (Non-destructive Evaluation System for the Inspection of Operation-Induced Material Degradation in Nuclear Power Plants) has received funding from the Euratom research and training programme 2014–2018 under grant agreement No 755330.

Institutional Review Board Statement: Not applicable.

Informed Consent Statement: Not applicable.

Data Availability Statement: The data are contained within the article.

Acknowledgments: Authors are grateful to Inge Uytdenhouwen (SCK CEN) for her valuable help in sample preparation and characterization. A508Cl.2 samples were prepared in SCK CEN, and the neutron irradiation of all samples was done in SCK CEN in BR2, together with the performance of the mechanical (destructive) properties for the DBTT. Microstructure picture of the A508Cl.2 sample was made also in SCK CEN. Partners performed NDE tests in the hot cell of SCK CEN.

Conflicts of Interest: The authors declare no conflict of interest.

References

1. Koutsky, J.; Kocık, J. *Radiation Damage of Structural Materials*; Elsevier: Amsterdam, The Netherlands, 1994.
2. Ferreño, D.; Gorrochategui, I.; Gutiérrez-Solana, F. Degradation due to neutron embrittlement of nuclear vessel steels: A critical review about the current experimental and analytical techniques to characterise the material, with particular emphasis on alternative methodologies. In *Nuclear Power—Control, Reliability and Human Factors*; Tsvetkov, P., Ed.; IntechOpen Limited: London, UK, 2011; ISBN 9789533075990. Available online: http://www.intechopen.com/articles/show/title/non-destructive-testing-for-ageing-management-of-nuclear-power-components (accessed on 14 May 2020).
3. Jiles, D.C. Magnetic methods in nondestructive testing. In *Encyclopedia of Materials Science and Technology*; Buschow, K.H.J., Ed.; Elsevier: Oxford, UK, 2001; p. 6021.
4. Tomáš, I. Non-destructive Magnetic Adaptive Testing of ferromagnetic materials. *J. Magn. Magn. Mater.* **2004**, *268*, 178–185. [CrossRef]
5. Tomáš, I.; Vértesy, G. Magnetic Adaptive Testing. In *Nondestructive Testing Methods and New Applications*; Omar, M., Ed.; IntechOpen: London, UK, 2021; ISBN 978-953-51-0108-6. Available online: http://www.intechopen.com/articles/show/title/magnetic-adaptive-testing (accessed on 8 November 2020).
6. Barroso, S.P.; Horváth, M.; Horváth, Á. Magnetic measurements for evaluation of radiation damage on nuclear reactor materials. *Nucl. Eng. Des.* **2010**, *240*, 722–725. [CrossRef]
7. Dobmann, G.; Altpeter, I.; Kopp, M.; Rabung, M.; Hubschen, G. ND-materials characterization of neutron induced embrittlement in German nuclear reactor pressure vessel material by micromagnetic NDT techniques. In *Electromagnetic Nondestructive Evaluation (XI)*; IOS Press: Amsterdam, The Netherlands, 2008; p. 54. ISBN 978-1-58603-896-0.
8. Vértesy, G.; Gasparics, A.; Uytdenhouwen, I.; Szenthe, I.; Gillemot, F.; Chaouadi, R. Nondestructive investigation of neutron irradiation generated structural changes of reactor steel material by magnetic hysteresis method. *Metals* **2020**, *10*, 642. [CrossRef]
9. Tomáš, I.; Vértesy, G.; Gillemot, F.; Székely, R. Nondestructive Magnetic Adaptive Testing of Nuclear Reactor Pressure Vessel Steel Degradation. *J. Nucl. Mater.* **2013**, *432*, 371–377. [CrossRef]
10. Rabung, M.; Kopp, M.; Gasparics, A.; Vértesy, G.; Szenthe, I.; Uytdenhouwen, I. Micromagnetic characterization of operation induced damage in Charpy specimens of RPV steels. *Appl. Sci.* **2021**, *11*, 2917. [CrossRef]

11. Vértesy, G.; Gasparics, A.; Szenthe, I.; Uytdenhouwen, I. Interpretation of Nondestructive Magnetic Measurements on Irradiated Reactor Steel Material. *Appl. Sci.* **2021**, *11*, 3650. [CrossRef]
12. Uytdenhouwen, I.; Chaouadi, R.; Other NOMAD Consortium Members. NOMAD: Non-destructive Evaluation (NDE) system for the inspection of operation-induced material degradation in nuclear power plants—Overview of the neutron irradiation campaigns. In *ASME 2020 Pressure Vessels and Piping Conference*; Non-Destructive Examination, V007T07A003, PVP2020-21512; ASME: New York, NY, USA, 2020; Volume 7. [CrossRef]
13. Uytdenhouwen, I.; Chaouadi, R. Effect of neutron irradiation on the mechanical properties of an A508 Cl.2 forging irradiated in a BAMI capsule. In *Pressure Vessels and Piping Conference*; Codes and Standards, V001T01A060, PVP2020-21513; ASME: New York, NY, USA, 2020; Volume 1. [CrossRef]
14. Available online: https://www.spectro.com/ (accessed on 26 August 2021).
15. Cullity, B.D.; Graham, C.D. *Introduction to Magnetic Materials*, 2nd ed.; Wiley & Sons: New York, NY, USA, 2009; ISBN 978-0-471-47741-9.
16. Bozorth, R.M. *Ferromagnetism*; IEEE Press: New York, NY, USA, 1993.
17. Altpeter, I.; Becker, R.; Dobmann, G.; Kern, R.; Theiner, W.A.; Yashan, A. Robust Solutions of Inverse Problems in Eletromagnetic Non-Destructive Evaluation. *Inverse Probl.* **2002**, *18*, 1907–1921. [CrossRef]
18. Szielasko, K.; Tschuncky, R.; Rabung, M.; Seiler, G.; Altpeter, I.; Dobmann, G.; Herrmann, H.G.; Boller, C. Early Detection of Critical Material Degradation by Means of Electromagnetic Multi-Parametric NDE. *AIP Conf. Proc.* **2014**, *711*, 1581.
19. Lo, C.C.H.; Jakubovics, J.P.; Scrub, C.B. Non-destructive evaluation of spheroidized steel using magnetoacoustic and Barkhausen emission. *IEEE Trans. Magn.* **1997**, *33*, 4035–4037. [CrossRef]
20. Kikuchi, H.; Ara, K.; Kamada, Y.; Kobayashi, S. Effect of microstructure changes on Barkhausen noise properties and hysteresis loop in cold rolled low carbon steel. *IEEE Trans. Magn.* **2009**, *45*, 2744–2747. [CrossRef]
21. Hartmannn, K.; Moses, A.J.; Meydan, T. A system for measurement of AC Barkhausen noise in electrical steels. *J. Magn. Magn. Mater.* **2003**, *254–255*, 318–320. [CrossRef]
22. Jiles, D.C. *Introduction to Magnetism and Magnetic Materials*; Chapman and Hall: New York, NY, USA, 1991.
23. Stresstech Group. Rollscan 350. Available online: https://www.stresstech.com/products/barkhausen-noise-equipment/barkhausen-noise-signal-analyzers/rollscan-350/ (accessed on 26 August 2021).

materials

MDPI

Article

Impact of Magnetization on the Evaluation of Reinforced Concrete Structures Using DC Magnetic Methods

Paweł Karol Frankowski [1,*] and Tomasz Chady [2,*]

[1] Doctoral School, West Pomeranian University of Technology in Szczecin, Piastow St. 19, 70-310 Szczecin, Poland
[2] Faculty of Electrical Engineering, West Pomeranian University of Technology in Szczecin, ul. Sikorskigo 37, 70-313 Szczecin, Poland
* Correspondence: frankowski@zut.edu.pl (P.K.F.); tomasz.chady@zut.edu.pl (T.C.); Tel.: +48-91-449-41-34 (P.K.F.)

Abstract: The magnetic method is the most promising method that can be used to inspect large areas of reinforced concrete (RC) structures. Magnetization is a crucial process in this method. The paper aims to present the impact of the magnetization method on the results in the detection of reinforced bars (rebars) and the evaluation of concrete cover thickness in reinforced concrete (RC) structures. Three cases (without magnetization, same pole magnetization, and opposite pole magnetization) were considered in the experiments. Results achieved in all the methods are presented and evaluated. Two different sensing elements were used in the measurements: a magneto-optical (MO) sensor and an AMR sensor. The advantages and disadvantages of both mentioned transducers are presented and discussed in the context of a large areas inspection. The new approach involves using various magnetization methods to improve measurement results for complex structures.

Keywords: nondestructive testing NDT; nondestructive evaluation NDE; magneto-optical (MO) sensor; anisotropic magneto resistance (AMR) sensor; reinforcement bars detection; rebars; concrete inspection; reinforced concrete

Citation: Frankowski, P.K.; Chady, T. Impact of Magnetization on the Evaluation of Reinforced Concrete Structures Using DC Magnetic Methods. *Materials* **2022**, *15*, 857. https://doi.org/10.3390/ma15030857

Academic Editor: Giovanni Bruno

Received: 3 December 2021
Accepted: 18 January 2022
Published: 23 January 2022

1. Introduction

1.1. Nondestructive Methods of Testing Concrete Structures

For over a century, reinforced concrete (RC) has been a dominant construction material for structures of every type and size. Usually, buildings of this kind are designed for 50–100 years of operating time. However, the remaining lifetime of a specific structure is challenging to estimate because many different factors have influences. Many structures built at the beginning of the twentieth century are still in service [1,2]. Therefore, in most countries, periodic inspections of old structures are required by a building code (usually once per five years). Even new construction acceptance tests are conducted to determine if the requirements of a specification or contract are met. The requirements may involve verification of the class, diameter, and arrangement of the rebars in the concrete.

Reinforced concrete could be tested in many different ways. The methods range from destructive, through semi-destructive (where the concrete is partially damaged), to utterly nondestructive testing (NDT). The NDT methods are usually cheaper and faster than methods of other groups. Unlike the destructive and semi-destructive, they can also be easily used in many points of the tested object. Therefore, they better reflect the actual state of the facility.

A full review of NDT methods used in construction diagnostics, along with their advantages and disadvantages, is given in [3]. The properties of a reinforced concrete structure which can be examined with NDT methods are presented in Figure 1.

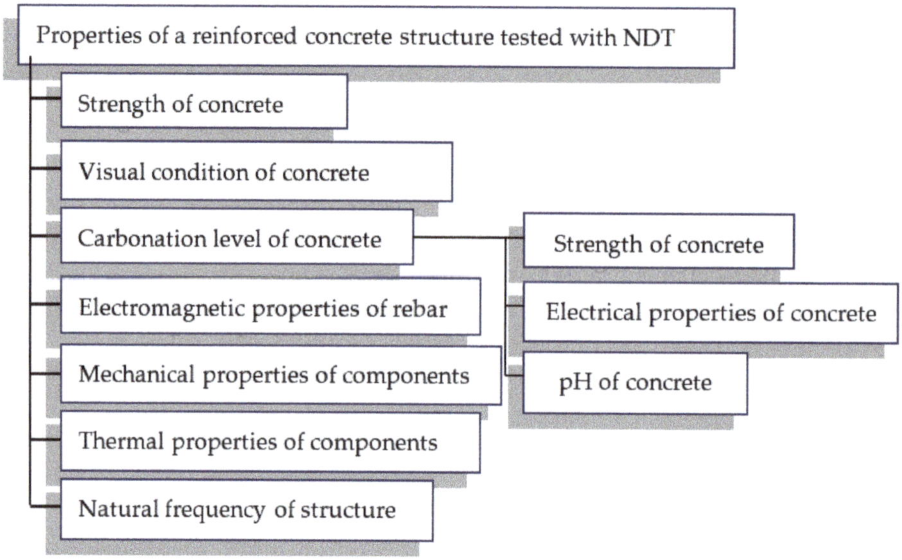

Figure 1. Properties of reinforced concrete structures that can be examined by NDT methods.

As described in [3] and presented in Figure 1, most of the NDT methods used in civil engineering are designed to evaluate concrete. Only methods that use an electromagnetic and mechanical wave can be effectively used for direct reinforcement assessment. The following methods can be distinguished in the mechanical group: high-frequency, active ultrasonic testing methods [3–5]; low-frequency-active mechanical methods [3,6], and passive-acoustic emission (AE) [7].

Electromagnetic methods are not universal, but on the other hand, they have many advantages over mechanical methods. The most crucial difference is that the results of the mechanical methods are affected by many factors because various phenomena may disturb the propagation of mechanical waves in complex structures. Therefore, electromagnetic and magnetic methods are preferred to assess reinforcement elements in concrete structures.

The electromagnetic methods may be used to localize rebars in the structure, precisely estimate basic structure parameters (such as the thickness of the concrete cover, the rebar's diameter, the rebars class [8–10], and detect corrosion or other flaws [11–13]). The most significant advantages of the methods from this group are the direct impact on reinforcement, the low damping of electromagnetic waves by concrete and the high spectrum of frequencies that can be used.

NDT electromagnetic methods can be categorized by the utilized excitation frequency (Figure 2). This frequency is crucial for all methods that use mechanical or electromagnetic waves. It affects resolution and an effective range. The same method may have good resolution and limited range (high frequency) or good effective range and low resolution (low frequency). In simplification, it can be assumed that the smallest size of the defect that can be detected is approximately comparable to the excitation frequency wavelength [14]. The penetration range depends on the frequency of excitation and magnetic permeability of concrete and steel. The fundamental division of NDT methods due to the frequency of excitation is shown in Figure 2.

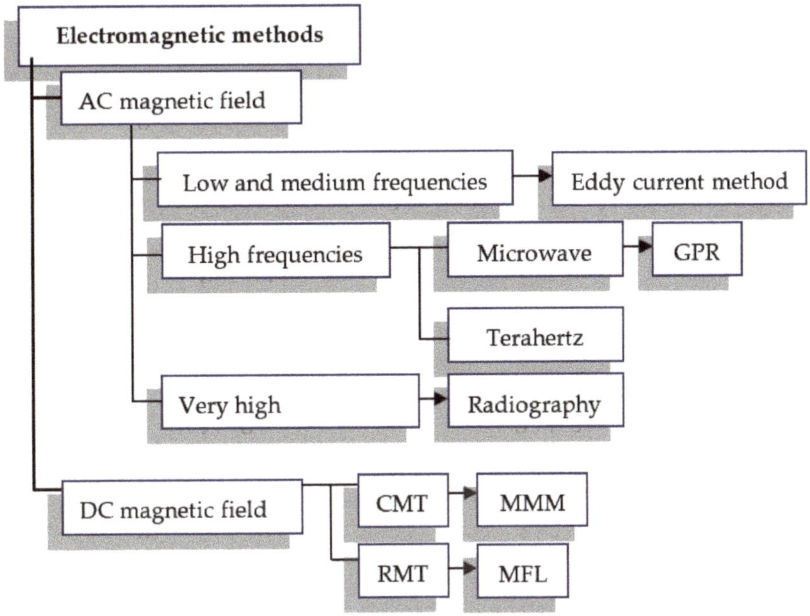

Figure 2. Classification of electromagnetic NDT methods according to the excitation frequency.

The most important AC magnetic field NDT method used in civil engineering is the eddy current (EC) method. In this method, the typical excitation frequency range is from 0.5 to 10 kHz (for reinforced concrete structures). The eddy current method can be used not only to detect the presence of rebars but also to determine the thickness of the concrete cover, the rebar's diameter, the alloy of reinforcing bars (due to different electrical properties), or even to detect corrosion of rebars [8–13]. The effective range of the eddy current method is from 0 to about 100 mm. Results can be really accurate and relatively easy to interpret. Lower excitation frequencies may be used in some versions of the magnetic flux leakage (MFL) and the magnetic force induced vibration evaluation MFIVE method [3] or in the method similar to MFIVE described in [6]. Both of these methods use low-frequency magnetic waves to induce rebar vibration. Natural frequencies of the reinforcement can be used to detect structure debonding, which is usually caused by corrosion.

Another important electromagnetic method is ground-penetrating radar (GPR). The standard operating frequency ranges from 100 MHz to 3 GHz. Rebars can be detected from the distance of several centimeters up to ten meters or more (when other electromagnetic methods have the maximum detection range not bigger than 200 mm). However, results are difficult to interpret and not very accurate [15,16]. The terahertz technique is rarely used due to the limited penetration in concrete, which is usually characterized by high water content, strongly damping electromagnetic waves at these frequencies. Higher frequencies are used in radiography, which can be very effective but, on the other hand, possess many limitations. The source and detector usually must be placed on both sides of the object. Moreover, this method generates risks for human health [3].

Inspection methods utilizing DC magnetic field can be divided into two categories: continuous magnetization techniques (CMT), also called active magnetic inspection (AMI) and residual magnetization techniques (RMT), called passive magnetic inspection (PMI). In the case of CMT, not only receiving devices but also excitation is required.

The leading representative of CMT is the magnetic flux leakage method (MFL). The method is commonly used in the inspection of ferromagnetic parts and components. However, currently, the adaptations of this method for civil engineering are also popular.

In the MFL method, the detector is usually placed between the poles of the magnets or electromagnet) to detect the leakage field. The relative permeability of concrete, stones, water, and the air is close to 1. Therefore they have practically no influence on the magnetic field distribution. The reinforced bars (rebars) made of steel as ferromagnetic materials concentrate the magnetic flux. In this way, the magnetic field is influenced by rebars and can be used to localize them in the concrete structure. The magnetic flux can be disturbed by discontinuities in the material, such as breaks or cracks [12]. The magnetic flux leakage caused by rebar inhomogeneity can be detected at a distance in the range of the typical concrete cover [17,18].

In some cases, the MFL method can be used to determine the material loss caused by corrosion [19–22]. Magnetic methods also allow to identify rebar diameter [23]. The magnetic flux leakage method can also be used for structural health monitoring [22]. Other active magnetic methods, such as Barkhausen emission (MBE), magnetoacoustic emission (MAE), stress-induced magnetic anisotropy (SMA), or magnetic powder method, usually are not used for the evaluation of reinforced concrete structures. The magnetic field is higher in the case of the active magnetic methods (CMT). However, the CMT methods also have disadvantages like longer measurement time, equipment deployment, and power consumption [3].

Residual magnetization methods are more economical and straightforward. The basic RMT is the magnetic memory method (MMM). The method can be used to detect abnormal conditions arising from changes in crystalline structures resulting from stress concentration, corrosion, or cracks. One of many versions of MMM is iCAMM (infrastructure corrosion assessment magnetic method). This method works through passive magnetic inspection under the effect of the Earth's magnetic field.

1.2. Novelty and Significance of the Research

Periodic evaluation of reinforced concrete structures is required by national law in most countries. However, in many cases, such inspection can be problematic. Standard 'in point' tests can be misleading (most of the structure is not checked). The point-to-point scans also cannot be used in large areas because tests of this kind are usually very time-consuming. The obvious solution is to use area tests. In such a way, the investigation time is significantly reduced and received results are reliable. However, currently, there is not even one method that can be used in that way on a large scale. Area tests potentially can also be used as a pilot or preliminary evaluation before applying other more precise methods. There are only a few methods that theoretically can be used for such evaluation. This group includes primarily visual testing, radiography, and thermography. Unfortunately, these methods have many limitations (e.g., thermography can be used only if the concrete cover is low [8,24]; radiography requires specialized equipment, generates risks for human health, and elements of the system must be placed on both sides of the object) and they are often insufficient. The full summary of the area testing methods is shown in [3]. The magnetic methods are not always considered to be good for area tests. However, this method possesses many advantages over others tests mentioned before. Tests executed with the magnetic method are cheap, the principle of operation is easy to understand and use, the used magnetic wave can avoid damping caused by concrete cover. The test showed that the magnetization method is crucial for the effectiveness of this method. The potential of the active and passive magnetic methods is presented in further sections of the paper.

1.3. The Article Outline

In the introduction of this paper, first, the importance of nondestructive testing (NDT) in periodic tests of reinforced concrete structures has been described. A brief overview of the NDT methods used in the construction sector is also presented. Next, the significance of the conducted research was indicated.

The Section 2 (Materials and methods) presents the tested samples and measurement systems. The section has much attention to magneto-optical (MO) sensors. The MO elements, one of the few magnetic field detectors, are designed for area testing. The evaluation of ferromagnetic objects remote from the sensor as much as in the case of reinforced concrete is an unusual issue for this type of sensor, which is intended and designed for surface testing. Therefore, before the tests, their accuracy in the case of reinforced concrete structure was doubtful. For more detailed investigations, AMR sensors connected in one matrix were proposed. In this section, examples of received results and algorithms of data processing are discussed.

The results of the measurements were placed in the Section 3. First, the entire section is briefly described. Next, results received for the MO sensor are presented. The experiments with the MO sensor show both the influence of magnetization on increasing the ability to detect rebars and the application potential of the MO-sensors.

In the other subsection, results received for three different samples and three different magnetization variants are presented. All experiments were conducted with the AMR sensor. The main point of the subsection is to show how significant the impact of the magnetization method on received results can be. The impact is even stronger for more complex samples. This part also presents the disadvantages of the passive method, which also becomes more significant during the tests on more complex samples.

The obtained results are summarized in the Section 4 In particular, the magnetization aspect is discussed in this part. The section 'Conclusions' discussed whether the magnetic method is finally suited for area testing and how the tests of this kind fall on the background of other methods. The two tested sensors are also compared in this part. The advantages and disadvantages of both systems are presented, and applications of the sensors have been proposed. In the section also plans for further research on the magnetic method for area testing are presented.

2. Materials and Methods

2.1. Measuring Systems and Samples

2.1.1. Test Samples

The main aim of the article is to investigate the influence of magnetization on the effectiveness of magnetic nondestructive testing methods in the evaluation of reinforced concrete structures. For this purpose, the three different samples are examined: S1—the sample with single rebar (Figure 3a); S2—the sample with two rebars, one placed 85 mm under the other (Figure 3b), and S3—the sample contains three rebars, all rebars placed one next to each other (Figure 3c). In the third sample, distances between rebars are 55 and 50 mm. The magnetic sensor was moved above the sample in a line perpendicular to the reinforcement bars. The distance between the rebar and the sensor (thickness of the concrete cover) is marked as h (Figure 3a). The results were obtained using an integrated AMR transducer that allows measuring three field components.

Configurations of the samples are presented in Figure 3. The magnetic transducer was moved along the x-axis, while rebars were positioned along the y-axis.

2.1.2. Systems for Active Magnetic Inspection

The measuring system consisted of four subsystems: excitation subsystem, positioning subsystem, magnetic field transducer, and data acquisition subsystem. The general block scheme of the system is presented in Figure 4. All subsystems are described in the following sections.

The simplest solution to magnetize reinforcement bars (rebars) can be achieved using permanent magnets. In the presented systems, two neodymium magnets in two different configurations were used for this purpose. The reference configuration was without any magnets, as shown in Figure 5a. In the second configuration, magnets have opposite poles facing the sample (Figure 5b). In the third configuration, the magnets were directed to the sample with the same poles (Figure 5c). The magnets were placed on both sides of the sensor at a distance of 500 mm.

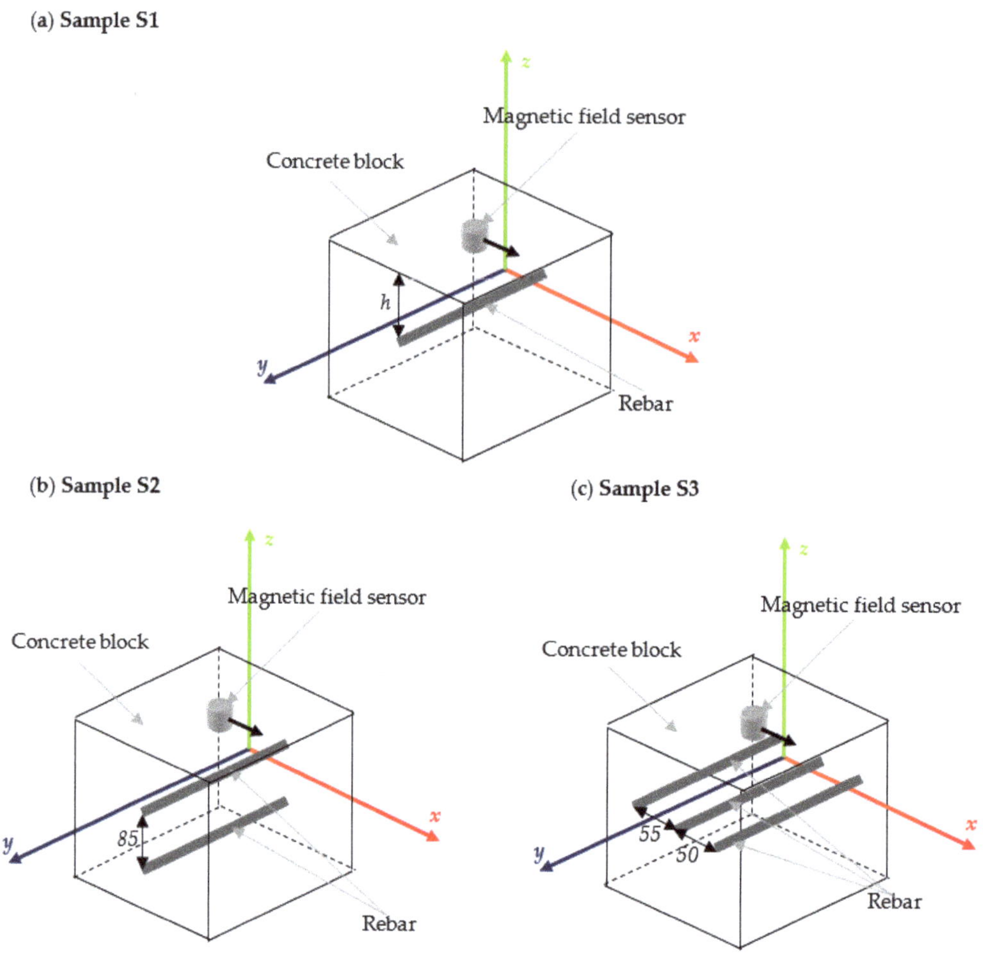

Figure 3. The samples used in the experiments; (**a**) sample S1 (with single rebar); (**b**) sample S2 (with two rebars one under the other); (**c**) sample S3 (with three rebars next to each other).

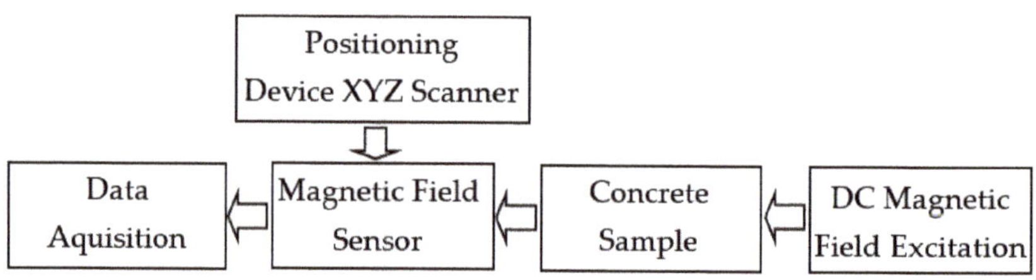

Figure 4. Block scheme of the measuring system.

In the experiments, a two-dimensional area over the sample surface was scanned. The area directly above the rebars is tested with a positioning system. The example of positioning subsystem is shown in Figure 6.

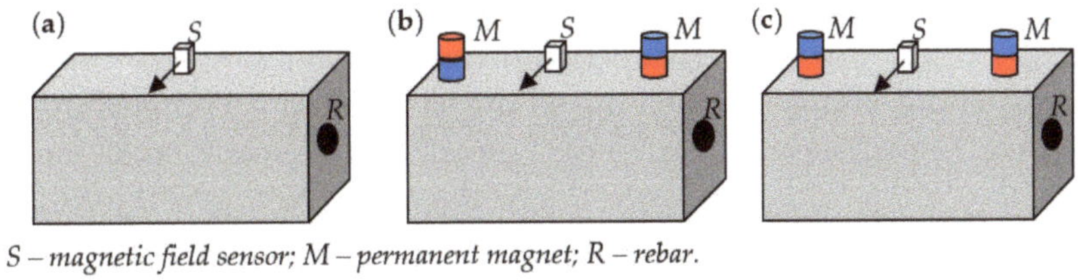

Figure 5. System configurations used in the experiments; (**a**) reference configuration; (**b**) configuration of opposite poles magnetization (OPM); (**c**) configuration of same poles magnetization (SPM).

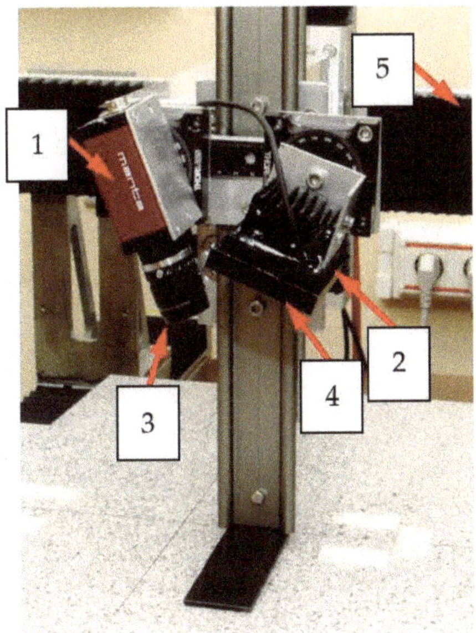

Figure 6. Photo of the MO transducer attached to the XYZ scanner: 1—camera; 2—the source of monochromatic light; 3 and 4—linear polarizers; 5—XYZ scanner.

The magnetic field sensor is an essential part of the system. Magnetoresistive (MR) and Hall effect sensors are of the greatest industrial importance among the magnetic field sensors. The Hall effect components account for approximately 85% of the world's production of magnetic sensors for DC and low-frequency applications. The MR sensors account for around 10%, and their market share grows [25].

The most used MR sensors are anisotropic magnetoresistors (AMR) and giant magnetoresistive effects (GMR) elements. The AMR and GMR sensors have high sensitivity and field resolution. Elements of this kind can operate even in the pT range. However, they can be permanently affected by strong magnetic fields and GMR sensors have a high hysteresis.

The Hall effect sensors have several advantages over MR elements. They show no saturation effects and can measure strong magnetic fields. For these reasons, the Hall effect sensors are preferably used at magnetic fields higher than 1 mT. They are the first choice in many industrial applications. However, large offset and relatively low sensitivity limit both the accuracy of the measurements and the minimum value of the magnetic field that can be measured. One of the issues examined in this research is testing non-magnetized reinforced concrete structures using magnetic methods. The MR sensors seem to be much better suitable for this purpose.

Most magnetic field sensors can measure the magnetic field at one point. The exception is magneto-optical (MO) sensors, which are well suited to constructing an area testing system. Therefore, magneto-optical (MO) sensors are preferable for testing large-scale reinforced concrete structures. The Faraday magneto-optical effect is used in MO sensors [26,27]. The main advantage of this solution is the immediate obtaining of the 2D field distribution over the sample surface.

2.1.3. Measuring System with Magneto-Optical Sensor

The Faraday magneto-optical effect is used in MO sensors. This effect describes an interaction between light and a magnetic field in a medium. The plane of polarization of linearly polarized light rotates parallel to the propagation direction of light waves passing through the magneto-optical medium. The mechanism of the Faraday effect is explained in Figure 7 [26,27].

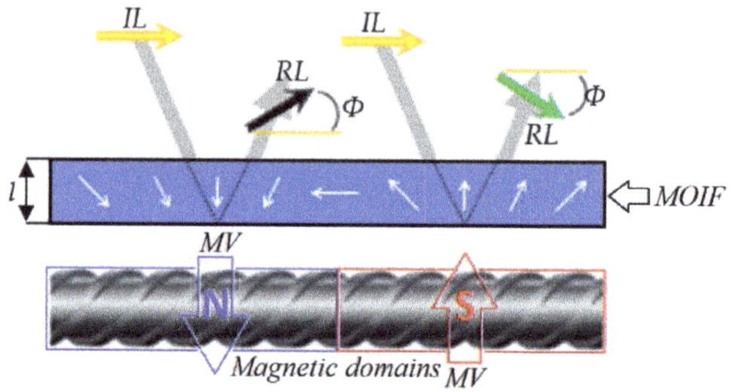

Figure 7. Operating principle of the MOIF. MOIF—Magneto-Optical Indicator Film (sensor); MV—Magnetic Vector; IL—Polarization Plane of Incident Light; RL—Polarization Plane of Reflected Light; Φ—Angle of Faraday rotation.

The MO sensor presented in Figure 8a consists of four layers, as shown in Figure 8b. Additional layers are necessary to improve the quality of the measurements. The mirror layer (for visible spectral range) is used to improve the sensor reflectivity. For mirror protection, the resistant material layer is used. The sensor also contains anti-reflection coated glass [26,27].

The most important advantage of the MO-sensor over other magnetic field sensors is the large area of observation of the magnetic field and the relatively high resolution. The most significant advantage of MO-sensor over other magnetic field sensors is the large area of magnetic field observation and relatively high resolution. The manufacturers offer sensors with diameters up to 3 inches. A few different types of MO transducers are used in many different applications [26,28]. Parameters and characteristics of the type A sensor used in the experiment are provided in Figure 9. The sensitivity of this MO-sensor is comparable to the Hall effect elements.

The type A sensor used in the experiments is an out-of-plane (OOP). The MO-sensors of this kind are generally more sensitive but have a smaller range and nonlinear characteristics. The hysteresis (Figure 9a) can also cause difficulties during measurements (in the case of less sensitive MO-sensors, there are no such problems). The A-type is chosen because of the lowest dynamic range (significant visible changes with minor magnetic field changes). An alternative to the A-type transducer in this kind of application is a D-type transducer. Sensors of this kind are more sensitive than A-type; field range is from 0.03 to 5 kA/m and can be used to test printed magnetic inks or steels alloys. The sensors are sensitive, but it also depends on the quality of the camera and other elements. There is another valuable property of the D-type element.

The D-type element can be working in two modes:

- Faraday: for applications without external excitation;

- Bias: for work in the environment of an external magnetic field. In this mode, performance is weaker, but other types of sensors would lose their performance entirely. This mode is used mainly with magnetically very soft materials, like inks.

The MO-transducers require a relatively complex setup. The block diagram of the system with MO-transducer is shown in Figure 10, and the setup photo is presented in Figure 6.

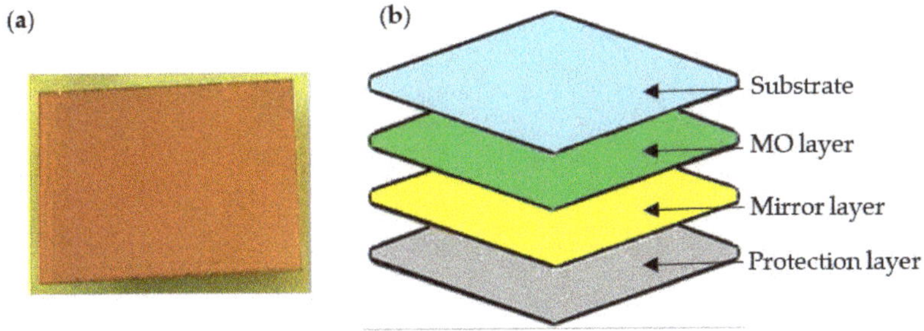

Figure 8. Magneto-optical sensor; (**a**) the photo of the A-type MO sensor in the protective packaging. (**b**) schematic showing the functional layers of the Magneto-Optical Indicator Film (sensor).

2.1.4. Measuring System with a Magnetoresistive Sensor

Systems based on MR sensors are less complex than these based on MO sensors. Moreover, the AMR sensors with three sensitivity axes are better suited for more accurate investigations of reinforced concrete structures.

AMR (anisotropic magneto resistance) elements belong to the MR group of sensors. The resistance of these elements decreases when a magnetic field is applied. This function is dependent on the direction of the magnetic force lines applied to the element (anisotropic). The material of the AMR element is an alloy of nickel, iron, and other metals (ferromagnetic). In these experiments, integrated transducer HMC5883L was used. The sensor has few advantages over GMR. The sensitivity is high, much higher than in the case of the MO sensor. Nevertheless, lower than it could be in the case of GMR [29].

(**a**) A-type

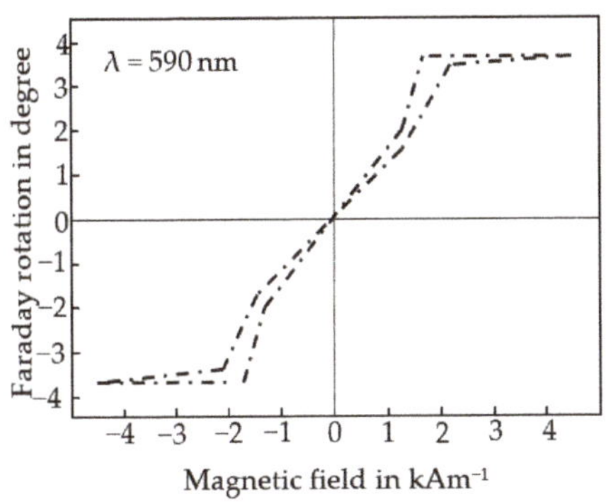

Manufacturer	Matesy
Type	A
Dimensions (mm)	15.5 × 20.5 × 0.5
Magnetic saturation	2 mT
Resistance to temperature changes	+10 to +50 °C
Working temperature range	+15 to +30 °C
Optical transmission range	$\lambda > 530$ nm
Optical resolution	1 to 25 μm
Faraday rotation angle (λ = 590nm)	1 to 10°

A-type. Field range: 0.05 ... 2 kA / m
Applications: magnetic stripe card, hard magnetic ink, steel inspection, audiotapes

Figure 9. *Cont.*

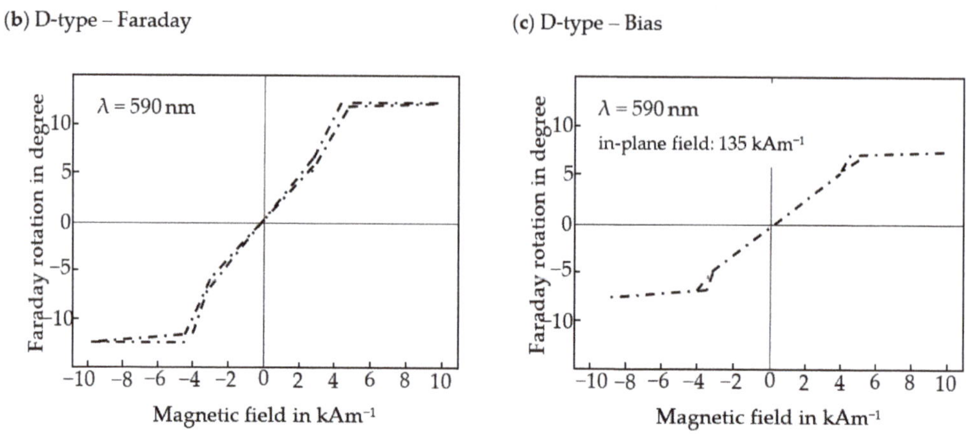

Figure 9. Parameters and approximate curves of characteristics-utilized MO-sensor; (**a**) A-type sensor: plot of magnetic field vs. Faraday rotation Φ (λ = 590 nm), and selected parameters of the sensor; (**b**) D-type Faraday version of the sensor: plot of magnetic field vs. Faraday rotation Φ (λ = 590 nm); (**c**) D-type bias version of the sensor: plot of magnetic field vs. Faraday rotation Φ (λ = 590 nm). (Based on materials received from the manufacturer Matesy).

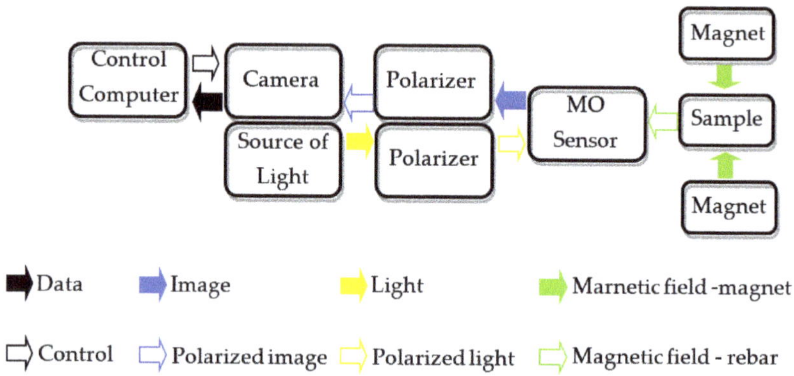

Figure 10. Block diagram of the system with the MO-sensor.

On the other hand, the sensitivity of GMR would be too high for this application. With the use of 'reset strap drive' the internal offset of the sensor and its temperature dependence is corrected for all measurements. This option could be helpful in the vicinity of large magnetic fields. In opposite to GMR, AMR sensors clearly indicate the results of the magnetic field direction. Because the positive and negative sides have symmetric characteristics, the same operation is performed even if the north and south poles of the magnet are reversed. This characteristic is used to improve the reliability and accuracy of the data. The sensor also has high linearity and low hysteresis.

2.2. *Methods of Processing the Results*

2.2.1. Measurement Results Processing in the System with MO Sensor

The results obtained from MO systems usually do not require complicated processing and are available in real-time. Nevertheless, in some cases, such as a high thickness of concrete cover h, even minor image changes have to be detected. Therefore the following algorithm of the image enhancement was implemented. First, the algorithm extracts the active area of the MO sensor from the image obtained from the camera. Then, since the axis of the camera lens was not perpendicular to the sensor surface, it was necessary to correct the perspective. The next step is to reduce the geometric distortions caused by the lens. Images processed in this way are saved in the system memory. Due to the relatively small size of the sensor area, the final image [A] consisted of several (5 to 7) images [A_n] taken at subsequent positions above the sample. The sensitivity of the MO transducer is not the same at different places on the sensor surface. Therefore, the images [A_i] are corrected using a coefficients matrix calculated from a uniform DC magnetic field measurement. In the cases of small (0–20 mm) or big (80–100 mm) thickness of concrete cover (*h*), it is also necessary to correct the non-linearity of the characteristic and hysteresis presented in Figure 8a. In order to remove noises, a 2D-median filter with a 5 × 5 mask is applied to the image [A]. In the last step, contrast and brightness were corrected. Effects of the processing are shown in Figure 11.

Action	Results
Step 1: RAW images.	…  …
Step 2: Perspective and lens corrections.	…  …
Step 3: Sensitivity corrections.	…  … … [A_{-2}], [A_{-1}], [A_0], [A_1], [A_2] ….
Step 4: Combining pictures.	… 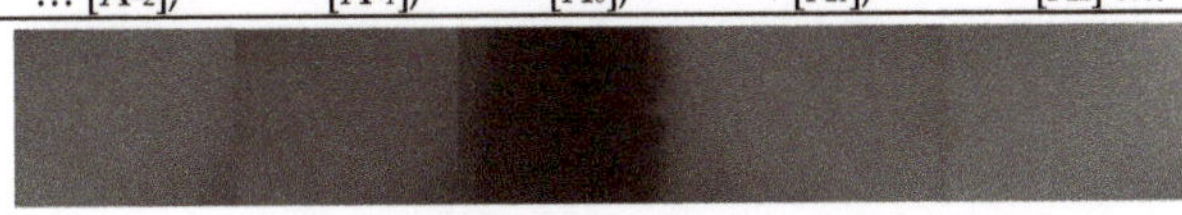 … … [A_{-2}] & [A_{-1}] & [A_0] & [A_1] & [A_2] …. → [A]
Step 5: Outliers removing using a median filter (mask 5 × 5).	…  …
Step 6: Contrast and brightness corrections.	…  …

Figure 11. The processing of images obtained with the MO sensor; same pole magnetization (SPM); Sample S1; *h* = 0.5 mm.

The MO sensors enable testing areas of objects under investigation without time-consuming point-by-point scans. Unfortunately, sensitivity, linearity, and repeatability are limited. Moreover, the images are noisy. The problems only to some degree, can be caused by hardware limitations (polarizers or video cameras). The MO sensors could be a solution for a preliminary evaluation.

2.2.2. Measurement Results Processing in the System with MR Sensor

MR systems are much more sensitive than MO systems. Moreover, systems of this kind can deliver information about three components of the magnetic field. In further investigation, measurements were taken by moving the transducer with a 1 mm step in the *x*-axis and 10 mm in the *y*-axis direction. The measurements were very time-consuming.

Examples of results received using opposite poles polarization for inspection of the sample S1 are presented in Figure 12.

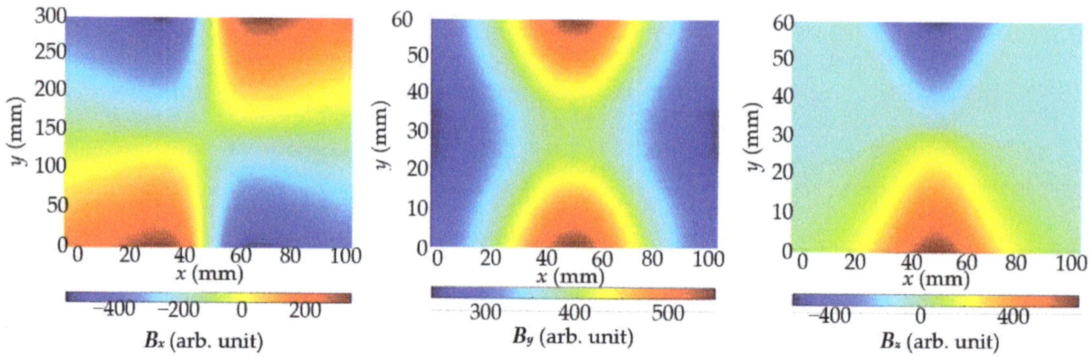

Figure 12. Magnetic field B_x, B_y, and B_z components measured in case of the same pole polarization and sample S1 with single rebar. Measurements were carried out with the AMR sensor; concrete cover thickness h = 20 mm.

3. Results

The results of the experiment are discussed in two subsections. The first part presents the magnetic field distribution measurements using MO-sensor and their application for rebars detection. The experiments can be assumed as preliminary studies. The measurements with the MO-sensor are carried out quickly, and they are easy to interpret. However, the sensitivity of the MO-sensors is lower than the AMR sensor, and there is no possibility to measure x, y, and z induction components. In this case, only sample S1 with single rebar is tested (all samples were tested with the AMR sensor as shown in the following subsection). The experiments with the MO sensor show both the influence of magnetization on increasing the ability to detect rebars and the application potential of the MO-sensors. The same pole magnetization (SPM) is used in this experiment.

In the following subsection, results received for three different samples and three different magnetization variants are presented. All experiments were conducted with the same magnets (having different orientations against the rebars). Therefore, the magnetization effect is weaker for samples with a bigger concrete cover thickness. In addition, always the same single AMR sensor was used. The main point of the experiments is to show the impact of the magnetization method on received results. Tests prove that the impact is even more significant for more complex samples. Experiments carried out on the samples simulating reinforced mesh (samples S2 and S3) showed that the CMT (Continuous magnetization techniques) were much more effective than RMT (residual magnetization techniques). Moreover, the SPM (same pole magnetization) allows identifying rebars more straightforwardly than OPM (opposite pole magnetization).

3.1. Experiments with the MO-Transducer

Experiments using the MO-sensor for sample S1 (with single rebar) were conducted to show the differences between CMT and RMT. Rebars are magnetized every time up to the same level and in the same orientation. The SPM was selected as a method of magnetization. As a reference, the same experiment was also conducted with the non-magnetized rebar. Experiments were taken with the step of 5 mm along the axis z (change of concrete cover thickness h), and 20.5 mm along the axis x (size of the sensor is 15.5 × 20.5). In this way, continuous measurements were obtained without any gaps. The thickness of the concrete cover h was changed in the range from 0.5 to 100 mm.

Predictably, experiments have shown that magnetized rebar can be detected with a much greater concrete cover than a non-magnetized. When the non-magnetized rebar is challenging to detect with a cover thickness above $h = 20$ mm, the magnetized rebar could be detected from a distance of more than 100 mm. However, the readability of the graphs for large cover thicknesses is limited. Examples of the measurements received for thick concrete cover are shown in Figure 13. Only half of the measurement results are shown (the other half is symmetrical).

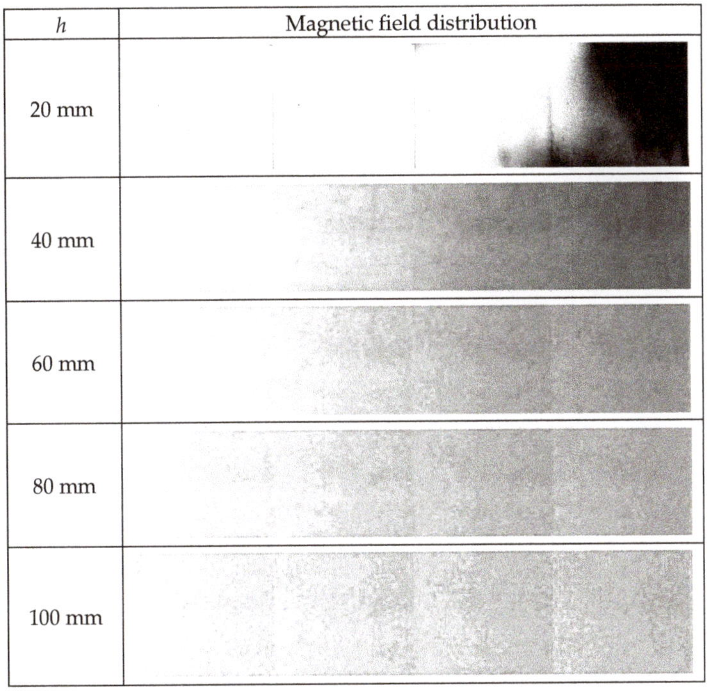

Figure 13. Magnetic field distribution measured with the MO-sensor for different concrete cover thicknesses; same pole magnetization; only half of the measurements are shown.

Plots showing the magnetic field distribution over the magnetized rebar vary depending on the thickness h of the concrete cover. Examples of such characteristics are shown in Figure 14. They are repeatable and unambiguous. Therefore, on their basis, it is possible to estimate the location of the rebar, the thickness of the concrete cover, and possibly other parameters of the structure, as was the case in [9,10].

In existing civil engineering constructions, the thickness of the concrete cover is usually between 10 mm to 50 mm over the reinforcing bars. When the reinforcing bars are not magnetized, the MO sensors are not sensitive enough to detect rebars from such distances. However, when the bars are magnetized, the efficiency of the MO sensors is sufficient. Thus, sensors of this type are suitable for the CMT and not for RMT. Examples of calculated signal to noise ratio (SNR) values are presented in Table 1.

Table 1. Signal to noise ratio SNR calculated for measurements obtained by MO sensor for different concrete cover thicknesses h.

h (mm)	0.5	20	40	60	80	100
SNR (dB)	29.7	26.9	24.8	23.0	22.5	22.0

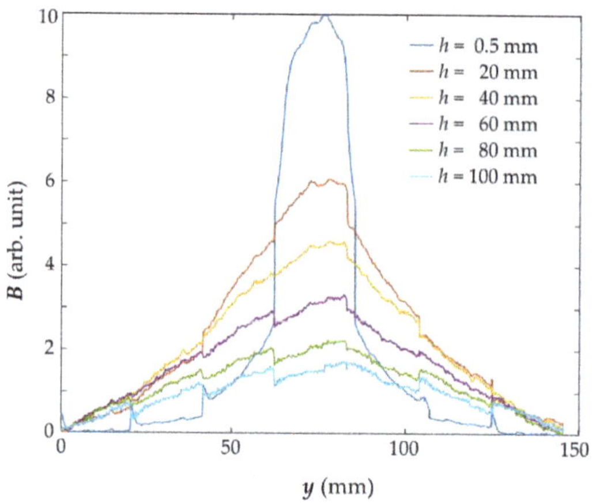

Figure 14. Impact of concrete cover thickness on the MO-sensor measurements. The average line profile of the magnetic field was measured using MO-sensor with the same pole magnetization SPM by moving the sensor along the *y*-axis.

The SNR was defined as:

$$\mathrm{SNR} = 20 \log \frac{A_{signal}}{A_{noise}}$$

3.2. *Influence of Rebars Magnetization Method on Magnetic Field Distribution*

All experiments in this section were taken with the step of 5 mm along the axis z (change of concrete cover thickness h), from 20 to 70 mm (typical concrete cover thickness).

The step along the axis y was equal to 20 mm and experiments were taken from −100 to 100 mm. Position 0 is a position in the middle of the rebar.

The step along the axis x was equal to 2 mm and experiments were taken from 0 to 98 mm. Rebars in S1 and S2 are placed in position 27 mm (axis x). In the case of S3, the middle rebar is placed in this position.

Magnets were moving together with the sensor and were placed on both sides of the sensor at a distance of 500 mm.

In the first set of experiments, the measurements were carried out for the sample S1 using different magnetization methods. The results for different thicknesses h of concrete cover are presented in Figure 15. The second set of experiments was carried out with three different samples S1, S2 and S3, shown in Figure 3. The measurement results were symmetrical concerning the rebar, and therefore the measuring range has been reduced nearly by half. Positions of the rebars were depicted on the plots in Figure 16 by dashed lines.

Figure 15 shows that the influence of magnets on the rebar decreases when the cover thickness h is increasing, and thus, the magnetic field measured by the sensor also decreases. The method of magnetization significantly influences the value of the magnetic field. Compared to the field measured for a non-magnetized bar, the use of magnetic excitation in any configuration of the magnets causes an increase in the field value. As a result, the magnetic field diagrams obtained for different cover thicknesses h differ significantly, which facilitates identification. The strongest field over the rebars was measured in the case of magnets directed towards the bar with homonymous poles (SPM), lower for magnets with opposite poles (OPM), and the lowest for the reference sample in which the rebar was not magnetized. One can observe that the maximum value of the magnetic field component B_z was similar in both magnetization methods.

In the case of non-magnetized rebar, the graphs representing the magnetic field along the x-axis perpendicular to the rebar did not change significantly with increasing cover thickness. Even the changes measured for thickness h above 50 mm are minimal.

(a) 2D plots of B components for concrete cover thickness h = 20 mm;

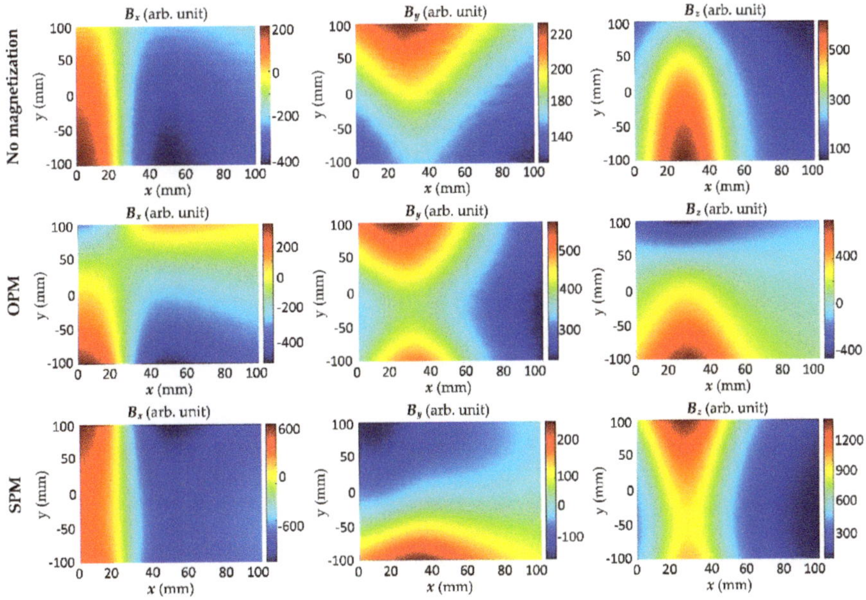

(b) 2D plots of B components for concrete cover thickness h = 30 mm;

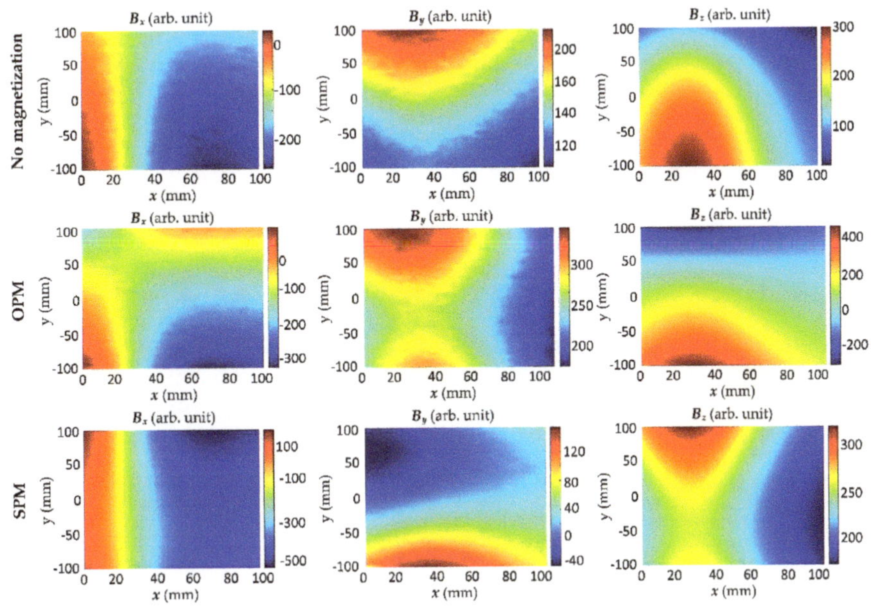

Figure 15. *Cont.*

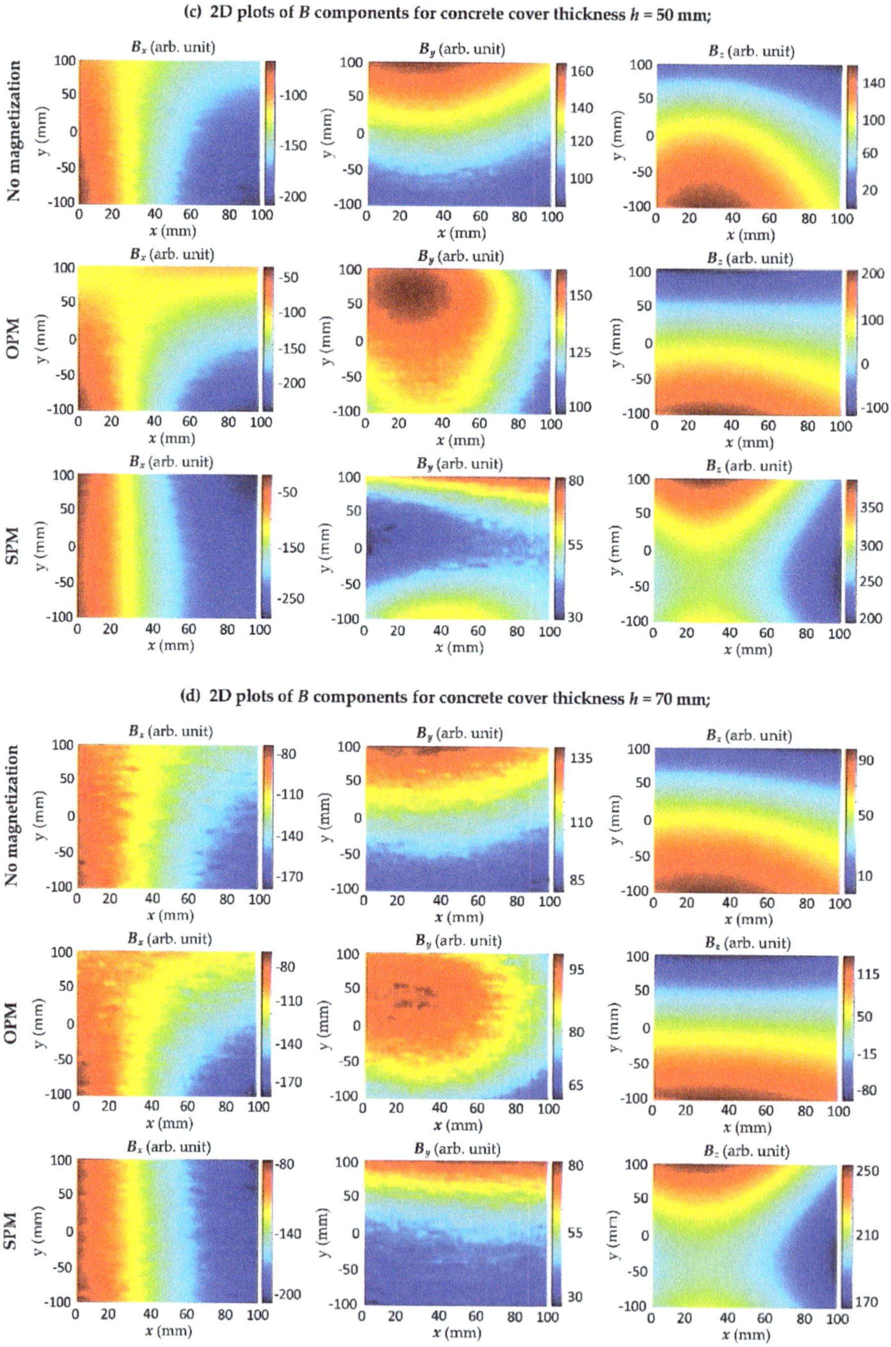

Figure 15. Results of 2D measurements using the AMR-sensor obtained for different variants of magnetization and different thicknesses h of concrete cover; experiment conducted for the sample S1 with single rebar; SPM—same pole magnetization, OPM—opposite pole magnetization; concrete cover thickness: (**a**) h = 20 mm; (**b**) h = 30 mm; (**c**) h = 50 mm; (**d**) h = 70 mm.

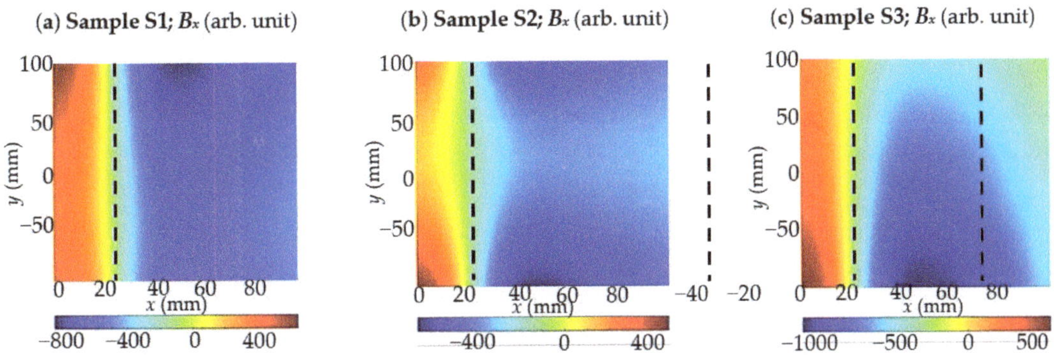

Figure 16. Selected results of 2D measurements of magnetic field component B_x with the depicted position of the rebar (dashed line); (**a**) sample S1—single rebar; (**b**) sample S2—two rebars one under the other; (**c**) sample S3—three rebars next to each other.

The results obtained with the SPM magnetization system are the easiest to interpret. The B_x component of the magnetic field is particularly interesting. It has a much larger value than the others and changes significantly with increasing cover thickness. Moreover, in contrast to magnetization OPM, the SPM looks similar, regardless of the measurement place in the y axis direction. The most important conclusion from the presented results is that the cover thickness h can be estimated based on the slope of the graph of the measured magnetic field (Figure 17).

The measurements show that the lack of magnetization causes a significant reduction of the magnetic field and, therefore, it may cause errors in the rebars identification. For example, in Figure 17, in the case of non-magnetized rebar, the B_y component takes very small values. The results of experiments show that the magnetization method can impact noise immunity. Signal to noise ratio calculated for different methods of magnetization and different thicknesses of concrete cover h is provided in Table 2.

Table 2. Signal noise ratio SNR (dB) calculated for measurements obtained using AMR sensor for different concrete cover thicknesses h, and different magnetization methods.

h (mm)		20	30	40	50	60	70
No mag.	B_x	35	31	25	23	21	21
	B_y	25	X	X	X	X	X
	B_z	38	36	33	29	27	27
SPM	B_x	48	45	37	38	35	34
	B_y	52	47	46	43	41	40
	B_z	58	56	56	55	56	53
OPM	B_x	45	44	44	36	31	29
	B_y	49	49	50	47	48	47
	B_z	49	51	51	50	49	47

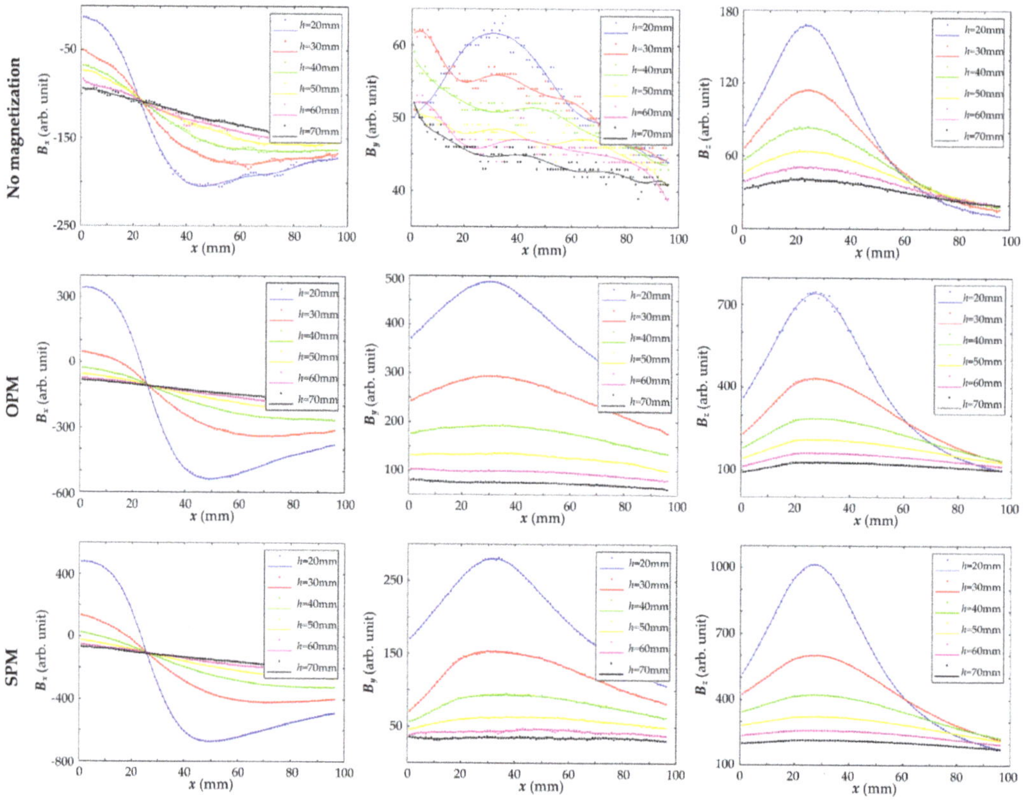

Figure 17. Plots of the magnetic field components as a function of sensor position x (mm) which are showing the influence of the concrete cover thickness h.

As mentioned, the signal to noise ratio (SNR) for y-component and non-magnetized rebar has very small values. Moreover, the presence of rebar nearby does not appear to be the dominant factor that formed this characteristic (due to the influence of external fields). Therefore, SNR is not calculated in that case. The impact of noise is much higher in the case of non-magnetized rebars. A slightly bigger SNR was achieved for the SPM magnetization compared to the OPM. However, in this respect, both methods are comparable. As it is not difficult to predict, the growth of the thickness of concrete cover has a negative effect on the SNR. The influence of h on the SNR is different for different magnetization methods and for different components. However, drawing conclusions based on Table 2 could be premature due to a small test attempt. In addition, in all cases, the impact of noise is moderate. It can be noted that the dominant influence on SNR has the maximum value of the obtained signal. The relationship between the signal value received from the AMR sensor and the thickness of the concrete cover h for different magnetization methods is shown in Figure 18.

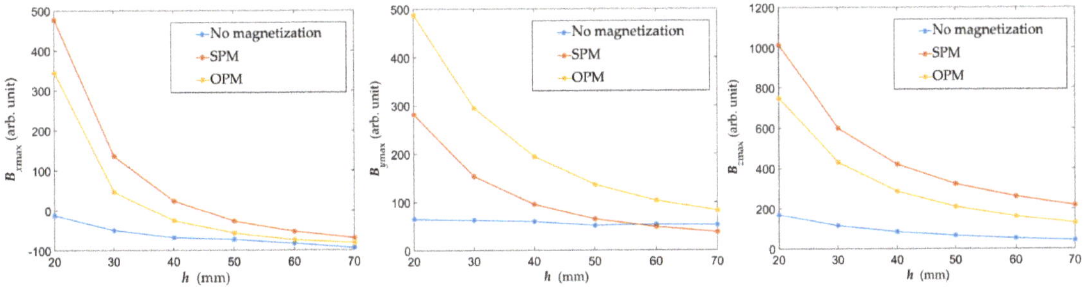

Figure 18. Graphs of the maximum value of the magnetic field components as a function of the concrete cover thickness *h* obtained for various magnetization methods.

The maximal value of signals presented in Figure 18 is greater in the case of the B_x and B_z for SPM (same pole magnetization). In the case of B_y, the biggest signal value is achieved for OPM (opposite pole magnetization). One can observe that the maximal signals are significantly smaller without magnetic excitation. Results of identification, in that case, are uncertain (nevertheless, detection of the reinforcement is possible). The problem of measurements conducted without magnetization is the low value of the received signals. This problem causes strong noise influence and the characteristics ambiguity. It is worth noting that the maximum amplitude of various components is significantly different. The maximal value of the signal obtained for B_z is much higher than for the two other components. This fact has a substantial impact on the SNR. The comparison of the characteristics for the non-magnetized rebars, magnetized with the SPM and magnetized with the OPM, is shown in Figure 19 (normalized curves). For non-magnetized rebars, there are differences in the shape of the characteristics caused by noise and the ambiguous polarity of the rebars. As a result, they are challenging to interpret, and the identification results could be inaccurate. Differences between maximum values of signals obtained for SPM and OPM are minor. Obtained results in these two cases are comparable.

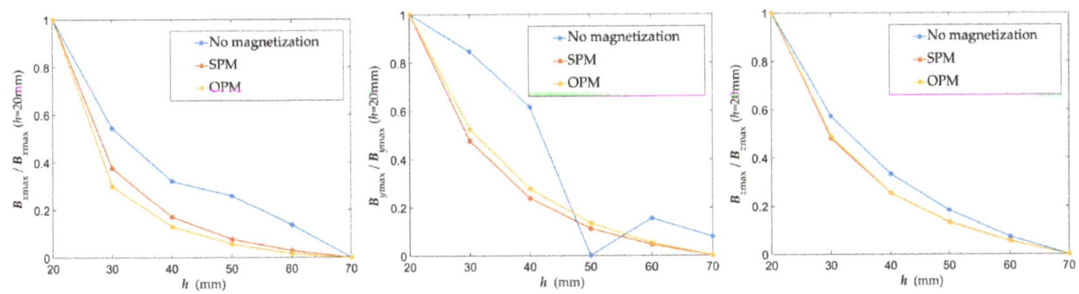

Figure 19. Graphs of the normalized maximum values of the magnetic field components as a function of the concrete cover thickness *h* obtained for various magnetization methods.

The type of magnetization method does not affect the steepness of changes in the measured linear profiles of the magnetic field components (Figure 20). However, also in this aspect, low SNR makes identification difficult in the case of lack of magnetization. The curves obtained for SPM and OPM are almost the same.

The following experiment was carried out to investigate the influence of magnetization on identifying the reinforcement mesh. Two kinds of specimens were tested: sample S2—(two rebars one over the other—Figure 15b) and sample S3 (three rebars are next to each other—Figure 15c) are considered in the tests and compared with measured earlier sample S1. The results are presented in Figure 21.

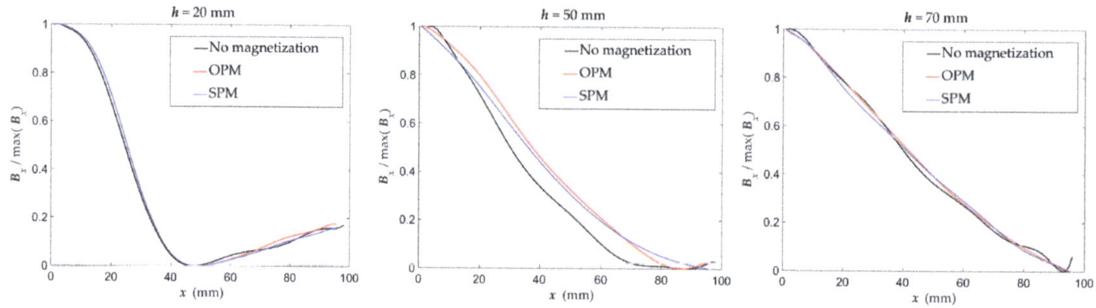

Figure 20. Graphs of the normalized value of the magnetic field component B_x (x) as a function of the concrete cover thickness h (mm) obtained for various magnetization methods.

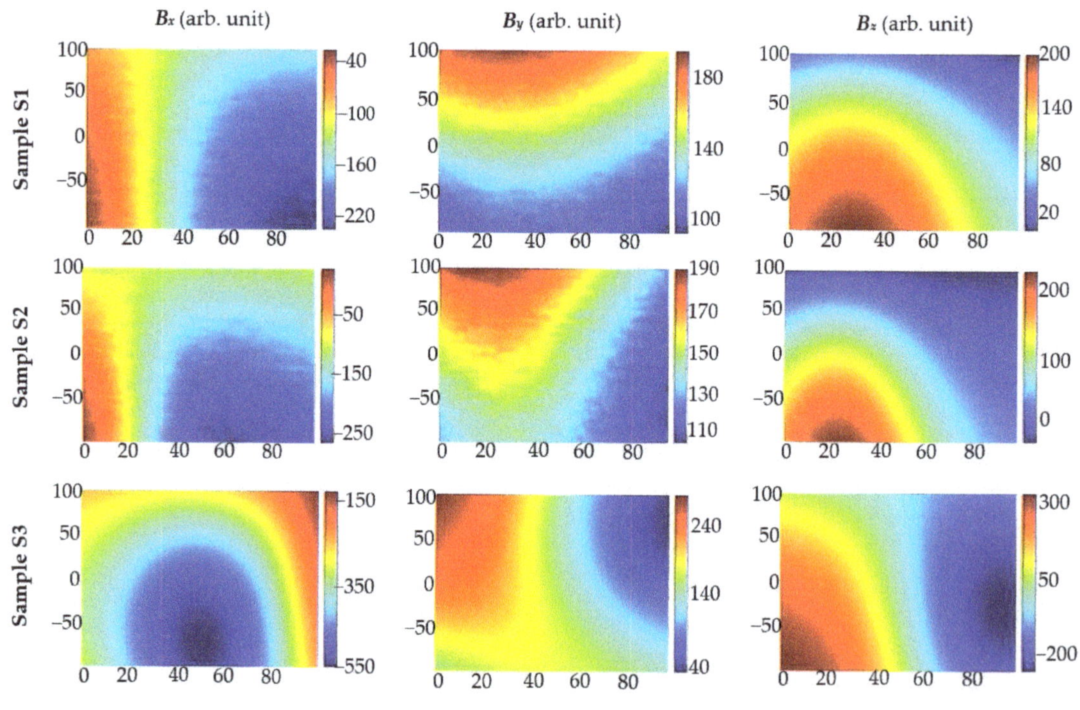

Figure 21. Results of 2D measurements using AMR sensor obtained for different samples (S1, S2, S3); without magnetization; thickness of the concrete cover h = 40 mm; x (mm), y (mm)—sensor positions.

The obtained results indicate that regardless of the method of magnetization or the lack of it, more complex structures containing several bars next to each other (samples S2 and S3) generate field distributions significantly different than in the case of a single bar (sample S1). Correctly-configured magnetic excitation creates opportunities to correctly identify complex structures, which are more similar to existing building structures.

There are many problems with testing reinforcement meshes, where more than one rebar strongly influences the sensor. In the case of concrete structures without magnetization (RMT) the most significant problem is a lack of knowledge about the residual magnetization of individual rebars. Another obstacle is that the rebars could be strongly magnetized during earlier operations (e.g., by a crane with an electromagnetic gripper) and the obtained results strongly depend on the magnetized rebars relative position as shown in Figure 22.

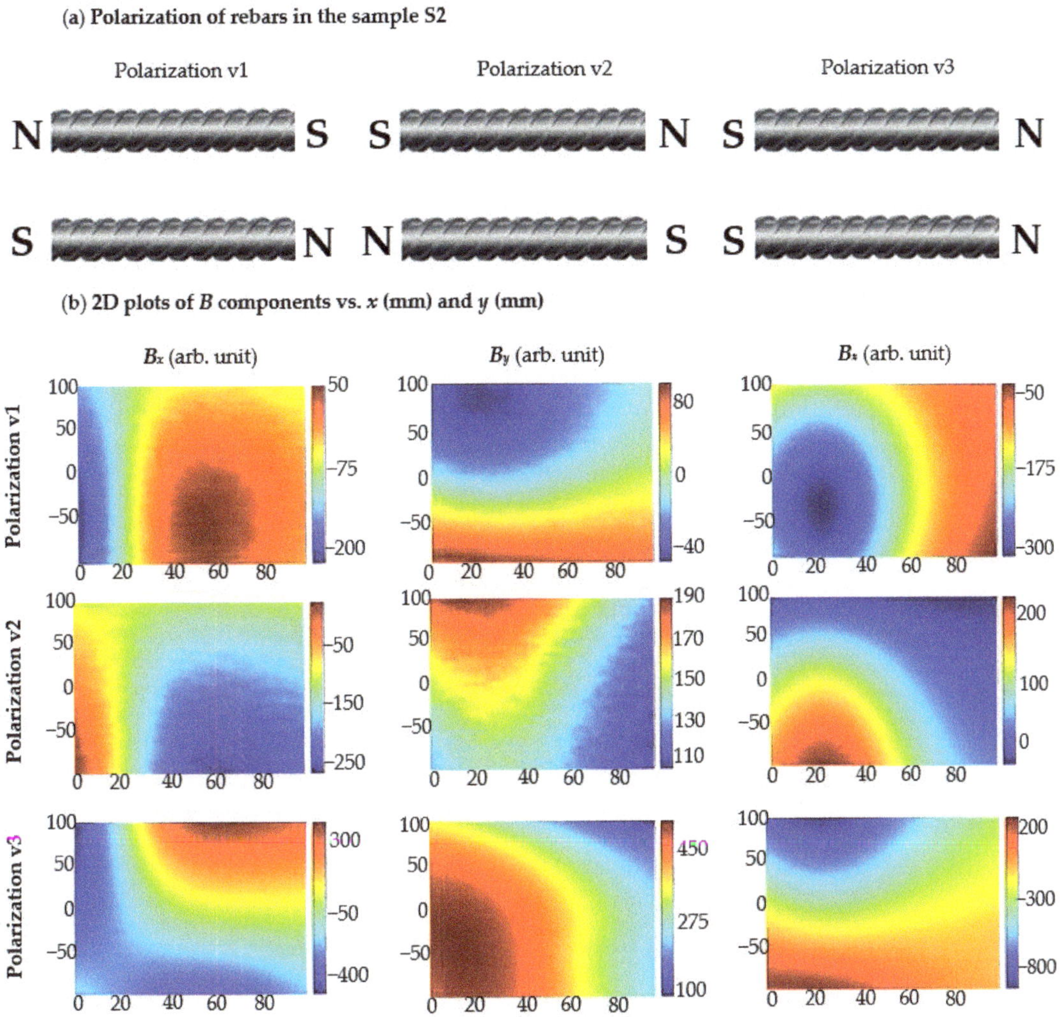

Figure 22. (**a**) Polarizations (residual magnetization) of earlier magnetized rebars in the sample S2; (**b**) Results of 2D measurements using AMR sensor received for different arrangements of earlier magnetized rebars; without external magnetization during measurements, the thickness of the concrete cover h = 40 mm; sample S2; x (mm), y (mm)—sensor positions.

The problem with unknown residual magnetization disappears when magnetic excitation is used. Moreover, the signal value is higher and the identification process is reliable. Next, experiments were conducted for three different samples with the use of different magnetization methods.

In the case of SPM, the value of the obtained signal is bigger than without the magnetization. Unfortunately, identifying the arrangement of the bars in the mesh is very difficult or even impossible. The shapes and maximal values of received characteristics are very similar for sample S2 and sample S3, as shown in Figure 23.

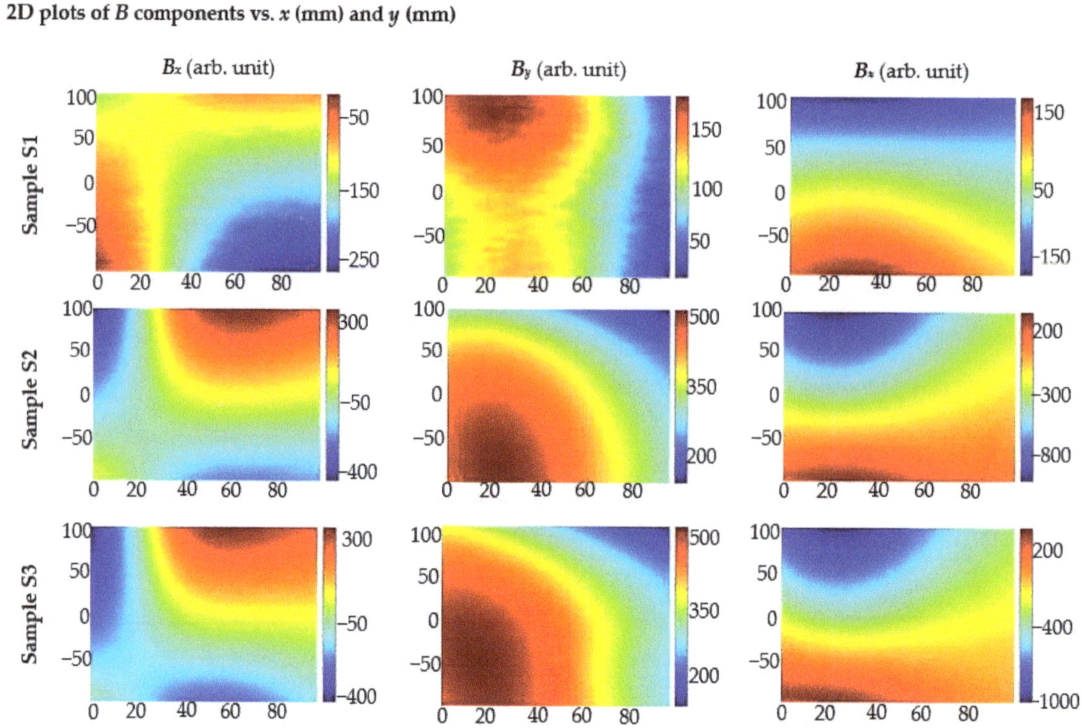

Figure 23. Results of 2D measurements using AMR sensor received for different samples (S1, S2, S3); the magnetization OPM; the thickness of the concrete cover h = 40 mm; x (mm), y (mm)—sensor positions.

Experiments prove that SPM is far superior to OPM in identifying complex structures. For sample S2, in which the bars are located one after the other, the characteristics are to some extent similar to those obtained for single rebar. However, their shapes and maximal values differ enough, and they are easy to distinguish. Therefore, it is possible to easily recognize this arrangement of rebars and even estimate the distance between them. In the case of sample S3, where the rebars are next to each other, the greatest signal values are obtained over the middle rebar (over which the magnets are placed). In addition, this case is easy to recognize. The SPM-results are presented in Figure 24.

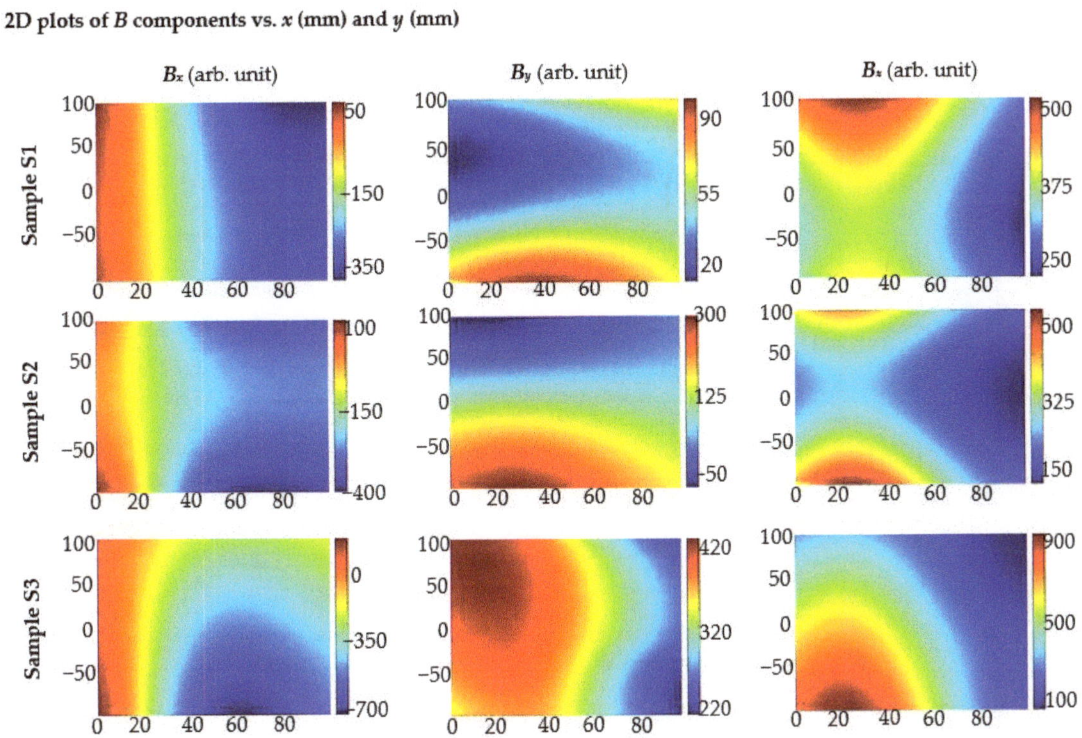

Figure 24. Results of 2D measurements using AMR sensor received for different samples (S1, S2, S3); the magnetization SPM; the thickness of the concrete cover h = 40 mm; x (mm), y (mm)—sensor positions.

4. Discussion

The use of magnetic excitation is crucial for the quality of the results in the magnetic evaluation of reinforced concrete structures. In the case of simple structures, where only one rebar is detectable, it affects noise immunity (Table 2) and the signal value. In addition, even a weak magnetic field makes the rebar's polarization predictable, which significantly facilitates identification. As shown in Figure 19, the results of measurements obtained without magnetization are challenging to predict and heavily dependent on residual magnetization (which can be unknown to the investigator). Generally, the identification of any parameters without magnetic excitation is a subject of significant uncertainty. However, it is possible to detect the rebar even without the magnetization. In the case of more complex structures (Sample S2 and S3), identifying the structure can be tricky when two or more rebars of unknown polarization affect the sensor.

The thickness of the concrete cover (h) can be estimated using the magnetic method. The relationship between the signal value and the h for different magnetization methods is shown in Figure 18. Potentially also different parameters of a reinforced concrete structure can be tested with this method (e.g., rebars diameter, rebars class, etc.). However, confirmation requires further investigations.

The magnetization method significantly impacts the results of measurements performed with the magnetic method. This aspect is often undervalued. In the case of sample S1, signal value and SNR depend on magnetization methods. Better results are received mostly for SPM (single pole magnetization). Moreover, in the case of the SPM, identification was more straightforward, as the results received for B_x are similar over the entire surface above the rebar (Figure 15). The magnetization method is even more critical in evaluating more complex structures. In the case of samples S2 and S3, it was possible to identify the structure only by using the SPM (Figures 23 and 24).

The MO sensors enable the evaluation of large areas of reinforced concrete structures in real-time. It is also helpful for fast pilot studies. In the case of greater concrete cover thicknesses, it is necessary to magnetize the rebars due to the moderate sensitivity of the MO-sensor. The signal to noise ratio (SNR) in the case of MO-sensors is much lower than in the case of the AMR sensor. Therefore, for more accurate tests, MO-sensors are not well suited. However, the quality of the results can be improved by hardware enhancement.

The AMR sensors enable effective testing of reinforced concrete structures without magnetization (with typical concrete cover thickness). However, when the concrete cover thickness is high, it is worth using even a small level excitation to improve the system's efficiency. This solution provides a stronger signal, easier to interpret and analyze the characteristics.

MR elements can be used for area testing. For this purpose, matrices of the sensors can be used. The experiments presented in the paper show that the elements of this kind are much more sensitive and resistant to noise than MO. Comparison is presented in Table 3. The MR sensors also allow the testing of particular spatial components B_x, B_y, B_z. However, these elements cannot be used if the magnetic field is out of range. Therefore, if the magnetic field can be stronger than 1 mT, it is recommended to use a proper MO-sensor or matrix of Hall-elements (Hall-elements possess all advantages of MR sensors, but the active range is much higher than in the case of AMR and the sensitivity is comparable with MO-sensors).

Table 3. Signal to noise ratio SNR calculated for measurements obtained by MO and AMR sensors for different concrete cover thicknesses h, and same pole magnetization.

h (mm)		20	40	60
MO		27	25	23
AMR	B_x	48	37	35
	B_y	52	46	41
	B_z	58	56	56

5. Conclusions

In the introduction of this paper, it has been shown that only two groups of nondestructive testing (NDT) methods enable direct and effective testing of the condition of reinforced concrete structures. A complete comparison of various NDT methods used in civil engineering is presented in [3]. Magnetic and electromagnetic tests are better suited for reinforcement testing than those that use a mechanical wave. Electromagnetic and magnetic waves affect mainly/only steel bars. Concrete is for such waves (almost) transparent. As shown in [3] magnetic tests can be used for very similar purposes as the eddy current (EC) method. However, the tested method has several significant advantages over electromagnetic evaluation (particularly the EC tests with which they can compete). The experiments show that the most significant advantage is the ability to perform area testing, which would be difficult to do with, e.g., EC tests.

Moreover, the excitation system in magnetic studies does not require advanced power electronic systems or even a power supply. This makes a magnetic method very cheap in implementing and universal in application. The next advantage lies in received data. Research results are relatively simple in interpretation (especially with a well-designed excitation system). Interpretation is even simpler than in the case of EC tests and much simpler than in the case of GPR. The last huge advantage of magnetic testing is the possibility to analyze particular spatial magnetic components, which, combined with the area test, creates unique possibilities which no other method gives.

The magnetic test also has limitations. Compared to EC testing, their spatial resolution is firmly limited. Compared to GPR tests, the effective range is small. Nevertheless, the possibility to performing the area tests in a simple way, cheap and straightforward hardware implementation, simplicity of the interpretation of results, and the ability to test for particular spatial components B_x, B_y, B_z, makes the method universal and useful in the evaluation of composite structures (in particular structures of reinforced concrete).

The results of the experiments presented in this paper prove that AMR sensors are well suited for area tests. The sensor, unlike MO enables the study of particular spatial components B_x, B_y, B_z. They are also more sensitive and more resistant to noise. In addition, they are linear and there is no hysteresis phenomenon. The disadvantage of matrixes of AMR sensors is the relatively small availability on the market. Moreover, in the case of AMR, BGA assembly is required, which makes such transducers challenging to build without proper equipment. However, with professional assembly, such sensors compete with the MO sensors. The MO sensors are not destroyed when a tested magnetic field is too strong, they have a high resolution and in many cases, measurement results do not require any processing. At this moment, the superiority of the MR sensor matrix over MO sensors cannot be clearly stated. Principles of operation of MO and MR sensors are completely different. Moreover, both sensors have some advantages. For example, at low h and relatively strong excitation, MO sensors ensure high resolution at relatively high (sufficient) SNR. For the same conditions, the AMR sensor can be damaged due to the too strong magnetic field. Simplifying, AMR matrices transducers are better for testing a weak magnetic field when the MO sensors are better suited to a strong field. Therefore, further comparative studies will be continued. A simple AMR sensor matrix has already been constructed for this purpose.

Experiments have shown that the efficiency of identifying the concrete cover thickness h may be in the case of magnetic methods similar to the efficiency of identifying with EC tests (very high for standard concrete cover thicknesses). In further studies, the possibility of identifying diameter and class (alloy from which rebars are made) will also be tested. Identification of such parameters is possible using EC Tests [8–11]. In the case of the EC system, the frequency and amplitude of the excitation are the main factors determining the efficiency of the method. Similarly, in the case of magnetic methods, the configuration of excitation magnets can be crucial for the identification of reinforced concrete structures. Moreover, component B_x, B_y, B_z analysis can be fundamental for more reliable evaluation.

Author Contributions: Concept of the method, T.C. and P.K.F.; the concept of the paper, P.F; software development, T.C. and P.K.F.; hardware development, T.C. and P.K.F.; preliminary experiments T.C.; final experiments and all measurements, P.K.F.; data curation, P.K.F.; formal analysis, P.K.F.; investigation, T.C. and P.K.F.; resources, T.C. and P.K.F.; visualization, P.K.F. and T.C.; writing—original draft preparation, P.K.F.; writing—review and extensive editing, T.C. and P.K.F.; scientific consultation, T.C.; project administration, T.C. and P.K.F.; Contribution of the Authors in %: P.K.F.—55%, T.C.—45%. All authors have read and agreed to the published version of the manuscript.

Funding: This research was partially funded by National Science Centre (in polish: NCN—Narodowe Centrum Nauki), program: Preludium, grant number 2021/41/N/ST7/02728 ("Smart support system for the Magnetic Force Induced Vibration Evaluation (M5)—an electromagnetic, nondestructive method designed for the evaluation of composite materials containing ferromagnetic and conductive elements.").

Institutional Review Board Statement: Not applicable.

Informed Consent Statement: Not applicable.

Data Availability Statement: The data presented in this study are available on request from the corresponding author. The data are not publicly available due to a complicated structure that requires additional explanations.

Conflicts of Interest: The authors declare no conflict of interest. The funders had no role in the design of the study; in the collection, analyses, or interpretation of data; in the writing of the manuscript; or in the decision to publish the results.

References

1. Wegen, G.; Polder, R.B.; Breugel, K. Guideline for service life design of structural concrete—A performance-based approach with regard to chloride induced corrosion. *Heron* **2012**, *57*, 153–168.
2. Neville, A.M.; Neville, A.M. *Properties of Concrete*; Pearson Education Limited: London, UK, 2011; ISBN 9780273755807.
3. Frankowski, P.K.; Chady, T.; Zieliński, A. Magnetic force induced vibration evaluation (M5) method for frequency analysis of rebar-debonding in reinforced concrete. *Measurement* **2021**, *182*, 109655. [CrossRef]
4. Mayakuntla, P.K.; Ghosh, D.; Ganguli, A. Nondestructive evaluation of rebar corrosion in concrete structures using ultrasonics and laser-based sensing. *Nondestruct. Test. Eval.* **2021**, 1–18. [CrossRef]
5. Ghosh, D.; Kumar, R.; Ganguli, A.; Mukherjee, A. Nondestructive Evaluation of Rebar Corrosion–Induced Damage in Concrete through Ultrasonic Imaging. *J. Mater. Civ. Eng.* **2020**, *32*, 04020294. [CrossRef]
6. Miwa, T. Non-Destructive and Quantitative Evaluation of Rebar Corrosion by a Vibro-Doppler Radar Method. *Sensors* **2021**, *21*, 2546. [CrossRef] [PubMed]
7. Vibration-Based Nondestructive Damage Detection for Concrete Plates. *ACI Struct. J.* **2021**, *118*, 118. [CrossRef]
8. Szymanik, B.; Frankowski, P.K.; Chady, T.; Chelliah, C.R.A.J. Detection and Inspection of Steel Bars in Reinforced Concrete Structures Using Active Infrared Thermography with Microwave Excitation and Eddy Current Sensors. *Sensors* **2016**, *16*, 234. [CrossRef]
9. Chady, T.; Frankowski, P.K. Electromagnetic Evaluation of Reinforced Concrete Structure. *AIP Conf. Proc.* **2013**, *1511*, 1355–1362. [CrossRef]
10. Frankowski, P.K.; Chady, T.; Sikora, R. Knowledge extraction algorithms dedicated for identification of steel bars in reinforced concrete structures. *AIP Conf. Proc.* **2014**, *1581*, 822. [CrossRef]
11. Frankowski, P.K. Corrosion detection and measurement using eddy current method. In Proceedings of the 2018 International Interdisciplinary PhD Workshop (IIPhDW), Swinoujście, Poland, 9–12 May 2018. [CrossRef]
12. Mosharafi, M.; Mahbaz, S.; Dusseault, M.; Vanheeghe, P. Magnetic detection of corroded steel rebar: Reality and simulations. *NDT E Int.* **2020**, *110*, 102225. [CrossRef]
13. Frankowski, P.K.; Sikora, R.; Chady, T. Identification of rebars in a reinforced mesh using eddy current method. *AIP Conf. Proc.* **2016**, *1706*, 090008. [CrossRef]
14. McCann, D.; Forde, M. Review of NDT methods in the assessment of concrete and masonry structures. *NDT E Int.* **2001**, *34*, 71–84. [CrossRef]
15. Solla, M.; Lagüela, S.; Fernández, N.; Garrido, I. Assessing Rebar Corrosion through the Combination of Nondestructive GPR and IRT Methodologies. *Remote. Sens.* **2019**, *11*, 1705. [CrossRef]
16. Tešić, K.; Baričević, A.; Serdar, M. Non-Destructive Corrosion Inspection of Reinforced Concrete Using Ground-Penetrating Radar: A Review. *Materials* **2021**, *14*, 975. [CrossRef]
17. Diederich, H.; Vogel, T. Evaluation of Reinforcing Bars Using the Magnetic Flux Leakage Method. *J. Infrastruct. Syst.* **2017**, *23*. [CrossRef]
18. Diederich, H.; Vogel, T. Break Detection in Reinforcing Bars Using the Magnetic Flux Leakage Method. In Proceedings of the International Symposium Non-Destructive Testing in Civil Engineering (NDT-CE), Berlin, Germany, 15–17 September 2015.
19. Sadeghnejad, A.; Valikhani, A.; Chunn, B.; Lau, K.; Azizinamini, A. Magnetic Flux Leakage Method for Detecting Corrosion in Post Tensioned Segmental Concrete Bridges in Presence of Secondary Reinforcement. In Proceedings of the Transportation Research Board 96th Annual Meeting, Washington DC, USA, 8–12 January 2017.
20. Zhang, H.; Liao, L.; Zhao, R.; Zhou, J.; Yang, M.; Zhao, Y. A new judging criterion for corrosion testing of reinforced concrete based on self-magnetic flux leakage. *Int. J. Appl. Electromagn. Mech.* **2017**, *54*, 123–130. [CrossRef]
21. Shams, S.; Ghorbanpoor, A.; Lin, S.; Azari, H. Nondestructive Testing of Steel Corrosion in Prestressed Concrete Structures using the Magnetic Flux Leakage System. *Transp. Res. Rec. J. Transp. Res. Board* **2018**, *2672*, 132–144. [CrossRef]
22. Sun, Y.; Liu, S.; Deng, Z.; Tang, R.; Ma, W.; Tian, X.; Kang, Y.; He, L. Magnetic flux leakage structural health monitoring of concrete rebar using an open electromagnetic excitation technique. *Struct. Health Monit.* **2018**, *17*, 121–134. [CrossRef]
23. Gobov, Y.L.; Mikhailov, A.V.; Smorodinskii, Y.G. Magnetic Method for Nondestructive Testing of Rebar in Concrete. *Russ. J. Nondestruct. Test.* **2018**, *54*, 871–876. [CrossRef]
24. Tran, H.Q. Passive and active infrared thermography techniques in nondestructive evaluation for concrete bridge. *AIP Conf. Proc.* **2021**, *2420*, 050008. [CrossRef]
25. Fermon, C. Introduction on Magnetic Sensing and Spin Electronics. *Nanomagn. Appl. Perspect.* **2017**, 1–18. [CrossRef]
26. Koschny, M.; Lindner, M. Magneto-optical sensors accurately analyze magnetic field distribution of magnetic materials. *Adv. Mater. Processes* **2012**, *170*, 13–17.
27. D'Silva, G.J.; Feigenbaum, H.P.; Ciocanel, C. Visualization of Magnetic Domains and Magnetization Vectors in Magnetic Shape Memory Alloys Under Magneto-Mechanical Loading. *Shape Mem. Superelast.* **2020**, *6*, 67–88. [CrossRef]
28. Matesy. Available online: https://matesy.de/en/products/materials/mo-sensors-magnetooptics (accessed on 13 November 2021).
29. Yang, S.; Zhang, J. Current Progress of Magnetoresistance Sensors. *Chemosensors* **2021**, *9*, 211. [CrossRef]

materials

MDPI

Article

Magnetic Investigation of Cladded Nuclear Reactor Blocks

Gábor Vértesy [1,*], Antal Gasparics [1], Ildikó Szenthe [1] and Sándor Bilicz [2]

[1] Centre for Energy Research, Institute of Technical Physics and Materials Science, 1121 Budapest, Hungary; gasparics.antal@ek-cer.hu (A.G.); szenthe.ildiko@ek-cer.hu (I.S.)
[2] Department of Broadband Infocommunications and Electromagnetic Theory, Faculty of Electrical Engineering and Informatics, Budapest University of Technology and Economics, Műegyetem rkp. 3., H-1111 Budapest, Hungary; bilicz.sandor@vik.bme.hu
* Correspondence: vertesy.gabor@ek-cer.hu

Abstract: The wall, made of ferromagnetic steel, of a nuclear reactor pressure vessel is covered by an austenitic (very weakly ferromagnetic) cladding. In this work, we investigated how the base material and the cladding can be inspected separately from each other by nondestructive magnetic measurements. It was found that with the proper choice of the magnetizing yoke, these two different materials could be measured independently of each other. The effect of the yoke's size was studied by the numerical simulation of magnetic flux, pumped into the material during magnetic measurements. Measurements were performed by two different sizes of yokes on pure base material, on base material under cladding and on cladding itself. Experiments verified the results of the simulation. Our results can help for the future practical application of magnetic methods in the regular inspection of nuclear power plants.

Keywords: magnetic nondestructive evaluation; nuclear reactor pressure vessel; austenitic cladding; steel degradation

Citation: Vértesy, G.; Gasparics, A.; Szenthe, I.; Bilicz, S. Magnetic Investigation of Cladded Nuclear Reactor Blocks. *Materials* **2022**, *15*, 1425. https://doi.org/10.3390/ma15041425

Academic Editor: Giovanni Bruno

Received: 19 January 2022
Accepted: 8 February 2022
Published: 15 February 2022

1. Introduction

In almost all major industrial countries worldwide today, nuclear power plants (NPPs) are used to generate electricity. However, the Fukushima Daiichi accident affected the nuclear energy renaissance and, since then, safety aspects have been significantly strengthened. For many existing NPPs, lifetime extensions to 40, 50, 60 or even 80 years have been requested [1]. The long-term operation of existing NPPs has already been accepted in many countries as a strategic objective to ensure adequate supply of electricity over the coming decades. Operating conditions that affect the design lifetime include neutron exposure (fluence), as well as the number and magnitude of temperature and/or pressure cycles in both normal conditions and hypothetical accidental conditions [2,3]. License renewal and periodic safety reviews (PSRs) are the two basic regulatory approaches that are required for an authorization of the long-term operation of NPPs [1]. Evaluation of these parameters during the PSRs allows for an estimation of the operational lifetime of NPPs [2].

As a result of the above arguments, the regular inspection of nuclear power plants is an extremely important task, because the mechanical properties of the reactor pressure vessel (RPV) wall are modified during its operation mainly due to the long-term and high-energy neutron irradiation [4]. In boiling water pressurized reactors, the most critical and most important part is the reactor pressure vessel, because it is not changeable during the whole period of operation. Apart from the standard destructive tests (mechanical Charpy impact testing [5]) several nondestructive electromagnetic methods have been suggested recently for determination of neutron irradiation-generated embrittlement of the pressure vessel steel material. The precise measurement of the Seebeck coefficient makes it possible to derive the neutron irradiation-induced embrittlement of RPV material [6,7]. An ultrasonic technique is also widely used in the inspection of NPPs [8–10].

The RPV material is ferromagnetic steel, so magnetic methods are useful for this inspection. An overview of nondestructive magnetic methods are given in References [11,12]. The magnetic Barkhausen noise (MBN) technique was developed for inspection of residual stresses, surface defects and microstructure changes [13–16]. Another method, magneto-acoustic emission, can also be frequently used for the monitoring of residual stresses [17].

A correlation exists between the modification of the microstructure of the material, generated by different effects, and the observed magnetic behavior if the material is influenced by a magnetic field. This phenomenon can be used to characterize the ferromagnetic materials via magnetic hysteresis measurement. Dislocation movement and domain wall motion are both affected by the microstructure of the material. In ferromagnetic materials, the correlation between mechanical and magnetic hardness is well-known and understood [18,19]. Magnetic methods are advantageous because they are not expensive, are technically simple, and they can be used easily, even on active materials. One of them, the so-called 3MA approach (3MA = micromagnetic, multiparameter, microstructure and stress analysis) applies several methods [20–22], and it was found suitable for the characterization of the damage in ferromagnetic materials like RPV steels and for monitoring the progress of materials.

Another method of magnetic nondestructive testing is the measurement of the magnetic hysteresis loops. For instance, the so-called magnetic minor loops power scaling laws (PSL) were developed, whereby different parameters of minor hysteresis loops are used for material characterization [23]. A similar method, magnetic adaptive testing, also measures systematically minor magnetic hysteresis loops. This is also a multi-parametric, powerful and sensitive method of magnetic inspection [24]. As a conclusion of these works, a reasonable correlation could be found between the destructively measured parameters and nondestructively measured magnetic characteristics through application of these methods. This fact makes possible the future potential use of magnetic methods in the inspection of RPVs' structural integrity.

In several previous works [16,20,25–27] irradiated Charpy samples were measured using magnetic methods, and results of magnetic measurements were compared with the destructively measured ductile to brittle transition temperature (DBTT). However, for the full inspection of nuclear reactors, blocks cut from RPV steel should be also measured, not only samples of Charpy geometry. This aspect of the inspection of a nuclear reactor's integrity has still not been extensively investigated. If the base material itself is directly measured, magnetic measurement is easy. However, in RPV, the base material (ferromagnetic steel) is covered by a cladding. This cladding, which is made of austenitic steel, is the integral component of a WWER 440-type nuclear reactor pressure vessel. Its role is to warrant an anticorrosive protection for the vessel material. Cladding is about a 10 mm thick stainless steel weld-overlay, which is deposited on the pressure vessel's inner surface. It shields the base metal of the pressure vessel from the corrosive environment produced by primary light water coolant [28]. Cladding in WWER 440 reactors is made by submerged arc welding technology by using strip electrodes. The surface of the cladding is either ground or machined roughly to ensure ultrasonic testing coupling.

For future inspection of nuclear reactors, a tool should be developed that is suitable for non-destructive evaluation of the embrittlement of the vessel wall. The final system should be capable of inspecting the degradation of the microstructure through the cladding. The first step has been made in this direction: cladded blocks were successfully measured even through the cladding [29] via the magnetic adaptive testing method [24]. In this work, cladded RPV blocks were investigated, which had been treated thermally by a step cooling procedure, which caused embrittlement of the material. It was demonstrated that the base material degradation could be followed by magnetic measurements even through the cladding. It was shown that a reliable, nearly linear correlation existed between magnetic parameters and DBTT, as expected.

In this type of investigation, when base material is measured through the cladding, the main problem for the magnetic measurement is that cladding means a thick, almost nonmagnetic layer between the magnetizing yoke and base material, which resulted in an extremely low and noisy probe response, as presented in Reference [29]. In other words, cladding causes serious difficulty for base material investigation. Nevertheless, through the proper choice of measuring parameters and suitable software, the measured signal could be successfully evaluated.

On the other hand, the impact of cladding on the reactor pressure vessel wall integrity has been investigated in a very limited way in spite of the fact that it can potentially be of great significance to RPV integrity. The reason is that plasticity and the elevated fracture toughness of cladding can provide additional strength to the pressure wall and this process can justify an extended reactor lifetime [30]. In addition, in thermal expansion coefficients, significant differences can be found with respect to pressure vessel base metals, which can cause a stress peak [31]. This is the so-called pressurized thermal shock and it is a potential risk of interfacial crack initiation and propagation. Safety analysis of this phenomenon has lately become a subject of interest for operators of nuclear power plants [32]. In a very recent work, the results of an experimental investigation were presented, aimed at the evaluation of microstructure and failure mechanisms of WWER 440 reactor pressure vessel austenitic cladding (made of stainless steel Sv 08Kh19N10G2B) [33].

The question is arising, as to whether the base material and cladding could be investigated independently of each other by magnetic measurements. The purpose of this paper was to study this problem and to give an answer to this question. Material of cladding is basically austenitic, but it also contains several percentages (2–8%) of ferrite (magnetic) phase. The existence of this ferrite phase gives a chance for successfully applying magnetic measurements to study the properties of cladding. However, the huge volume of highly ferromagnetic base metal, close to the weakly ferromagnetic cladding material, causes difficulty for cladding material investigation. In this work, we will show that the characterization of base and cladding material can be separated from each other by a suitable technique of measurement.

The idea is to choose the proper size of magnetizing yoke—different sizes of yokes can be used for the characterization of base metal and for the characterization of cladding. To make a qualitative interpretation of the experimental results, the effect of the yoke dimension is calculated by numerical simulation of the magnetic flux distribution in the sample.

2. Materials and Methods

2.1. Materials

As a first sample, a cladded block was investigated. Chemical composition of the 15H2NMFA base material can be seen in Table 1. The block was cut from the forged ring according to Figure 1. This block is shown in Figure 2. The size of the block was 110 mm × 77 mm × 278 mm. The 10 mm thick cladding is clearly seen on the top of the block.

Table 1. Chemical composition of 15H2NMFA base metal (wt%).

C%	Mn%	Si%	S%	P%	Cr%	Ni%	Mo%	V%	Cu%	Co%	Sb%	Sn	As%
0.16	0.42	0.29	0.08	0.012	1.97	1.29	0.52	0.12	0.12	0.06	0.001	0.003	0.003

2.2. Magnetic Measurement

Permeability of the material was measured by attaching a magnetizing yoke on the surface of the sample. The yoke itself was made of Fe-Si laminated sheets. An exciting coil, a wound on the leg of magnetizing yoke, was used for producing magnetizing field *F* in the sample.

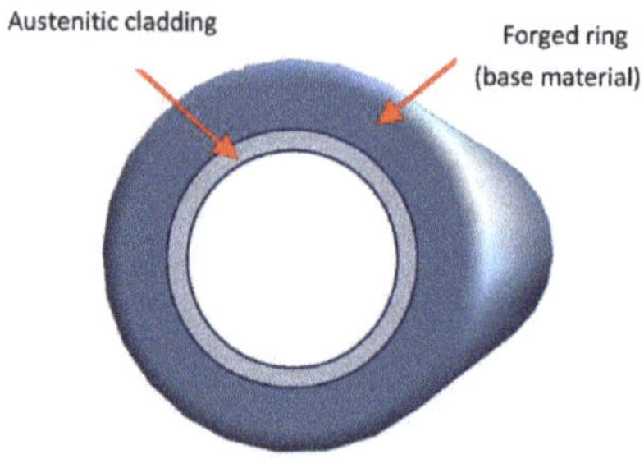

Figure 1. Cutting of the cladded block from a forged ring.

Figure 2. Photograph of the cladded block.

Another sample was also prepared. This was only a piece of cladding, which was cut from the top of a cladded block. The photograph of this sample is shown in Figure 3. The size of the sample was 114 mm × 50 mm × 12 mm.

Figure 3. Photograph of the cladding itself (cut from the top of another cladded block).

Before starting the measurement, the sample was magnetized close to saturation by applying a magnetizing current in the exciting coil. Then, the value of magnetizing current was decreased, linearly by time t, to zero and then increased again in the opposite direction up to saturation with the same slope. The slope of the magnetizing current was 0.125 A/sec. Due to the time variation of the effective field in the magnetizing circuit, a signal is generated in the pick-up coil, which is wounded also on the magnetizing yoke. As long as the magnetic field (or magnetizing current) is sweeping linearly with time, t, the U signal voltage in the pick-up coil is proportional to the differential permeability, μ of the magnetic circuit.

$$\mu = \text{const} \times U(\mathrm{d}F/\mathrm{d}t) = \text{const} \times \partial B(\mathrm{d}F/\mathrm{d}t)/\partial t = \text{const} \times \mu(\mathrm{d}F/\mathrm{d}t) \times \mathrm{d}F/\mathrm{d}t \quad (1)$$

In future parts of the text, magnetizing current, I, will be used instead of magnetizing field to describe the magnetization of the sample, because in an open magnetic circle the real value of magnetizing field inside the sample is never known due to the dissipation of the magnetic field into the air. It means that in non-uniform magnetic circuits, it is not possible to speak about the signal *U* as proportional to the differential permeability of the material, but we use an effective differential permeability values of the existing circuit. The current values also characterize the magnetic state of the investigated samples well.

The magnetizing yokes with different dimensions, Yoke A and Yoke B, used in our measurements, can be seen in Figure 4. The two (driving and pick-up) coils, wound on the legs of yokes, are seen well in the photos, especially on the right side photo (Yoke B).

Figure 4. Photographs of two different size magnetizing yokes (**Yoke A** and **Yoke B**) on the top of the same block.

The signal of the pick-up coil can be seen well in figures below in Research and Discussion section. The magnetizing current value at the maximal value of permeability was used as the characteristic parameter for the magnetic behavior of the investigated samples. During measurements, the steel side of the block (down in Figure 2), the cladded side of the block (up in Figure 2) and the cladding (Figure 3) were measured by applying two different size magnetizing yokes.

The schematic drawing of the sole of magnetizing yoke is given in Figure 5, and Table 2 presents the dimensions of the two yokes numerically. In this table, the heights of the yokes are also given.

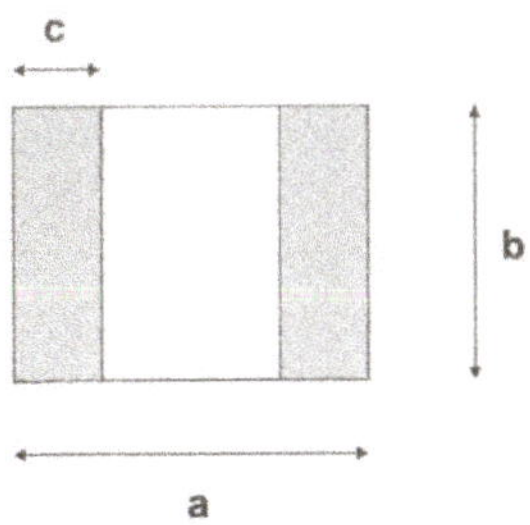

Figure 5. Schematic drawing of the sole of magnetizing yoke. a: total length of the yoke, b: width of the yoke, c: width of the leg.

Table 2. Dimensions of the magnetizing yokes.

	a (mm)	b (mm)	c (mm)	Height (mm)
Yoke A	62.0	19.0	16.0	55.0
Yoke B	11.5	12.5	4.5	13.0

2.3. Numerical Simulation

The idea behind applying two different sizes of magnetizing yokes for measurement of cladded blocks was that by doing this, we can separate the magnetic signal from the base material and from the cladding. Numerical simulation of the distribution of the magnetic flux was performed for both Yoke A and Yoke B. The ferromagnetic base material was characterized by a nonlinear $B(H)$ curve with saturation around 2 T and initial relative permeability of $\mu_{rel,0}$ = 1200. Calculations were performed for three different values of relative permeability of the cladding, μ_{rel} = 1, 5 and 10, respectively.

The calculations have been performed by 3D Finite Element Method (FEM), using the COMSOL Multiphysics software. The partial differential equations of the stationary magnetic field have been formulated for the magnetic vector potential [34]. The model domain was closed by an artificial boundary on which the normal component of the magnetic flux density was set to zero. The exciting coil has been modelled as an equivalent surface current density on the yoke's surface. The coil on Yoke A has 150 turns, whereas Yoke B has 40 turns. The exciting current was set as 0.45 A and 0.3 A, respectively, which approximated the exciting current at maximum differential permeability in the experiments. The nonlinear system of equations resulting from the FEM-discretization has been solved iteratively by the software.

In the post-processing step, two magnetic fluxes were calculated: Ψ_1 is the flux together in the cladding and in the base material, while Ψ_2 is the flux only in the base material, both evaluated at the symmetry plane of the model. The quotient Ψ_2/Ψ_1 characterizes the relative magnetization of the base material. The numerical results for the two yokes and three different values of the relative permeability of the cladding, are given in Table 3.

Table 3. Numerical result of simulation of the magnetic flux for the two different yokes and for three different values of relative permeability of the cladding.

	μ_{rel}	Ψ_1 (Wb)	Ψ_2 (Wb)	Ψ_2/Ψ_1
Yoke A	1	4.55×10^{-6}	4.24×10^{-6}	0.931
Yoke A	5	1.49×10^{-5}	1.43×10^{-5}	0.956
Yoke A	10	2.54×10^{-5}	2.43×10^{-5}	0.955
Yoke B	1	3.35×10^{-7}	1.96×10^{-7}	0.586
Yoke B	5	1.12×10^{-6}	5.88×10^{-7}	0.526
Yoke B	10	1.95×10^{-6}	9.86×10^{-7}	0.505

It can be seen that in the case of Yoke A, this was a large value (>93%), while in the case of Yoke B, it was only around 50%. In this latter case, Ψ_2/Ψ_1 depended more on the relative permeability of the cladding.

The distribution of the magnetic flux is shown for the three values of the relative permeability of cladding (μ_{rel} = 1, 5 and 10) for both yokes in Figures 6 and 7.

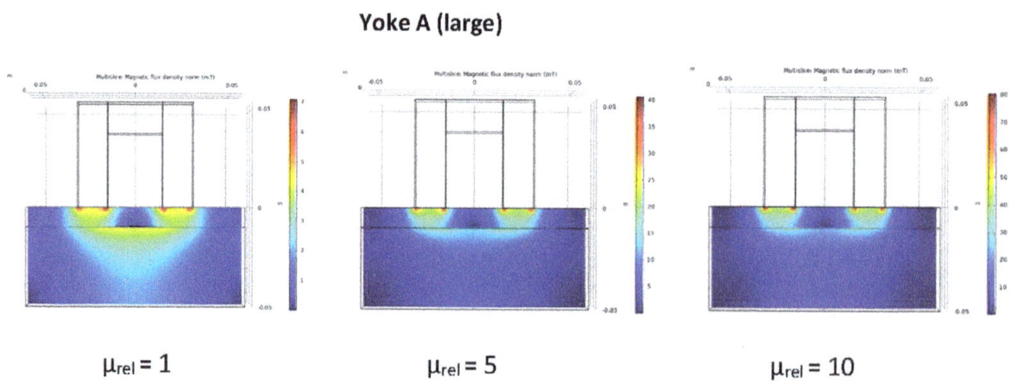

Figure 6. Distribution of the calculated flux density in the cladded block for three values of relative permeability of the cladding if the large magnetizing yoke (Yoke A) is applied.

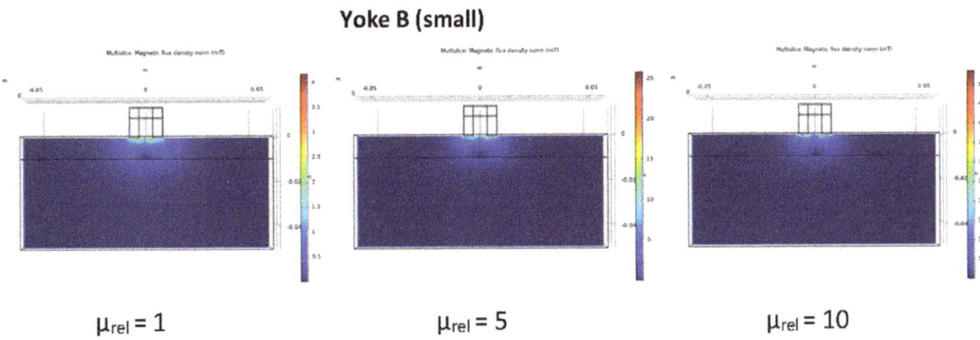

Figure 7. Distribution of the calculated flux density in the cladded block for three values of relative permeability of the cladding if the small magnetizing yoke (Yoke B) is applied.

The result of numerical simulation revealed that by using the large magnetizing yoke, the base material could be magnetized enough even through the cladding. On the other hand, if a small-sized magnetizing yoke is applied, it is sensitive only to the region of cladding, while the magnetic influence of the base material below cladding is very limited, almost negligible. In the next section, it will be shown how the real measurements verified the result of simulation.

3. Results and Discussion

3.1. *Yoke A*

The first measurement was performed on the base material (bottom of block, shown in Figure 2), by applying the larger yoke (Yoke A). The signal of the pick-up coil (proportional to the permeability of the material according to Equation (1)) as a function of the magnetizing current is presented in Figure 8. The error of the magnetizing current at the top permeability is also given in the figure. The sample magnetically was saturated before measurement by a negative current, then the value of the magnetizing current was linearly decreased to zero, then increased with the same slope of current to positive saturation.

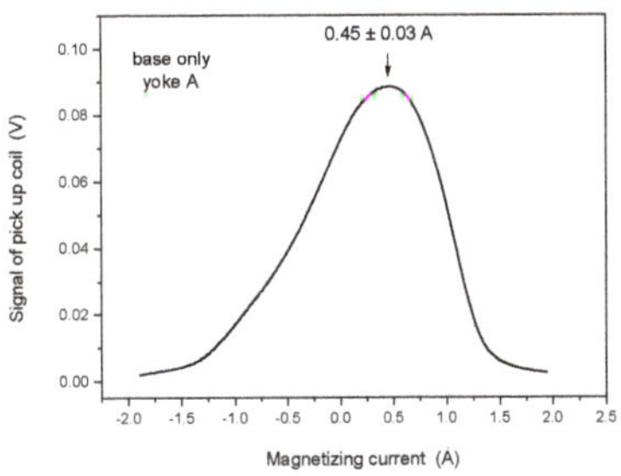

Figure 8. Signal of the pick-up coil as a function of the magnetizing current, measured on base material by applying Yoke A.

The value of the magnetizing current at the maximal value of permeability was chosen to characterize the magnetic behavior of the measured material. This value, which was 0.45 A in the case shown above, does not depend on the actual parameters of the measurement. Evidently, only results of those measurements can be compared with each other, which were performed by the same magnetizing yoke.

The next measurement was performed on the top of the block (shown in in Figure 2), again by applying Yoke A. In this case, cladding and base material were measured together. In principle, the magnetic behavior of both somehow influence the measured signal. However, as concluded from the result of simulation, a relatively high amount of flux was pumped into the ferromagnetic base material, so it is expected that mainly the high permeability base material determines the measured signal. The result of this measurement is shown in Figure 9. The values of the magnetizing currents at maximal permeability were very close to each other (0.45 A and 0.47 A), within the error of measurement. This means that base material was also detected when measurement is performed through the cladding. The influence of cladding was almost negligible in this case. This result verifies our previous measurements on thermally treated cladded blocks [31].

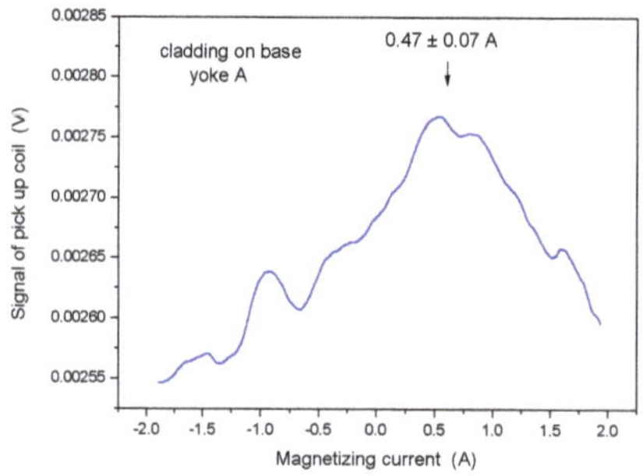

Figure 9. Signal of the pick-up coil as a function of the magnetizing current, measured on the top of cladded block by applying Yoke A.

It should be mentioned, however, that the registered curve is rather noisy if measurement is performed through cladding. For better presentation, the curve of Figure 9 was smoothed. Smoothing, made by adjacent averaging of measured points, decreases the scatter of points, but it has no influence on the value of magnetizing current at the maximal permeability. Measurements were repeated five times after each other, removing and placing the magnetizing yoke back. Practically no difference was found in the registered curves by this repetition of measurement.

In order to study the situation better, measurement by Yoke A was done also on pure cladding (see sample in Figure 3). This result is shown in Figure 10. In accordance with our expectation, a weak maximum can be seen in permeability, but at different values of magnetizing current, compared to measurements made either on base material or on cladding above base material. The maximum of the curve appeared at I = 0.47 A if the base material was under the cladding and appeared at I = 0.78 A in the case of pure cladding. This difference cannot be explained by any experimental error, only by the difference in the magnetic behavior of the cladding and base material. The magnetic behavior of cladding is due to the small ferrite content of cladding, as mentioned already in the Introduction. The signal was very low and noisy, but a maximum definitely existed. The two signals, measured on pure cladding and on the cladding above the base material, can be compared if the two curves are presented on the same scale, as done in Figure 11.

3.2. Yoke B

The same series of measurements as described above was performed on the samples by applying the small-sized yoke. Results are shown in Figures 12–14. Note that the values of magnetizing current at maximal permeability are not comparable with similar current values in the previous section because different yokes were used in the two measurements.

As defined above, the characteristic parameter of the base material was 0.30 A (see Figure 12). When the same measurement was performed, but on the cladding, the magnetizing current at maximal permeability was 0.90 A (see Figure 13). This parameter was very far from the base metal, so it can be considered as characteristic for the cladding. This statement is confirmed by the result of the measurement, performed on pure cladding (see Figure 14), where the maximum was observed at 0.87 A magnetizing current.

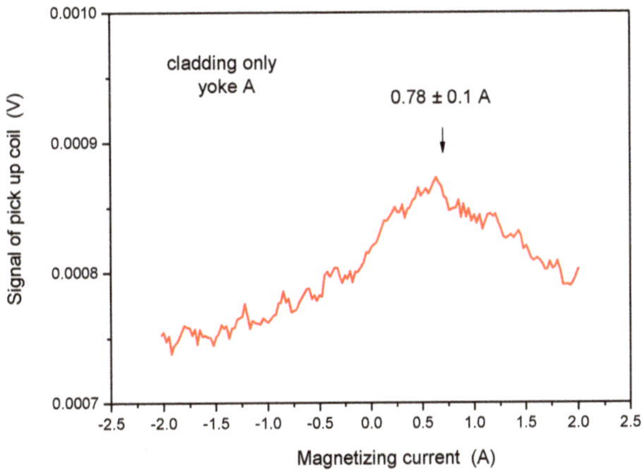

Figure 10. Signal of the pick-up coil as a function of the magnetizing current, measured on the pure cladding by applying Yoke A.

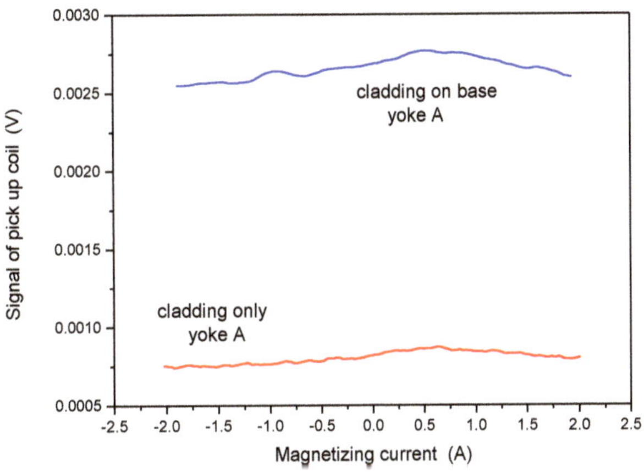

Figure 11. Signal of the pick-up coil as a function of the magnetizing current, measured on the pure cladding and on the cladding above the base material by applying Yoke A.

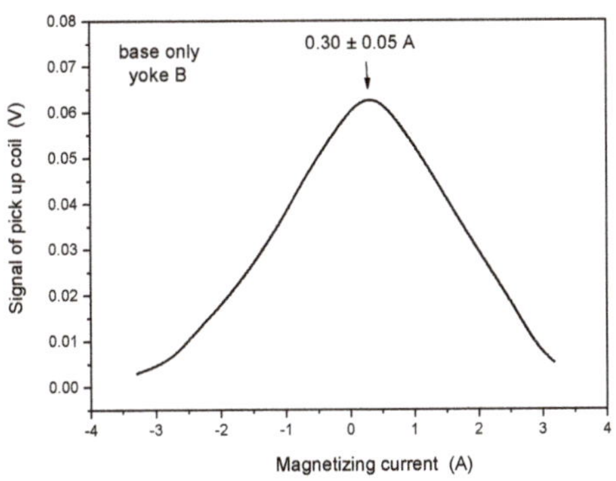

Figure 12. Signal of the pick-up coil as a function of the magnetizing current, measured on the base material by applying Yoke B.

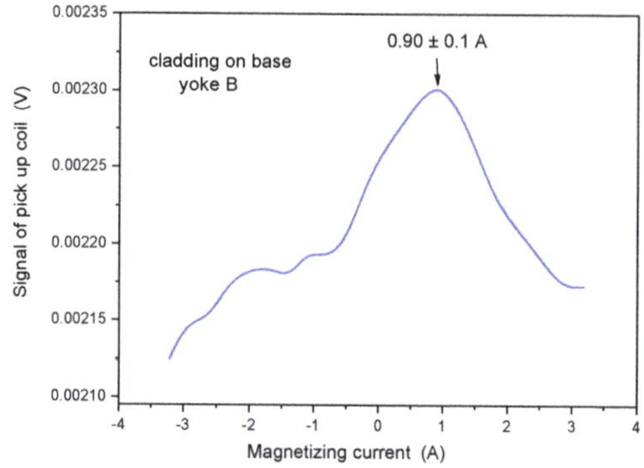

Figure 13. Signal of the pick-up coil as a function of the magnetizing current, measured on the top of the cladded block by applying Yoke B.

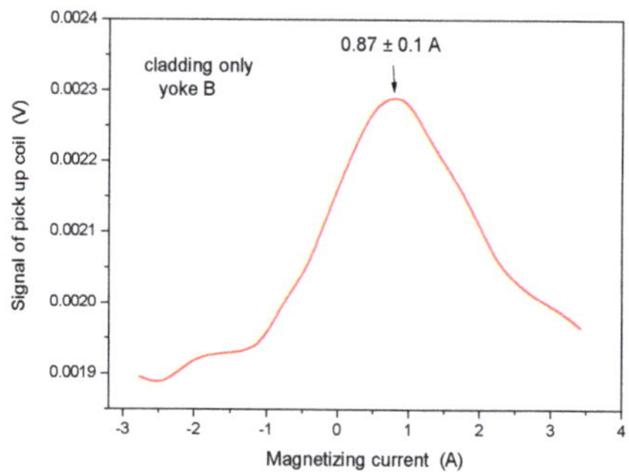

Figure 14. Signal of the pick-up coil as a function of the magnetizing current, measured on the pure cladding by applying Yoke B.

When the small size yoke was applied, the influence of the base material could not be detected, but the magnetic behavior of cladding could be excellently measured, even in the presence of highly ferromagnetic base metal.

Application of magnetic nondestructive methods can be important for the future inspection of austenitic steel degradation. As is known, originally paramagnetic steel became more and more ferromagnetic under stress, due to the appearance of bcc α′-martensite. In our previous work, titanium stabilized austenitic stainless steel, 18/8 type, was studied [35]. Stainless steel specimens were cold-rolled at room temperature. The compressive plastic deformation of the material increased its hardness. It was found that this change could be followed by a nondestructive magnetic method with substantially higher sensitivity and reliability than the traditionally used destructive hardness measurements.

In another work, austenitic stainless steel SUS316L was also plastically deformed by a tensile stress [36]. In contrast to the compressed samples, the tensile deformation did not introduce such a large percentage of the ferromagnetic phase into the deformed samples. Nevertheless, magnetic indication of the strain values was possible, and the method was also able to reflect anisotropy induced into the material by the stress.

4. Conclusions

A magnetic method was developed by which the cladded blocks of a nuclear pressure vessel can be characterized by a nondestructive technique.

It was demonstrated, by applying two magnetizing yokes with different dimensions, that the two types of very magnetically different components (ferromagnetic base metal and almost austenitic cladding) can be investigated separately from each other. To our best knowledge, this way of measurement is new.

If a large-sized yoke is applied, the ferromagnetic base metal can be measured. This yoke is not suitable for investigation of cladding on the top of base material, because the signal from the ferromagnetic part suppresses the signal of weakly magnetic cladding.

With the application of a small-sized yoke, the cladding itself can be measured, even in the presence of highly ferromagnetic steel. The magnetic flux is closed in the cladding and cannot penetrate into the base material.

The effect of the different yoke dimension was determined based on simulation of the magnetic flux distribution in the given geometry. The results of measurement correlate very well with the suggestions of simulation.

By using our results, the possible material degradation of austenitic cladding on the pressure vessel—due to different effects, like neutron irradiation, thermal treatment, etc.—can be inspected by a simple and nondestructive magnetic method. This way of investigation implies the ability to monitor the integrity of the various layers of the reactor walls. Furthermore, this approach of using two different yoke sizes to test a bilayer material can be applied in general in other areas, where a highly ferromagnetic material is covered by another weakly ferromagnetic layer. This would make the work more meaningful than in connection with the testing of one particular wall. This seems to be possible, since the depth of magnetic permeability testing should increase with the size of the yoke.

Based on our results, in the future, it will be possible that material degradation of cladding generated by any effects (neutron irradiation, thermal shock, plastic or elastic deformation) could be inspected by magnetic hysteresis measurements, mainly by magnetic adaptive testing. The measurement can even be done directly on the reactor pressure vessel wall.

Author Contributions: Conceptualization and original draft preparation, G.V.; investigation and methodology, G.V. and A.G.; numerical simulation, S.B.; project administration, A.G. and I.S. All authors have read and agreed to the published version of the manuscript.

Funding: This research was carried out in frame of the "NOMAD" project. This project (Non-destructive Evaluation System for the Inspection of Operation-Induced Material Degradation in Nuclear Power Plants) has received funding from the Euratom research and training program 2014–2018 under grant agreement No 755330.

Institutional Review Board Statement: Not applicable.

Informed Consent Statement: Not applicable.

Data Availability Statement: The data are contained within the article.

Conflicts of Interest: The authors declare no conflict of interest.

References

1. *The Economics of Long-Term Operation of Nuclear Power Plants*; Nuclear Energy Agency and Organisation for Economic Co-Operation and Development: Paris, France, 2012.
2. Ballesteros, A.; Ahlstrand, R.; Bruynooghe, C.; von Estorff, U.; Debarberis, L. The role of pressure vessel embrittlement in the long term operation of nuclear power plants. *Nucl. Eng. Des.* **2012**, *243*, 63–68. [CrossRef]
3. Al Mazouzi, A.; Alamo, A.; Lidbury, D.; Moinereau, D.; Van Dyck, S. PERFORM 60: Prediction of the effects of radiation for reactor pressure vessel and in-core materials using multi-scale modelling—60 years foreseen plant lifetime. *Nucl. Eng. Des.* **2011**, *241*, 3403–3415. [CrossRef]
4. Koutsky, J.; Kocık, J. *Radiation Damage of Structural Materials*; Elsevier: Amsterdam, The Netherlands, 1994.
5. Ferreño, D.; Gorrochategui, I.; Gutiérrez-Solana, F. Degradation due to neutron embrittlement of nuclear vessel steels: A critical review about the current experimental and analytical techniques to characterise the material, with particular emphasis on alternative methodologies. In *Nuclear Power—Control, Reliability and Human Factors*; Tsvetkov, P., Ed.; IntechOpen Limited: London, UK, 2011; ISBN 9789533075990. Available online: http://www.intechopen.com/articles/show/title/non-destructive-testing-for-ageing-management-of-nuclear-power-component (accessed on 20 October 2019).
6. Niffenegger, M.; Leber, H.J. Monitoring the embrittlement of reactor pressure vessel steels by using the Seebeck coefficient. *J. Nucl. Mater.* **2009**, *389*, 62. [CrossRef]
7. Niffenegger, M.; Reichlin, K.; Kalkhof, D. Application of the Seebeck effect for monitoring of neutron embrittlement and low-cycle fatigue in nuclear reactor steel. *Nucl. Eng. Des.* **2005**, *235*, 1777–1788. [CrossRef]

8. Seiler, G. Early detection of fatigue at elevated temperature in austenitic steel using electromagnetic ultrasound transducers. In Proceedings of the Seventh International Conference on Low Cycle Fatigue: LCF7, Deutscher Verband für Materialforschung und-prüfung e.V. (DVM), Aachen, Germany, 9–11 September 2013; pp. S359–S364.
9. Smith, R.L.; Rusbridge, K.L.; Reynolds, W.N.; Hudson, B. Ultrasonic attenuation, microstructure and ductile to brittle transition temperature in Fe-C alloys. *Mater. Eval.* **1993**, *41*, 219–222.
10. Dobmann, G.; Kröning, M.; Theiner, W.; Willems, H.; Fiedler, U. Nondestructive characterization of materials (ultrasonic and magnetic techniques) for strength and toughness prediction and the detection early creep damage. *Nucl. Eng. Des.* **1995**, *157*, 137–158. [CrossRef]
11. Jiles, D.C. Magnetic methods in nondestructive testing. In *Encyclopedia of Materials Science and Technology*; Buschow, K.H.J., Ed.; Elsevier: Oxford, UK, 2001; p. 6021.
12. Blitz, J. *Electrical and Magnetic Methods of Nondestructive Testing*; Adam Hilger IOP Publishing, Ltd.: Bristol, UK, 1991.
13. Lo, C.C.H.; Jakubovics, J.P.; Scrub, C.B. Non-destructive evaluation of spheroidized steel using magnetoacoustic and Barkhausen emission. *IEEE Trans. Magn.* **1997**, *33*, 4035–4037. [CrossRef]
14. Kikuchi, H.; Ara, K.; Kamada, Y.; Kobayashi, S. Effect of microstructure changes on Barkhausen noise properties and hysteresis loop in cold rolled low carbon steel. *IEEE Trans. Magn.* **2009**, *45*, 2744–2747. [CrossRef]
15. Hartmann, K.; Moses, A.J.; Meydan, T. A system for measurement of AC Barkhausen noise in electrical steels. *J. Magn. Magn. Mater.* **2003**, *254–255*, 318–320. [CrossRef]
16. Barroso, S.P.; Horváth, M.; Horváth, Á. Magnetic measurements for evaluation of radiation damage on nuclear reactor materials. *Nucl. Eng. Des.* **2010**, *240*, 722–725. [CrossRef]
17. Augustyniak, B.; Chmielewski, M.; Piotrowski, L.; Kowalewski, Z. Comparison of properties of magnetoacoustic emission and mechanical barkhausen effects for P91 steel after plastic flow and creep. *IEEE Trans. Magn.* **2008**, *44*, 3273–3276. [CrossRef]
18. Devine, M.K. Magnetic detection of material properties. *J. Min. Met. Mater. JOM* **1992**, *44*, 24–30. [CrossRef]
19. Kronmüller, H.; Fähnle, M. *Micromagnetism and the Microstructure of Ferromagnetic Solids*; Cambridge University Press: Cambridge, UK, 2003.
20. Dobmann, G.; Altpeter, I.; Kopp, M.; Rabung, M.; Hubschen, G. ND-materials characterization of neutron induced embrittlement in German nuclear reactor pressure vessel material by micromagnetic NDT techniques. In *Electromagnetic Nondestructive Evaluation (XI)*; IOS Press: Amsterdam, The Netherlands, 2008; p. 54, ISBN 978-1-58603-896-0.
21. Altpeter, I.; Becker, R.; G Dobmann, R.; Kern, W.A.; Theiner, A. Yashan: Robust Solutions of Inverse Problems in Eletromagnetic Non-Destructive Evaluation. *Inverse Probl.* **2002**, *18*, 1907–1921. [CrossRef]
22. Szielasko, K.; Wolter, B.; Tschuncky, R.; Youssef, S. Micromagnetic materials characterization using maschin learning—Progress in Nondestructive Prediction of Mechanical Properties of Steel and Iron. *Tech. Mess.* **2020**, *87*, 428–437. [CrossRef]
23. Takahashi, S.; Kobayashi, S.; Kikuchi, H.; Kamada, Y. Relationship between mechanical and magnetic properties in cold rolled low carbon steel. *J. Appl. Phys.* **2006**, *100*, 113908. [CrossRef]
24. Tomáš, I. Non-destructive Magnetic Adaptive Testing of ferromagnetic materials. *J. Mag. Mag. Mat.* **2004**, *268*, 178–185. [CrossRef]
25. Vértesy, G.; Gasparics, A.; Uytdenhouwen, I.; Szenthe, I.; Gillemot, F.; Chaouadi, R. Nondestructive investigation of neutron irradiation generated structural changes of reactor steel material by magnetic hysteresis method. *Metals* **2020**, *10*, 642. [CrossRef]
26. Tomáš, I.; Vértesy, G.; Gillemot, F.; Székely, R. Nondestructive Magnetic Adaptive Testing of Nuclear Reactor Pressure Vessel Steel Degradation. *J. Nucl. Mater.* **2013**, *432*, 371–377. [CrossRef]
27. Rabung, M.; Kopp, M.; Gasparics, A.; Vértesy, G.; Szenthe, I.; Uytdenhouwen, I. Micromagnetic characterization of operation induced damage in Charpy specimens of RPV steels. *Appl. Sci.* **2021**, *11*, 2917. [CrossRef]
28. Gillemot, F. Overview of Reactor Pressure Vessel Cladding. *Int. J. Nucl. Knowl. Manag.* **2010**, *4*, 265–278. [CrossRef]
29. Vértesy, G.; Gasparics, A.; Szenthe, I.; Gillemot, F. Magnetic nondestructive inspection of reactor steel cladded blocks. *Glob. J. Adv. Eng. Technol. Sci.* **2019**, *6*, 1.
30. Gillemot, F.; Horváth, M.; Úri, G.; Fekete, T.; Houndeffo, E.; Acosta, B.; Debarberis, L.; Viehrig, H.W. Radiation Stability of WWER RPV Cladding Materials. *Int. J. Press. Vessel. Pip.* **2007**, *84*, 469–474. [CrossRef]
31. Trampus, P. Pressurized Thermal Shock Analysis of the Reactor Pressure Vessel. *Procedia Struct. Integr.* **2018**, *13*, 2083–2088. [CrossRef]
32. Chen, M.; Yu, W.; Qian, G.; Shi, J.; Cao, Y.; Yu, Y. Crack Initiation, Arrest and Tearing Assessments of a RPV Subjected to PTS Events. *Ann. Nucl. Energy* **2018**, *116*, 143–151. [CrossRef]
33. Štefan, J.; Siegl, J.; Adámek, J.; Kopřiva, R.; Falcník, M. Microstructure and Failure Processes of Reactor Pressure Vessel Austenitic Cladding. *Metals* **2021**, *11*, 1676. [CrossRef]
34. Comsol Multiphysics. AC/DC Module User's Guide. Available online: https://doc.comsol.com/5.4/doc/com.comsol.help.acdc/ACDCModuleUsersGuide.pdf (accessed on 10 January 2022).
35. Vértesy, G.; Tomáš, I.; Mészáros, I. Nondestructive indication of plastic deformation of cold-rolled stainless steel by magnetic adaptive testing. *J. Magn. Magn. Mater.* **2007**, *310*, 76–82. [CrossRef]
36. Vértesy, G.; Ueda, S.; Uchimoto, T.; Takagi, T.; Tomáš, I.; Vértesy, Z. Evaluation of Plastic Deformation in Steels by Magnetic Hysteresis Measurements. In *Elecstromagnetic Nondestructive Evaluation (XIV)*; Takagi, T., Udpa, S.S., Chady, T., Gratkowski, S., Eds.; IOS Press: Amsterdam, The Netherlands, 2011; pp. 371–378.

MDPI
St. Alban-Anlage 66
4052 Basel
Switzerland
Tel. +41 61 683 77 34
Fax +41 61 302 89 18
www.mdpi.com

Materials Editorial Office
E-mail: materials@mdpi.com
www.mdpi.com/journal/materials